GENERAL AND APPLIED MECHANICS

MECHANICAL ENGINEERING AND APPLIED MECHANICS

V. Z. Parton, Series Editor

VOLUME 1: GENERAL AND APPLIED MECHANICS, A. Yu. Ishlinsky, N. N. Krasovsky, V. V. Rumyantsev, and V. N. Rubanovsky

In Preparation

VOLUME 2: FLUID MECHANICS, G. A. Liubimov, L. I. Sedov, and G. G. Chernyi

VOLUME 3: MECHANICS OF SOLIDS, N. Kh. Arutiunian, I. F. Obraztsov, and V. Z. Parton

VOLUME 4: APPLIED MECHANICS AND TECHNOLOGY, V. M. Kudinov, N. A. Makhutov, and K. V. Frolov

GENERAL AND APPLIED MECHANICS

Edited by
A. Yu. Ishlinsky
N. N. Krasovsky
V. V. Rumyantsev
V. N. Rubanovsky

Series Editor
V. Z. Parton
Institute of Chemical Engineering
Moscow, USSR

HEMISPHERE PUBLISHING CORPORATION
A member of the Taylor & Francis Group
New York Washington Philadelphia London

GENERAL AND APPLIED MECHANICS

1 2 3 4 5 6 7 8 9 0 BRBR 8 9 8 7 6 5 4 3 2 1 0 9

This book was set in Times Roman by Edwards Brothers, Inc. The editor was Janine Ludlam.
Cover design by Sharon M. DePass.
Braun-Brumfield, Inc. was printer and binder.

Library of Congress Cataloging-in-Publication Data

General and applied mechanics / edited by A. Yu. Ishlinsky . . . et al.
p. cm.—(Mechanical engineering and applied mechanics ; v.
1)
Translated from Russian.

1. Mechanics. 2. Mechanics, Applied. I. Ishlinskiĭ, Aleksandr
Iul'evich, date. II. Series.
TJ7.M36 vol.
[TA350]
621 s—dc20
[620.1] 89-15506

ISBN 0-89116-724-2

CONTENTS

FOREWORD

Whether considered as the study of forces and motions or as the knowledge of machinery, mechanics has always been closely associated with all areas of human activity, from an individual to a national level. It is mechanics which underlies the design and construction of machines, instruments, technological systems, and structures with varying orders of complexity.

Perhaps the most characteristic features of modern mechanics are its close links with physics and chemistry, its deep penetration into biology, medicine, and geology, and its wide use of numerical methods based on modern high-speed computer technology. Today's mechanics make it possible to investigate the mechanical properties of new materials and constructions, and to study their behavior under extreme load and temperature conditions, as well as under various and hostile environments. We are now witnessing the increasing introduction of mechanics into the theory of technological processes and into the life and reliability analysis of structures. The development of optimization methods and the automation of design are also among the subjects of present-day mechanics.

Russian and, subsequently, Soviet scientists have always played an important and sometimes leading role in the development of mechanics. The works of Chebyshev, on applied mechanics, and Lyapunov, on stability, form the principal classic reference to their respective subjects, as also do the works of Zhukovskii and Chaplygin on hydrodynamics, and of Krylov on ship theory. The works of Muskhelishvili on the elasticity theory are widely known, as are the remarkable results obtained by Keldysh in a number of fields. The fundamental achievements of Soviet mechanical science enjoy worldwide recognition.

The processes of restructuring now underway in the Soviet Union are posing new and demanding tasks for science in general and mechanics in particular, placing emphasis on the intensification of research and the accelerated introduction of scientific results into production use. The reconstruction and the intensive development of mechanical engineering is one of the top priorities of Soviet mechanical science.

The articles in the present collection have been written by the leading Soviet scientists and devoted to the most topical problems of mechanics as they stand today. The collection is divided into four volumes: 1. General and applied mechanics; 2. Fluid mechanics; 3. Mechanics of solids; and 4. Applied mechanics and technology. The collection opens with a review article by Academician Frolov, Vice-President of the USSR Academy of Sciences. The papers are based on the talks given at the Sixth All-Union Congress on Theoretical and Applied Mechanics, Tashkent, September, 1986. The forum, which collected about 3,000 scientists and researchers from across the USSR, made it possible to analyze the achievements of Soviet mechanics in the preceding five years and to indicate the principal direction to be taken in future work in light of the decisions of the XXVII CPSU Congress.

MECHANICAL PROBLEMS IN MODERN MACHINE BUILDING

K. V. Frolov

Soviet mechanical science held its Sixth Congress at a time when the 27th CPSU Congress had already adopted its magnificent program [1] and the process of all-out change was well underway in our society and in the scientific community as well.

Scientific research provides the basis for technological development in various and diverse fields—and in machine building in particular—and mechanics, perhaps needless to say, plays a particularly important role among other scientific disciplines allowing, as it does, to penetrate deep into the mysteries of the physical world and, on the other hand, to solve many practical engineering problems.

Modern electronics offers limitless possibilities for further development and exploitation of the research methods of classical mechanics. To draw examples from different fields, one only has to think of new manufacturing techniques, breath-taking space programs, robots, aircraft, gigantic engineering projects, tailor-made materials, and advanced power generation methods.

"It may be rightfully said," Mikhail Gorbachev points out, "that the science and technology of today beautify our life and enable all-round development of every human being on the Earth. But," he goes on, "the same science and technology, these amazing creations of the human mind, are now endangering the very existence of the human race. Absurd, isn't it? We consider that science should no longer serve two masters, life and death; it must serve life alone" [2].

Over the period since the 5th Theoretical and Applied Mechanics Congress, Soviet mechanical science has made major advances beneficial both for national welfare and national security. To describe the role of mechanics in machine building and to review the most important achievement of, and the problems faced by those involved in the field is, of course, a demanding task. I find it appropriate to begin by considering some general results pertaining to machine building [3], deferring until the concluding sections a detailed discussion of the subject.

1. GENERAL AND APPLIED MECHANICS

Soviet science undoubtedly plays a leadership role in the development of variational methods in mechanics. A number of highly effective mathematical methods now being used in machine building have resulted from the extension of classical variational methods, either in their differential or integral form, on systems with nonholonomic nonlinear constraints or distributed parameter systems.

The method of Lyapunov functions, so fruitful in the stability studies of nonlinear systems, has been and is continuously being developed and has already found applications in fields as diverse as equilibrium stability of nonlinear elastic bodies; stationary motions of compound engineering structures; flutter; heat conduction and chemical processes; hydrodynamic and magnetohydrodynamic flows.

The Soviet Union's preeminence in the theory of mechanical vibrations is an old tradition stemming from the pre-revolutionary pioneering works on stability and nonlinear vibrations and is associated with such renowned names as Lyapunov, Chetaev, Krylov, and later Bogolyubov, Mandelshtam, Papaleksi, Andronov, Witt, Malkin, and Kamenkov. The last two decades or so have seen many developments—and the appearance of totally new research areas—in the field of nonlinear vibrations.

In particular, much work has been done on computer-oriented algebraic constructions for finding asymptotic and exact solutions of nonlinear systems. It has been shown that the theory of Markovian processes and the asymptotic methods of nonlinear mechanics can be effectively applied to the study of random vibrations in viscoelastic systems perturbed by white-noise random influences. Methods have been developed for treating nonlinear vibrations in free-surface liquids and in solids with liquid-containing cavities; a theory has been devised to account for the effect of external vibrations on nonlinear mechanical systems. A great deal of work has been done on vibration transport, limited-excitation oscillations, engineering damping, distributed dry friction systems [4], vibration conductivity, vibrorheology, and the effects of vibrations on pulps and suspensions. These theoretical investigations have resulted in quite a number of new technologies, many of which are already being used in industrial mills, crushers, etc. [5].

One further development is the synchronization theory of dynamic systems, capable of describing among other things, the self-synchronization of unbalanced moving rotors. The vibration machine with a self-synchronizable vibration exciter [6]—a practical outcome of this theory and the last word in vibration technology—has already been licensed in the USA, Japan, and several other countries. It should be noted, however, that synchronization theory has yet to realize its full potential in machine building.

Considerable progress has also been made in understanding rotor systems. One major result obtained in this field is the generalization of Laval's self-balance principle. If a frequency of rotation exceeds the maximum natural frequency of a given mechanical system, it turns out that unbalanced rotors linked to the linear oscillatory system tend to reach a self-balanced state; for frequencies lower than the minimum natural frequency, there is observed a build-up of vibrations; whereas at intermediate frequencies either situation may occur, depending on the specific value of the frequency.

Parametric oscillations present one further area where a number of significant results have been obtained.

The achievements listed above have been covered in the six-volume reference book "Vibrations in Technology" issued by the "Machinostroyeniye" ("Machine Building") publishers in the years 1978–1981. (The reader may also be referred to [6] for information on the subject). Speaking in terms of technical applications, however, it should be admitted that some factors exist which still prevent full advantage being taken of these theoretical achievements.

The development of complex navigational systems has led to an upsurge of interest in the mechanics of gyroscopes and gyroscopic systems, and calls for a dependable theory of their performance in the Earth's gravitational field.

Much progress has been made—but much more is still to be done—in studying the dynamics of compound systems and developing methods for the identification of their unknown characteristics.

Soviet scientists engaged in general and applied mechanics are expected to, and surely will, make further contributions into the development of modern mechanical engineering. The lines of activity which seem to be of most importance in this respect are as follows.

Extensive theoretical research in the fields of information, control, and vibration dynamics should provide a scientific basis for developing effective on-process diagnosis methods for machines and structures operating under complex loading conditions. This is particularly important for a number of unique machine complexes now under development or already at work in transportation, power engineering, and other fields.

Computer—man interactive systems and in particular computer graphics techniques open up new possibilities for organizing design and manufacturing processes. Information on the current state of a technical object can be displayed on a video screen either in a numerical form, when the position of the object relative to a chosen coordinate system is shown by certain sets of numbers or, more conveniently, in the form of a point or figure moving in this coordinate system. Assuming a certain mathematical model for the object, it proves possible to effectively use a computer for monitoring and analyzing the dynamic, thermal, force and deformation variables at the most critical parts of the machine.

Prognosis of the vibrational state of large-scale structures is a major concern in both conventional and nuclear power engineering, necessitating further development of the theory of nonlinear vibrations in mechanical systems. Special emphasis should be placed on nonlinear vibration studies in multiphase systems, with potential applications in oil, gas, and nuclear power industries. There are reports [7] that first steps have been made toward the application of wave mechanics to the problems of this kind. The feeling is that new fundamental ideas are needed to ensure further technological progress in this area.

And finally, there is the problem of the seismic resistance of mechanical systems, of particular and growing importance for atomic power engineering. Activity in this field has been steadily increasing over the last years in many scientific institutions across the nation (including, quite understandably, the Central Asian Republic of Uzbekistan [8]), but, unfortunately, the results obtained thus far do not appear to be of much practical significance.

2. FLUID MECHANICS

In recent years both fundamental and applied research on fluid mechanics has received considerable attention in this country. Methods of high-speed hydrodynamics are applicable to highly important problems of flowing free-surface liquids, both for surfaces formed by the motion of a body (cavitation) and for those existing independently of the motion (hydroplaning, undersurface wings, etc). It has become increasingly evident that many practical applications of high-speed hydrodynamic methods depend on the adequate understanding of the fundamentals of free-surface motion, and a number of results achieved in this field have thrown considerable new light on the subject of hydrodynamics as a whole and permitted some highly efficient engineering solutions in a number of areas.

A result of fundamental importance for hydraulic machine engineering has been the explanation of the mechanism by which macromolecules and other inclusions interact with turbulent perturbations occurring in a fluid. Of no less importance is a turbulence theory of diluted polymer solutions, capable of reliably indicating a polymer additive necessary for reducing the hydrodynamic resistance of pipes and of bodies moving in liquids.

A field which is rapidly extending its application area is gas dynamics [9]. Numerous industrial processes (particularly those occurring in power plants [10]) are being successfully modeled by employing fluid dynamic approaches, and advanced numerical methods make it possible to remove many of the simplifications used of necessity in the earlier models (such as a plane-grid representation of a turbine blade rim). Among the most interesting achievements I would like to mention are

1. methods for determining the integral and local characteristics of gas-turbine air vents and nozzles, including the case of the three-dimensional shock-wave flow past an air vent;

2. mathematical simulation of transient processes in, and flow stability analysis of, gas-turbine power installations;
3. accurate and effective methods for calculating flows in turbine blade rims.

Of particular importance is the work of Davidov et al. [11] on the simulation of non-stationary processes taking place in internal combustion engines. The large-particle method employed by the authors provides a design engineer with: (1) velocity projection distributions in various radial sections and (2) the equal concentration lines of combustion products, the information being presented in an easy-to-read graphic form for different times in the course of the gas exchange process and hence for different positions of the moving boundaries (i.e., pistons). Another major benefit of the method is the information on the composition of combustion products throughout the cylinder, also for different times during the process. The concentration line patterns obtained by Davidov and co-workers support the earlier suggestions that the gas mixture within an engine is nonuniform in concentration at the final stage of gas exchange and that convective mass transfer plays an important role in the combustion process.

There has been much work in recent years on developing numerical or analytical-numerical methods for determining nonstationary loads acting on blades and other turbine elements [12] and for treating the aeroplasticity problems arising in the operation of these elements (as exemplified by blade vibrations or blade motions relative to the guiding system).

Aerodynamic performance of flying vehicles is yet another major concern of fluid dynamics specialists [13]. Some of the recent successes in this field are as follows:

- Three-dimensional hypersonic nose design minimized for aerodynamic resistance;
- Solution of the contour optimization problem for a planar asymmetric thrust-lift nozzle;
- Planar and axisymmetric nose profiles maximizing the critical Mach number M* for the ongoing flow (which means no local hypersonic regions and a zero wave resistance factor in the $M < M^*$ regime);
- Contour optimization of the sub-, super-, and hypersonic parts of the Laval nozzle for the purpose of producing the maximum thrust for a fixed fuel consumption rate and certain size restrictions.

The non-ideality of liquids and gases, and primarily their viscosity, are factors which must necessarily be taken into account in determining the aerodynamic characteristics of flying vehicles and their propulsion systems as well as in the effectiveness evaluation of compressors, turbines, pipelines, heat exchangers and other industrial systems. This implies that the corresponding computation methods must be sufficiently sophisticated that highly nonlinear and high-order systems of equations can be treated (or even non-closed ones if it comes to turbulent flow problems).

The progress in fluidics would be impossible or at least too slow without experimental techniques capable of providing necessary information about a system being studied. Extensive research conducted by the Academy of Sciences in cooperation with leading industrial institutes has resulted in a number of new experimental methods of this kind [15]. Some successes, in particular, have been made in the important field of flow vizualization. A very satisfying fact is the worldwide recognition of the experimental work by Kochanov with co-workers [16] on the stability and laminarity-turbulence transitions in boundary layers.

Recent advances in numerical methods and computer technology have opened new opportunities and are being increasingly used in fluid studies. The importance, indeed necessity, of numerical aerodynamics is accentuated by the fact that in many problems of interest it is simply impossible to model experimentally the flow under study. One cannot experiment, for example, with a hypersonic real gas flowing past and being heated by a spaceship descending in the Earth's atmosphere.

Application of numerical methods to the study of viscous fluids has increased considerably in recent years. An important aspect of this activity is the use of asymptotic methods for treating processes occurring at large Reynolds numbers. These methods are extremely useful in investigating the physical nature of fluid phenomena (such as flow detachment) and in determining the diagnostic variables or similarity parameters of a flow, and I feel safe in saying that Soviet scientists are well ahead of their western colleagues in developing this approach. In particular, the asymptotic methods have been applied to such problems as flow stability, the aerodynamic heating due to hypersonic gas flow, and the self-induced detachment of a boundary layer at sub- and supersonic velocities.

The problems of hydrodynamic instability [18] and related turbulence problems are fundamental to modern hydrodynamics because most flows occurring in practice are due to instabilities arising in simpler types of flow. Some of the state of the art results achieved by Soviet researchers in recent years are as follows.

Experimental work on stability, detachment, and laminarity-turbulence transitions in boundary values has revealed a new LTT variety involving the excitation of higher harmonics and non-classical resonant three-wave interaction. This result suggests a possibility of using small perturbations as a means for controlling the detachment and turbulization of a boundary layer.

It has been shown that an unusual type of vibration-induced convection may arise in a thermally nonhomogeneous medium subjected to rapidly changing mass forces. This convection mode may be realized under weightlessness conditions and is likely to be useful for controlling outer-space technological processes.

A new method has been put forward, and theoretically justified, for stabilizing distributed mechanical systems by slightly varying boundary conditions in step with (sensor-monitored) changes in one of the system's parameters. It has been shown experimentally that the method allows considerable increases in load-carrying capability (as witnessed by an eightfold limiting load increase in bar experiments) and enhances the effectivity of continuous chemical reactors (in hydrocarbon oxidization, 25 to 30 percent in net product, and several hundred percent in gross product). The method is also advantageous in that it requires little or no redesign if some modifications are to be introduced into a technological process or a structure.

Turbulence is one of the most complicated physical phenomena [19], and its role in engineering and applied research can hardly be overestimated. Turbulent flows determine atmospheric and oceanic processes, and are always there in aviation and rocketry, in power engineering, and in chemistry [20]. It is unfortunate that the poor coordination of academic and industrial research in the field slows down the development of state-of-the-art investigation methods for theoretical and experimental turbulence studies. Experiment is of primary importance in this area but, regrettably, our industry fails to provide scientists with instruments for measuring velocity pulsations and for automating the spectrum and correlation analysis of high-speed turbulent processes.

In the field of hydrodynamics, many important practical problems have found their solutions in recent years. In particular, much work has been done, both theoretically and experimentally, on the use of an electromagnetic field for controlling separation and convection processes in MHD systems. Acceleration of such processes finds its uses in metallurgy and in some applications in other fields where a separation of mixed components is needed; suppression may be useful, for example, in the production of composite materials or in laboratory simulations of space technologies such as the growing of large and nearly perfect single crystals. Mention should also be made of the stability studies of the quasi-homogeneous electric current distribution on MHD electrodes. The electrode corrosion caused by current localization effects makes these studies extremely important for the design of MHD generators and similar heavy-current systems.

Many scientific institutions throughout the country are engaged in active research on percolation and other types of motion of underground fluids like water, oil, or gas, with obvious application to fuel and agricultural watering issues. Much research effort is being devoted to the very complicated problem of simultaneous motions of surface and underground waters. The level movement in rivers, lakes, and underground waters is generally a rather demanding problem involving nonlinearities, singularities, complicated initial and boundary conditions, and necessitating computer simulation and careful analysis of mathematical correctness in each particular case. This direction of research is of vital importance for those concerned with the rational use of national water resources, and the recent government resolutions concerning these issues oblige scientists to double their research efforts in the area.

In underground oil and gas hydrodynamics, emphasis is being placed on problems related to the intensification of oil/gas production. These are inherently mechanical problems, and it is from this viewpoint that they have been and are being treated. In particular, some investigations have been carried out on the possibility of increasing oil production by affecting the oil-pool face zone in some way or another (by applying vibration, for example).

Despite some undeniable successes, the pace of progress in this area is too slow, however. Lack of experimental data narrows the usefulness and possibilities of the existing models. Far too little has been done to develop experimental methods, especially for studying the microscopic mechanisms of multiphase flows in porous media. Only rudimentary studies exist in the wave technology and the hydraulic seam fracture technology, both of which are powerful enough to allow deposit exploitation even in low-permeability rocks. We are still far from a full understanding of multiphase flows in oil wells operating under gas-expansion drive; the improvement of the method would drastically increase well efficiency and service life.

Much support has been given in recent years to the theoretical and experimental studies of the inflammation and combustion of gaseous and liquid fuels in turbulent air flows. Results of such studies are of great importance for controlling combustion processes occurring in supersonic flows in the combustion chambers and in the vicinity of the surface of flying vehicles [21]. We now have effective methods for calculating the basic characteristics of such flows. It has been shown [22] that in many fire/explosion safety problems one should be able to determine detonation criteria for the situation under study. Such criteria have been found experimentally for a number of mixtures. In rough-surface pipes, it has been found that so-called quasi-detonation regimes of combustion propagation are possible. Much research work has been done on combustion processes in heterogeneous systems.

3. MECHANICS OF DEFORMABLE SOLIDS

This subject is finding increasing applications in major mechanical engineering problems such as strength, service life, stiffness, durability, and material saving [23]. There has been a great deal of recent interest in dynamic, mixed, and nonclassical problems of elasticity theory as well as in problems involving anisotropy and nonhomogeneity or piecewise homogeneity. Major advances have been made in generalizing elasticity theory models to include thermal and moment-force stresses, and electromagnetic, diffusion and other fields. New methods have been devised, and classical ones improved, for both theoretical and experimental studies of deformable media (potential parameter expansions; integral equations and integral transforms; homogeneous solutions and superpositions; experimental techniques employing polarization and holography effects; etc.). Solutions in the form of series or quadratures have been found for a number of elasticity and thermoelasticity problems, and necessary computer software has been developed for and applied to many important practical problems including, in particular, the elasticity theory optimization problem for objects subjected to interacting physical and mechanical fields and to operating conditions of various natures.

Much research has been carried out on the strength properties of natural and, especially, artificial elastic materials when subjected to various fields and operating under various dynamic loading conditions [24, 25]. It proved necessary in doing this to measure, or to correct, the elastic characteristics of many such materials. An effective method has been devised—and provided with the necessary software—for handling two-dimensional, nonlinear, elasticity problems arising at finite deformations of intricately shaped nonhomogeneous three-dimensional bodies.

It can be said with confidence that Soviet mechanics is second to none or even that it plays a leadership role in such areas of elasticity theory as contact and mixed problems; three-dimensional problems for bodies of complex geometry and for those having stress concentrators; and elasticity of nonhomogeneous, anisotropic, and prestressed bodies [26]. It would be leaving too rosy a picture, however, if some of the shortcomings were not mentioned.

The theory of coupled fields (thermo-, electro-, and magnetoelasticity) is definitely lagging behind practical needs, even though a number of fundamental results may be claimed in the area. In the field of nonclassical models and deformed elastic bodies, adequate models are needed for describing—and automated instruments for measuring—the mechanical properties of new materials like ceramics, cermet, and metallic polymers. Because of the lag in computer technology, we are not yet at the "state of the art" in hydro-aeroelasticity, dynamic problems, and (geometrically and/or physically) nonlinear problems.

The experimentally verified principles underlying the theory of non-elastic processes have provided a basis for a number of major developments and sometimes for entirely novel possibilities in the theory of plasticity. In particular, new approaches have been developed for certain types of processes, and comprehensive analysis of their mathematical formulations and of the required solution methods has been given.

A result of fundamental importance for the analysis of nonelastic stress-strain states is the establishment of a physical truth criterion for the corresponding boundary value problem, and there is every reason to believe that before very long the functionals to be used in the general theory of plasticity will be found.

Major advances have also been made in applied research. A recently developed theory of surface plastic flows is of wide use today in the analysis of metal working processes; the design of critical structures is provided with stability study methods suitable for bodies undergoing elastoplastic deformations. The general theory of plasticity has been extended to include anisotropic media. Exact solutions have been obtained for creep and plasticity problems involving twisted nonheterogeneous tapered pipes. Governing equations have been derived for polycrystalline materials for large curvature loading trajectories with angular points. A theory of thermoplasticity capable of describing nonelastically deformed metals subjected to intermittent non-isothermic loading regimes has been developed. Of wide use is the elastoplastic stability theory developed for complex loading regimes.

Deformation theory and flow theory, two traditional approaches to plasticity problems, have found wide application in recent years in the optimization of technological processes and in structural strength design owing to the availability of the computer software developed for these applications in Moscow, Novosibirsk, Perm', Gorkii, Tashkent, and other scientific centers throughout the country.

A subject of considerable importance in a variety of applications is the rheology of solid bodies depending on the rate of deformation for their properties. Basic research in this area is aimed at the development of finite deformation thermomechanics for elastoviscoplastic bodies susceptible to considerable loosening and/or damage accumulation effects in the process of loading. Although there are encouraging results in this field, only a few of the hereditary elasticity models thus far proposed find their justification for loading histories other than trivial (constant in time) ones. The development of general theories in this area of research depends on new ideas and further efforts for its success.

Increasingly more engineering design problems are being solved today by applying the concept of creep. A common practice is to employ simplified creep theories, the choice of a specific version depending on the ranges of the parameters of the particular process. Such an approach clearly places strong emphasis on the collecting and storing of the relevant experimental material.

Of the recent advances in the field of nonclassical models, the theory of growing viscoelastic bodies is perhaps one of the most important. The theory allows for phase transitions and aging processes in bodies under study and, apart from numerous technical fields, has applications in biology and medicine. The main difficulty encountered in the development of the theory was to establish appropriate conditions at the growth surface and at the first-kind transition boundary. A major conclusion to be drawn from the application of the theory to a number of technological problems is that the formation of a stressed state in a growing body is highly sensitive to preload growth rate and perhaps other parameters of the growth process.

Encouraging results have been obtained in the study and technological application of electroplasticity, superplasticity, and shape memory, even though no adequate theory of these effects has yet been developed. It is satisfying that a new academic institute has been set up (in Ufa, Bashkiria) specifically for studying these phenomena.

There is every indication that a general theory will be available in the near future capable of describing mechanical fields in solids in their interaction with thermal and electromagnetic fields. A theory of electroelasticity of piezoceramic materials has already been completed and is being employed in electronics and radio industry applications.

There are currently many research groups in the Soviet Union concentrating on fracture mechanics and on the strength properties of structural materials. During the past five years or so, a special emphasis has been placed on the behavior of structural materials under extreme loading conditions characterized by very high (or very low) temperatures; nonstationary thermal and/or force influences; sharp temperature variation; gaseous or liquid corrosive agents; radiation; intense and sustained loads, constant or varying in time. These factors may drastically change such strength-related characteristics as the number and size of stress concentrators (including cracks), the form of the stressed part, temperature stresses, and initial stresses (including those due to manufacturing processes), and clearly neither reliability nor design optimization requirements can be met without an adequate understanding of the physical mechanism of such changes. Over recent years valuable theoretical results have been obtained in the area and useful experimental methods developed for their verification, so that quite a number of sophisticated testing techniques are now available, some of which are unique and most automated and computer-controlled.

Of great practical interest are strength studies performed on materials and structural elements under conditions closely simulating operating regimes [27]. The focus of attention in this work was on high-stressed structural elements operating under hostile environments and responsible for the strength and reliability of an entire chemical, transportation, or power engineering installation. A survey of the results obtained is given in the series of monographs, titled "Fundamentals of Machine Design," which provides a design engineer with operational stress characteristics of machines and structures; parameters for calculating the service life of machinery; strength calculations performed for sustained, static, low- and high-cycle; and instantaneous fracture criteria. The series also provides calculation methods for mechanical engineering applications in power industry, transport, agriculture, metallurgy and other fields.

It is a characteristic feature of fracture studies that they are heavily oriented to industrial application. An important event worth mentioning in this connection is the establishment of methodological instructions for crack resistance measurements at various loading conditions [28, 29], a titanic work that took several years of sustained effort by many researchers in the Academy, State Standard Committee, and other institutions. It has been realized, in particular, that it is imperative in most strength problems to estimate as closely as possible the limiting states for crack formation and propagation processes with respect to viscous, brittle, and low-cycle fatigue criteria. The energy and deformation criteria for viscous, brittle, and quasi-brittle fracture have been appreciated as the principal ones for heavy-duty life and strength design.

The application of fracture mechanics has proved exceptionally fruitful in many diverse areas of machine engineering, particularly in power industry. Of particular importance to further progress in this field and undoubtedly a very successful form of cooperation between all those involved are, in my opinion, the conferences on the theoretical and applied problems of crack resistance, long-term strength and service life, regularly held under the joint aegis of the Academy, the Standard Committee, the Science & Technology Committee, and leading industrial enterprises of our country. Recent studies in fracture mechanics have provided a basis for the reliability, service life and strength evaluation of power engineering structural elements with a known defect concentration. Indeed the reconstruction, updating, and dismantling programs adopted by the Ministry of Power Industry and Electrification for the period up to the turn of the century are largely based on the strength data obtained for major housing elements from their defect concentrations and from their creep and crack-resistance properties; as well as on the system of primary and auxiliary criteria developed for the in-process estimation of jet-turbine service life.

Changing to a somewhat less optimistic note, it is certainly detrimental to the field of strength mechanics and to material science in general that scientific and applied research in this area is being conducted on the intra-sectorial or, shall we say, parochial basis. At a time when reliability and trouble-free service problems have become the order of the day, it is increasingly evident that rapid progress in fundamental and applied continuum mechanics, as well as the development and introduction of technological diagnostics, depends largely on the combined efforts of research scientists and engineers drawn from different branches of industry [30].

In the theory of plates, shells and rod systems, the major lines of activity over the past five years have been as follows: dynamics; geometrically nonlinear phenomena (including stability problems); complex-structure shells; and the effects of various environments and of interacting internal fields.

In the field of dynamics, progress has been made in the study of vibrations and nonstationary processes occurring in thin-walled constructions interacting with deformable obstacles and deformable media. Some data have been obtained concerning the effects of physical and geometrical nonlinearities on the behavior of such constructions under complex combinations of external influences and complex inter-component intractions. It has proved possible, further, to determine the manner in which the eigenfrequency spectrum depends on the shape and area of the middle surface, as well as on the perimeter and the fixing conditions of the shell (plate) being studied.

In the study of the stability properties and of the nonlinear behavior of shells, previously known stability criteria have been developed and generalized. This enabled the formulation of a number of new problems in which a complex rheology of a material and a highly intensive influence may be allowed for. These problems rely on the aid of computers for their solution and, consequently, new computational techniques of various levels of accuracy have been developed based on the final-element method and final-difference approaches.

Further developments in this area include:

- the derivation of governing equations for interacting fields of various natures (mechanical, thermal, piezoelectric, and electromagnetic fields);
- new mathematical models for complex shell structures exhibiting local features of various kinds (cracks, cuts, reinforcement, lamination, perforation, etc.);
- theories for treating piezoceramic and magnetoelastic shells;
- a number of results concerning the behavior of soft shells in fluids.

It is of course a well-recognized fact that technological progress in mechanical engineering, aviation, rocketry, ship-building, and civil engineering is to a large degree controlled by the materials available, and that new materials must combine very high mechanical properties, such as strength and stiffness, with low weight and low cost. These conflicting requirements are often met by the so-called composite materials utilized in the form of thin-walled structures of properly chosen geometry. Such structures usually include cavities or reinforcements of some kind and must possess spatial periodicity to satisfy structural, technological, and weight requirements. Because of the inherent non-homogeneity of such systems, it is usually impossible to apply numerical methods directly to their study, and to circumvent this difficulty an asymptotic method has been developed using the ratio of the spatial period to the dimension of the body as a small parameter. The method averages processes occurring in periodic structures and makes it possible to reduce a three-dimensional thermoelasticity/heat conduction problem to a "smeared out" shell (or plate) model whose elastic and thermophysical characteristic are subsequently found by solving certain local problems on the scale of the periodicity cell. Models thus obtained prove very efficient for determining stress, strain, and temperature fields in composite shells and plates with different geometries and reinforcements (stiffened and orthogonally stiffened plates and shells; grids; perforation). The method is sufficiently accurate to yield the "microscopic" structure of stress and temperature fields in the material and so provides a solid basis for local-criterion strength design.

Among substantial contribution to the field of composite technology also are the theories of plasticity and nonlinear viscoelasticity of composite materials, and numerical computational methods for (three-dimensional) elastic, viscoelastic, and elastoplastic problems involving these materials [31]. These results have permitted the analysis of stress formation processes in these materials with attention to the chemical and thermal shrinkage of a binder; the initial tension of the reinforcing layers; and the filtration of the binder through the filling agent. Using statistical fracture models (accounting for damage accumulation processes) and applying methods available for the stability studies of such systems, a theory of the optimum design of load-carrying elements has been developed, as have engineering methods for strength stiffness, stability, and vibration design.

Stochastic analysis and optimization methods as means for improving the reliability and performance quality of structures and machinery is one further important line of research in the mechanics of deformable media. One of the main objectives in this field is to master the mathematical simulation and experimental modeling of engineering structures in terms of strength, stability, stiffness, reliability, service life, material consumption, and similar characteristics, and it is safe to say that in fulfilling this objective Soviet researchers are at the state of the art or even ahead of their foreign colleagues in some respects. It should be noted, however, that progress in this field could be more rapid if better computer hardware and more advanced experimental techniques were available. And, of course, there is increasing demand in this field—as well in many others, for that matter—for a closer cooperation between various research centers, both in the Academy and in industry.

4. APPLICATION OF MECHANICS TO TECHNOLOGICAL PROBLEMS

In addition to the three traditional fields we have thus far discussed, it has been found necessary by the previous Congress to set up one further: the fourth, aimed at dealing specifically with technical applications of mechanical science. As is the case with building mechanics, hydrodynamics, and some other departments of mechanics, technological mechanics has already taken its shape and has its own scope of investigation and its own mathematical models. Indeed it represents now a major line of research in mechanics without which little if any technological progress seems to be possible. The newly established section is expected to rely on fundamental and applied research in the field of mechanics in the search for solutions of the top-priority development issues in the nation's industry, primarily in mechanical engineering and modern technologies. Over the past several decades, it thas been recognized both nationally and internationally that further technological advance crucially depends on such major problems as the reliability, strength, longevity, and safety of structures and machines; development of new materials; industrial use of composite materials; control of mechanical motions involved in chemical processes; robotization of production processes. The progress in these fields affects engineering design on both a small and large scale and, in fact, controls the effectiveness and the possibility itself of the incorporation of new technologies in mechanical engineering.

Reliability is quite a sophisticated concept dependent not only on longevity and trouble-free performance but also on maintainability and keeping ability, that is, on the properties which determine the effectivity of a technology and the preservation of the product's quality under given operating conditions. Unfortunately, our industry has suffered for too many years and is in fact suffering now from costly and unremitting downtime due to the sudden failures and subsequent repair of many species of mass-produced machinery, caused primarily by inadequate maintenance. The reliability, longevity, and safety of machinery are becoming increasingly dependent on the skill and judgement of personnel and the performance of control systems. It is a bitter truth of our time that an error of a human operator in charge of a large engineering system may affect not only the production process but, by far more important, the health and life itself of millions of people—now and in generations to come. Experience teaches us that the human factor and ergonomics should be given more and more attention now in dealing with new technologies, from the design stage to the moment of its introduction in operation [32].

It has been estimated that in the U.S. navy and missile troops upwards of 50 percent of military equipment failures can be blamed on the human factor; in the air force and civil aviation, crewmen are responsible for more than 70 percent of the total of flight accidents. It is a result of the sad statistics that 150 U.S. companies have now employed ergonomics specialists for the study of "man-machine-environment" relationships, an initiative that Soviet industrial enterprises might do well to consider emulating.

The following seem to be the top priority directions to be taken in the field of the reliability, service life, longevity, and safety of complex technical systems:

- the development of the scientific basis, early diagnostics methods, and technical facilities for minimizing the probability of catastrophic breakdowns of one-of-

a-kind machines and structures, particularly in nuclear and thermonuclear power stations;
- the development of the physics and mechanics of advanced reinforcing techniques;
- the development of standards and other technical documentation on design methods and laboratory, bench, and full-scale tests, on the basis of research institutions, design offices, and the Intersectorial Science & Technology Complex "Reliability of Machinery";
- further extension and widening of intra-COMECON cooperation in the field.

As of now, the principle features of R & D activities in the fields of service life and reliability are as follows:

- standardization of operating conditions and attention to all structural and technological aspects of a problem as necessary requirements in the reliability/safety design of heavy-duty and critical machinery;
- mathematical simulation and optimization of technological processes (plastic metal working, heat treatment, electrophysical machining methods, wave technology, etc.);
- service life design and longevity prognosis at the damage accumulation and crack propagation phases;
- statistical approach to the reliability evaluation and service life control for individual machines and for systems of machines;
- multicriterial approach to the examination and design of the load-carrying capacity of welded metallic constructions, of corrosive and mechanical strength, wear resistance, and contact endurance.

Major practical applications of the results obtained are in the design of flying vehicles, nuclear power reactors, agricultural machinery, and chemical and metallurgical equipment.

One major area of research in recent years has been the technological mechanics of composite material structures. The main aspects of the problem are: (1) the development of models for the formation of (initial) technological stresses and strains; (2) the description of limiting state surfaces, and (3) the search for advanced quality control methods for use at the manufacture stage and under operating conditions. The efficiency of ribbon-filter reinforced composite materials has been demonstrated both theoretically and experimentally. Solutions have been found for a number of important reinforcement optimization problems, and new design methods have been developed for composite-material structures of complicated geometries (including meshed ones).

There are some areas, however, in which research work does not meet the present-day requirements in terms of reliable methods and computer software available for calculating technological and operational stress and strain fields and for optimizing technological and structural parameters. Such is the case, for example, with normalization of defect concentrations permissible in typical composite-material structures at various static and dynamic loading regimes. Nor is the situation satisfactory with fatugue fracture problems or with the problem of choosing a suitable composite structure for materials interacting with physical fields of various nature when under operating conditions. The reason for this is that R & D work on composite materials is so far divided between numerous research centers and different branches of industry. It is the imperative today that results of individual research be integrated and cooperative efforts initiated on large interdisciplinary problems whose solution is essential to composite materials engineering. Research-and-production associations, with their powerful research departments and their pilot production facilities seem to offer the most efficient way to speeding up the progress in this field.

Mechanics of robots is one further active topic in today's mechanical science [33]. Research work in this area provides the basis for designing flexible manufacturing systems and robotized complexes, whose functional characteristics such as precision, speed, reliability, and cost effectiveness are in large part determined by the mechanical properties and by control systems of the robots involved. A robot's precision, for instance, depends very sensibly on its elastic properties, type of drive, friction forces, and the nature of dynamic processes involved. The investigation of robot dynamics in terms of the above characteristics is particularly important for the development of the so-called precision robot, which is a high-speed, low-metal content and, as the name suggests, high-precision device attracting much research effort today in Japan, the United States, and West Germany.

In the Soviet Union, much attention is being given to the study, modeling, and motion control of robots with elastic elements. In particular, an effective general method for analyzing elastic-element mechanical systems has been developed based on motion-separation and motion-synthesis asymptotic methods. A set of programs is available for the computer simulation and dynamic precision evaluation of elastic-element manipulators. A method for evaluating elastic properties of industrial robots has been proposed and successfully tested. Useful data have been obtained concerning the effects of external vibrations or robot dynamics and robot precision. Methods have been developed for employing resonant and energy-recuperation drives, for damping elastic vibrations, and for enhancing robot precision.

The application of optimum control theory to electromechanical robots of various kinematic designs has resulted in a number of control laws optimized for speed and power requirement. Experimental tests have shown that these laws considerably reduce robot operation time, and that their realization gives hope of increasing the efficiency and cost effectiveness of microprocessor-controlled industrial robots currently in production use.

In the field of instrumentation, new techniques for measuring mechanical parameters have been developed, enabling robot motion control and the interaction of machines in the framework of flexible manufacturing systems. New principles of developing robotic instrumentation have been suggested and partly realized by employing currently available robots equipped with microprocessors and with devices for measuring mechanical parameters.

Practical realizations of the above achievements in the field of robotics include pilot robots and pilot robotized systems for use in welding, assembly, machining, rotary conveyors, and in instrumentation applications.

One recognizes four main research areas in robot mechanics which are in need of intensification. These are:

- the search for new application areas for robotic systems, primarily in complex technological processes;
- the enhancement of the reliability of industrial robots and improvement of their performance in terms of precision, speed, power, and material savings;
- the development of new control systems, effective software, and robot sensing systems;
- the upgrading of research into the mechanics and control of walking machines.

It follows from the above discussion that further progress in both basic and applied mechanical research requires that some new directions be taken in the field, such as the development of integrated approaches in continuum mechanics, and the numerical simulation of mechanical processes and systems on the basis of the fundamental physical, mechanical and chemical laws. Problems of ever-increasing complexity are arising in biomechanics in connection with the growing role of "man-machine-environment" aspects of modern life [32]. Entirely new practical approaches should be developed for the problems of teaching mechanics in educational institutions and of effectively using computer machinery potentialities.

The technological aspects of mechanical science are receiving considerable application in material research aimed at obtaining high-quality materials with "tailor-made" structures and properties, with special emphasis being placed on valuable materials like single crystals and certain biological substances usually produced in small quantities.

Although solid-state properties of most materials of interest have been understood fairly well by now, we are still far from a full understanding of the mechanisms which govern defect formation processes occurring during the production of a material from its liquid or gaseous form. This means that special attention should be given to hydromechanical and heat/mass exchange phenomena concomitant to material production processes, which is, of course, a formidable task in view of the diversity in material production methods, and physico-chemical and thermophysical properties of materials, and their intended end uses.

This line of research is particularly important in electronics, where single crystals and semiconductor structures of utmost homogeneity and purity are required for computer, laser, telecommunication, and other applications; in biotechnology, for separation purposes in work with medicinal and biological substances; and in continuous rolling, casting, and in numerous special-purpose metallurgical processes involved in the production of polymers and composite materials. The motions involved in the above processes classify into two categories, forced motions and natural (or convective) motions. In the former case they are caused either by some mechanical action (rotation, vibration [34], directional flow in a channel of some type, jet motion, mixing process) or by a magnetic field. Convective motions may be due to aerodynamic lift, or to surface-tension gradients arising in a non-isothermal and non-homogeneous (in composition) medium, liquid or gaseous. The velocities involved may vary from the speed of sound well into the ultrasonic region.

Problems of such complexity depend mainly on physical modeling and mathematical simulation for their answer, and accordingly a number of effective simulation and modeling methods have been developed which are being successfully applied to crystal growth technology design.

It will be noted that further progress in the study of hydrodynamics and heat/mass transfer in the context of materials research is hardly conceivable without a thorough understanding of such things as material formation processes; vibrations; electromagnetic, gravitational and thermocapillary forces; plasma phenomena; and perhaps some other factors which, separately or in concert, influence the quality and properties of the final product.

Special attention should be given, in my opinion, to the application of mechanical science to the ever-developing field of pulse technology. It seems worthwhile mentioning in this connection the theory of shock wave compression, the theory of high-speed deformation, and cumulative theory [35], which combine achievements of fluid dynamics, hydrodynamics, and continuum mechanics and represent remarkable examples of interdisciplinary research. Owing to recent advances in basic and applied research, superhigh pressures, speeds, and temperatures have become more or less a matter of routine in modern material-manufacturing and material-machining technologies, and although intensive pulse loads may be identical to ordinary static actions in some respects (in terms of pressure or temperature, for example) they should still be approached from an entirely different viewpoint if they are considered—as they must be—in the context of shock wave propagation picture.

What makes the difference between static pressures and shock waves arising from detonation, magnetic field pulses, electrohydrodynamic effect, etc., is that with a shock wave, it takes only a few ten-billionths of a second for the medium to assume a pressed state, and the whole shock-pressure state lasts for no longer than a few millionths of a second. This results in the atoms of the material drastically intensifying their motions, and if the material was originally a crystalline one, its crystal lattice strongly deforms, giving rise to a large concentration of defects (dislocations, etc.) and probably to a greater strength and lower plasticity of the material.

It is a well known fact that many physical and chemical processes are slowed down by the low rate of diffusion process. This latter is drastically accelerated by the intense deformations and enormous defect concentrations produced by an impact load, and so are, of course, physical and chemical reactions.

Some processes, like the formation of a new phase structure (diamond synthesis), may take place during the short period of the impact compression, that is, as rapidly as in a few millionths of a second or so. Some other processes among those stimulated by a shock wave actually occur after the wave has gone, as a consequence of transformations it caused in the material. Some metal cutting and metal welding techniques, for example, are based on the principle that a high-speed collision of two metallic bodies may under certain conditions result in a liquid-like flow of a metal.

The above features of impact load make it possible to change the properties of a material in a prescribed direction and to realize material transformations unheard of in the framework of conventional techniques. Advances in explosion physics [36], in gas dynamics and hydrodynamics, in metal physics and the mechanics of deformable media have provided the scientific basis for such innovations as explosion techniques for welding, cutting, stamping, and strengthening metals; explosion coating techniques; pulse extrusion and pulse flaring techniques for critical parts; etc. The power source employed in each particular case (explosives, powders, gas mixtures, pulse magnetic fields, intensive electrical discharges, high-speed strikers, etc.) determines the features of the design and the equipment required (blast chambers strong enough to allow the on-floor use of explosion technologies; detonation guns; powder presses and hydrodynamic presses; and a wide variety of electric charge machines for stamping, welding, casting, purification and so on).

The subject of impact interaction has already been applied to numerous practical and situations ranging from space projects (meteorite issue in the Veha mission [37]) to traditional technologies like oil production or cotton picking, and gives every indication of great future potential. Again, it is essential for the progress in this field that emphasis be placed on the mathematical simulation of complex multifactor phenomena like deformation, degradation, heat release, and structural transformations resulting from pulse loads of various nature.

One of the most important areas of research in applied mechanics is the hydrodynamics of chemical production processes. It is becoming increasingly clear today that progress in many of the leading national industries crucially depends on the development and introduction of large-scale technological processes. But it is clear on the other hand that there is much room for improvement in many of the large-scale production technologies that already exist, and one way of doing this is by using the achievements of aerodynamics and the mechanics of nonhomogeneous media. In many of the chemical reactors now in operation, the flow of the raw material through the catalyst layer is strongly uneven in terms of velocity, which is greatly detrimental to the efficiency and productivity of the reactors. A kinetic theory of dispersive and biphase systems, developed within the framework of the Academy of Sciences, has provided a basis for a number of practical recommendations concerning possible improvements of certain chemical technologies. In particular, a new method of arranging catalyst granules has been advanced, not only reducing hydrodynamic resistance (by a factor of about ten), but also increasing the yield of a reaction. The method is now being used in a number of chemical plants across the country in the production of anhydrides, carbolactam, and methane.

The application of classical mechanical methods to the analysis of motion in dispersive and biphase media has provided useful insight into the behavior of large systems of hydrodynamically interacting particles and bubbles, and made it possible to bring out some of the determining effects in a number of important technological processes [38]. In particular, the results obtained point to the possibility of some productivity increases in barbotage, fermentation, and ore-dressing flotation technologies.

The scientific results obtained in this country in the fields of linear [39] and nonlinear [40] vibrations are being used in one way or another in many of the vibration machines and technologies based on wave principles.

To end this section, it must be admitted, however, that there is a definite lack of cooperation between research groups engaged in technological application of mechanics, and it would be too optimistic therefore to say that the achievements I have thus far outlined are being introduced and utilized as rapidly and effectively as they should.

5. MECHANICS IN MACHINE BUILDING

It hardly needs repeating that advances in the machine-building industry are of vital importance to the technological progress and welfare of a society, and it is only too natural that the "major directions in the economic and social development of the USSR for the period of 1986–1990 and up to 2000" place a strong emphasis on the development and upgrading of machine building in our country.

In the last few years remarkable progress has been made in the industry as witnessed by the ever-growing importance of automated manipulators and by the introduction of a large number of new high-technology machines and processes.

There have been many impressive developments in recent years in such fields as the nuclear power industry; high-output power-generating installations; gas-pumping systems; electroslag remelting and vacuum refinement technologies: high-speed rolling mills; flexible manufacturing systems; numerical control machine tools; automated machine tool lines; and electron-beam, laser, and plasma microprocessor-controlled techniques.

State of the art machinery is being produced for the agro-industrial complex, and the chemical, transport, light, and food industries. The service life of motor cars and of metal-cutting machine tools has increased by a factor of about 1.5 to 2 over recent years.

Among the recent developments that are finding increasing application in machine building, mention should also be made of high-technology methods for improving the reliability of machinery, notably laser, ultrasound, explosion, electron-beam and some other welding techniques. New and effective corrosion protection methods have been devised for a number of metals, and new corrosion-resistant steels and alloys have been developed.

Automated and Optimum Design

The present-day design engineer is faced with problems which are by far more complicated than those his predecessors of the not so far past had to solve. The development of new-generation machines [41] makes the designer keep in close touch with major achievements in science, design, and technology, and makes economic, ecological, and man-protection factors [42] major considerations in design work. The problem of quality should already be addressed at the design stage of a project, when it is still possible—at least in principle—to consider all the aspects of the future machine and to satisfy as many design requirements as possible. What complicates the task of a designer is that many of the design requirements conflict with each other (minimum weight and high reliability, high speed and minimal dynamic load, low cost, longevity, etc.) and it is therefore necessary to choose structural, kinematic, dynamic, and operation design variables in an optimum way with respect to some predetermined criteria. The most obvious—and traditional—way to attack this problem is by performing necessary calculations for different sets of initial (input) parameters, which, however, is a totally impractical approach for a large number of such parameters. If, say, we have 10 parameters, each assuming 10 values in the process of calculation, there will be 10^{10} different variants to consider, a problem which is beyond the reach of even the most powerful computers available.

To choose the optimum design parameters it is necessary to have computation formulas or, more conveniently, a computer program capable of describing the behavior of the machine and to evaluate its quality characteristics for any sets of the parameters of interest.

The majority of high-tech machines operate in dynamic, often very intensive, regimes. This is true both of technological innovations (flying vehicles, gyroscopes, vibration machines) and traditional technologies (mining machines, rolling mills, turbines, internal combustion engines, etc.) and implies high speeds, lower weight, random variation of resistant forces, and some other "heavy-duty" conditions.

Quality criteria are calculated for a given set of the object's parameters using a computer program. Each calculation is termed a trial. The choice of trial points in the multi-dimensional parameter space is carried out by a generator programmed in such a way that the point's projections on the coordinate axes are different from each other and distributed quasi-uniformly on the axes. If the number of trial points is N, then each of the parameters is made to take N different values (located uniformly over the range of values of the parameter), so that the total number of the trial values of each parameter is equal to the number of trials. This is a highly effective approach in the sense of guaranteeing a uniform survey of the entire parameter space even for a relatively small number of trials. In the trial tables constructed for all the values of each criterion, the quality criteria are placed in descending order, that is, the first of the models comes as the best one with respect to a given criterion.

Experience shows that no machine or structure can be optimized with respect to all the criteria simultaneously, and the analysis of the trial table makes it possible to choose a compromise solution to the problem.

The approach we have outlined thus provides the designer with a basis for a correct formulation of a multi-criterial optimization problem and enables him to consider simultaneously many conflicting criteria in solving the problem. As a result, a qualitatively new design method has been developed, based on the designer-computer dialogue principle.

The performance of the system being projected depends on r variables α_1, α_2, $\ldots$, α_r, which represent the inertial, dissiputive, elastic and other parameters of the system and may be thought of as coordinates of the point $\alpha = (\alpha_1, \alpha_2, \ldots, \alpha_r)$ in the r-dimensional parameter space. These coordinates enter the differential or final-difference equations governing the dynamics (kinematics) and perhaps some other aspects of the system's functioning. The constraints of the problem usually fall into two categories: the parametric constraints

$$\alpha_j^* \leqslant \alpha_j \leqslant \alpha_j^{**}, \quad j = \overline{1, r} \tag{1}$$

and the functional constraints

$$c_l^* \leqslant f_l(\alpha) \leqslant c_l^{**}, \quad l = \overline{1, t} \tag{2}$$

where the quantities α_j, α_l^{**}, and c_l^{**} are specified *a priori*, and the $f_l(\alpha)$ are the functional constraints dependent on α.

The parameters of the system are not allowed to vary in the process of optimization, and the problem to be solved is to find the parameter D of the set of allowable points. The analysis of the parameter space proceeds in three steps, as shown in Fig. 1.

The above method allows designers and customers to consider as many local criteria as is necessary in the design process and holds particular promise for progress in mechanical engineering design.

The growing diversity of available equipment, materials, and technologies makes modern machinery design, and particularly the optimization of design parameters, an increasingly complicated and labor-consuming process. It has been estimated, for example, that the volume of scientific and technical documentation is doubling every 8 years; the nomenclature of products, every 10 years; and the complexity of machinery in terms of the number of their parts is doubling every 15 years. These and some other factors related to the design process account for the fact that the volume of design work is now increasing about tenfold every ten years, and this trend, not unnaturally, stimulates continuous improvement of design methods, collection and organization of normative and reference documentation [43], and more distinct specialization and labor division in the design process.

It must be admitted that the current state of the design art lags far behind what is necessary in today's mechanical engineering. The average time spent for

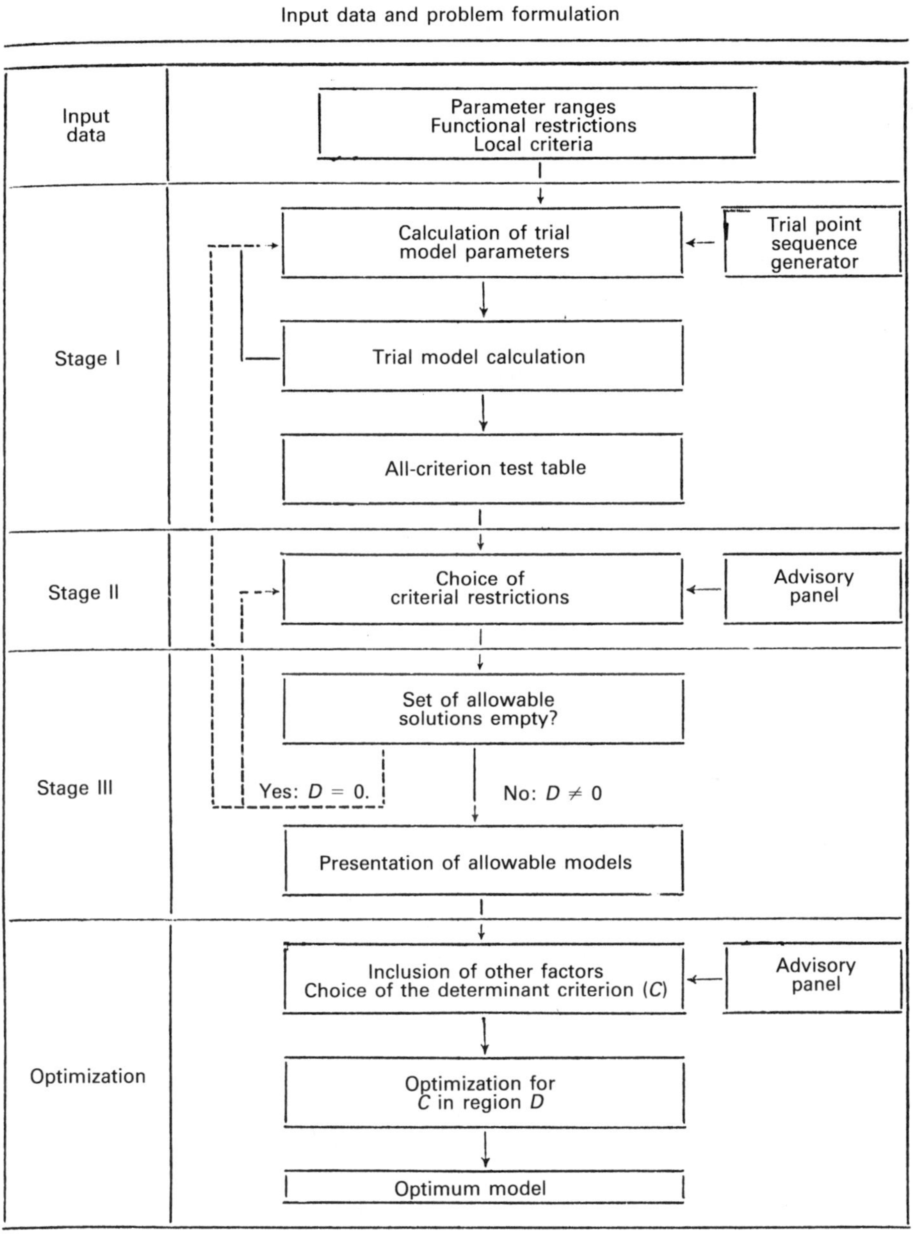

Figure 1 Flow chart of the optimum design process (interactive formulation/solution system)

designing an average machine is continuously increasing, and as time goes, more and more people are becoming involved in design work. Furthermore, only some 10% of the total engineering design time is being spent on what may be called creative work; the reminding 90% account for search for necessary information, standard calculations, recording of the results, and similar non-creative work which can equally well—or better—be done by computers and computer-based automated devices capable of processing symbolic and graphic information. What is needed is, of course, a wide-ranging modification and automation of design technology. There are five major directions in which this task is being carried out:

1. Development of both hardware and software for automating the operations of input, output, transfer, processing, and distribution of symbolic, graphic, and perhaps of some other types of design information [44];
2. Automation of design calculations, including optimization problems as discussed above. Automated calculations will have to rely on man-machine interface based on the use of dialogue regime and informal euristic procedures (which implies large computers, of course), and it can be safely predicted that engineering thought in this area will always go ahead of the possibilities of computer technology;
3. Development of integrated Computer-Aided Design systems [45] capable of combining large computers, for solving a problem proper, and smaller computers, for automated processing of design information. A CAD system should be only a part of a wide and ramified computer network whose size will be dictated by the complexity of a design problem being solved;
4. Development and realization of fundamentally new CAD principles, such as [46]:
   - integration of the entire design process, starting from the description of the projected system and ending by issuing design documentation in the form of process sheets, sketches, and CNC machine programs;
   - wide use of the module structure principle for facilitating the possible modification and build-up of the projected system;
   - sophisticated operating systems, appropriately organized design language hierarchy, and extensive use of the data base concept;
   - control systems capable of choosing and generating special-purpose applied programs best suited to a given design problem;
   - simulation and control at various design stages;
   - copying ability and easy change-over to another computer type;
5. Revision of all traditional design methods in the light of new possibilities offered by computer technology and design automation techniques.

CAD systems based on standard complexes or, to use the accepted term, on the concept of a, workstation [47], may differ either in the capabilities of the base computer and its peripheral devices, or in the number and possibilities of applied programs developed by or for users. Linking such a system to large computer complexes greatly increases its potentialities and makes it possible, in particular, to effectively solve optimum design problems.

The main features of an integrated CAD system are the automation of engineering design, automation of the pre-production preparation process, and data bank. The control of the design process is carried out by the computer operating system or, more precisely, by its monitor program, or monitor for short.

The introduction of workstations and similar automated design equipment has made it possible to solve a number of important design automation problems. The top priority problems to be addressed today by CAD specialists may be divided into the three following groups [48]: (1) gradual formation of what may be called a design softwave store, by further developing the concept of unified information at both the data level and the programming language level; (2) development of cheaper, more powerful, and more reliable multi-access hardware, and (3) development of the theoretical aspects of design automation. Those engaged in the development of new technological processes should be provided with high-volume information-and-reference systems for the purposes of calculating time rates, machining regimes, loading regimes, etc., and with algorithms for optimizing technological characteristics of the parts involved and for evaluating the design with respect to technological criteria. One of the most complicated and urgent problems in the field—of particular importance for one-of-a-kind and small-batch production—is the adequate description of the geometry of the object being designed, the subsequent transformation of this information, and its transfer to automatic drawing devices and to systems engaged in the development of CNC programs.

Robots and Flexible Manufacturing Systems

Of what has been achieved in mechanical engineering in the past two decades or so, much has to be attributed to the wide introduction of CNC machine tools [49].

The use of CNC, primarily in low-volume production, has made it possible to automate the replacement of tool bits, the movements of machine tool actuators, and some other aspects of part machining processes, but truly unattended production also implies an automated retooling for a new type of product, the high reliability of the technological equipment, and productivity control.

The geometric adjustment of a machine tool can be automated by using an adaptive control system based on the correction of the control program of the machine. The machine tool is equipped with a measuring instrument and a measuring head (Fig. 2), and its control system is provided with a correction block. At the first stage of the adaptation procedure a trial part is machined ("trial run"), and then the measurement of the machined surface of the part provides the information necessary for correcting either the control program itself or the geometry of the cutting tool.

In an alternative approach to adaptive control the technological tooling process becomes the object of automation. In this case the information on the dynamics of the machine tool is provided for the control system by a set of sensors

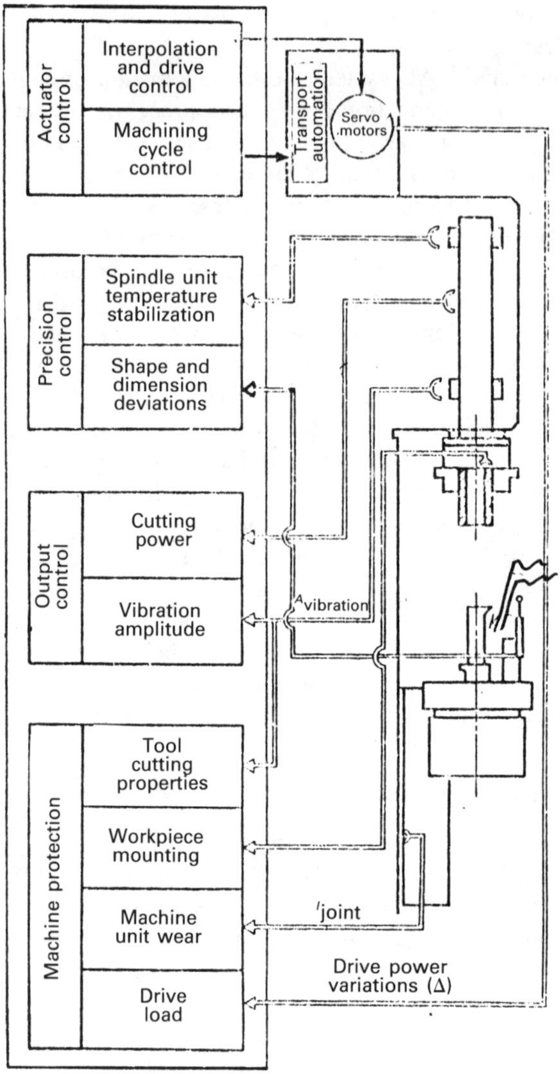

Figure 2 Flexible manufacture processing cell

measuring torque moments, vibration, and similar characteristics. The machine tool performance is diagnosed with respect to such parameters as the temperature of the lubricating oil, dynamic errors in servo drives, temperature regime in moving joints, etc. Machine tool diagnostics is a well-developed field which has already found many applications in industry. Many adaptive systems are now at the pilot testing stage or have already started being applied to manufacturing processes.

Robots and Manipulators

The introduction of robots and robotic systems is rapidly changing the face of manufacturing throughout the world. Combined with advanced machining technologies, robotics offers extremely high potential in increasing productivity and reducing direct labor in production processes, and represents in fact one of the major directions in modern technological progress.

Over the decade of the 1970's, the world's robot population grew by a factor of about 40 and by 1980 there were about 170,000 industrial robots at work worldwide. Marked increase has also occurred in the range of applications of robot technology. Although stamping, casting and machining have been and remain the areas of the most intensive robot activities, gradually the robot starts finding its way into welding, spray painting, circuit manufacturing, part inspection, material handling, and other value-added or non-value added processes in low-, medium-, and high volume batch production.

The majority of today's industrial robots have no means by which to accept information about their environment, nor about the object of manipulation in particular. So a robot must operate—of necessity—in strict accordance with some prescribed set of commands and, no less important, the environment itself must be arranged as rigidly as possible to confirm with the robot's motion, both in terms of space (accurate positioning) and time (precise synchronization). Accordingly, a changeover to a new type of production often implies a need for special resetting equipment and for some period of resetting time which, of course, makes the robot less attractive in terms of flexibility.

An important step toward overcoming this disadvantage can be taken by equipping the robot with special sensors that provide information about the progress of the robot's interaction with the environment and make the robot readjust its program automatically to external conditions to be able to perform its task properly. This type of robot is known as the adaptive (intellectual, "smart") robot, a device capable of functioning not only in a predetermined but also in a constantly changing situation.

A necessary part of every industrial robot is a built-in computer whose function is to process the information coming from the system of sensors. The algorithms it employs in doing so may vary in complexity depending on the specific task of the robot and the volume of the information in question. The algorithms may be very simple if the robotic mechanisms are actuated by a few one-bit (or yes-no) signals. In most cases, however, a robot computer processes huge amounts of multilevel information (including, for example, machine vision signals) and must be taught to reject low-valued information, to analyze and correct the mathematical models of the robot's mechanical system, and to perform a vast amount of calculations and information transformations before finally producing control signals on the output. An important point to be made is that for the robot to function properly, clearly its mechanical system must follow the control system, not the other way around, so the latter must operate (at least) somewhat faster than the former ("real time" control). That adaptive control is the future of robotics is a fact which is well appreciated by the world engineering community, and over the last two decades much work has been done on the development of this concept and its introduction into practice.

By 1990, it is projected that about a half of the total number of industrial robots will have adaptive elements, of which number no less than 50% will be equipped with computer vision. Since robots are mechanical devices after all, clearly very much attention must be given to purely mechanical problems in carrying out adaptive control research. Indeed, the entire information network of the robot and the methods by which sensor information is processed and transferred, are, in the final analysis, only a means of achieving a proper functioning of the robot mechanical subsystem. The introduction of adaptive control poses serious requirements on both the mechanical precision and the information-processing abilities of robot systems. Much work must also be directed to the development of the robot drive systems, where reduced power requirement, change to the electrical drive, and minimization of size are among the main objectives.

The production of robot sensors of various types is now growing at a rather high 10% annual rate throughout the industrial world; for computer vision systems the figure is as high as 50%. Increasingly more attention is also being paid to microprocessor-based control technology. By 1995, an estimated 96% of all industrial robots will have computers in their control loops. This, of course, implies much research effort, and large-scale capital investments on computer software development programs, but the prospect of labor cost savings leaves no doubt about rapid payback not only in machining industry, but also in aerospace technology, agriculture, light industry, oceanography, and other fields.

In recent years, much research has been done in academic and industrial scientific centers on both conventional and adaptive robot technologies. The concept of the dynamic separation of motions is perhaps the most noteworthy achievement in the field of robot mechanics. Ordinarily, the dynamics of the robot actuator is governed by a coupled multidimensional system of nonlinear equations. Use of the separation concept reduces this system to a number of one-dimensional subsystems, thus considerably simplifying the control of the robot, and hence reducing power requirement, and enhancing the accuracy of the robot's performance. Important results have also been obtained in solving what is known as the inverse kinematic problem. The term refers to a situation where the current position and orientation of the robot gripper must be recalculated into generalized coordinates to be fed into the control system. From an analysis of typical actuator configurations, a class of configurations has been found for which an explicit solution of inverse kinematics is possible. This result greatly simplifies real-time robot control and is therefore of great significance to the development of adaptive control computer software.

Results obtained from the study of closed mechanisms, including parallel topology systems, may be very helpful in improving the accuracy of robot technology in assembly and similar high-precision robot applications, and provide a basis for designing both robot devices and robot-testing facilities.

Work is also continuing on the development of conventional (non-adaptive) robot technology. The two major directions in this field are the design of regime-optimizing computer software and the calculation of manipulator dynamics for robotic structural elements with known elastic properties. Among the achievements in this area I would like to mention new (prototype-ready) high-speed low-power requirement cycle robots and the regime optimization scheme (for speed) obtained for a number of robot systems currently in use in domestic industry.

Regrettably, the USSR's robot technology has not yet achieved the state of the art level in terms of precision and reliability, which are, of course, the two main factors in determining the efficiency and effectiveness of the future robotized industry. To overcome this problem, use should be made of the robot testing, robot diagnosis and robot inspection techniques that have been developed in recent years. Another important direction should be the application of computer-aided design (CAD) technology. The introduction of domestic computer graphics systems has intensified robot-oriented CAD research work and led to a number of tangible results in such areas as manipulator systems, robot hydropower drive system, and the robot/environment interaction problems in assembly and machining applications.

Use of robot labor in this country is at present limited to cycle- or position-type robots and to relatively simple operations such as stamping, machine loading, and spot welding. The equipment of robots with adaptive control is, to risk repetition, the only way to widen the range of applications for robot technology. Application of robots to assembly, arc welding, spray painting, and other operations requires more sophisticated robot control systems, advanced computer software and the incorporation of tactile, force, force/torque, vision, and similar sensing capabilities into robot systems. Purely mechanical issues are, of course, of central importance in solving these problems, particularly in the development of tactile and force/torque sensors.

By now, a theoretical basis has been laid for developing sensor devices and control systems, and necessary software is to a large extent available; the development and application of the multiprocessor concept is perhaps the most impressive success in this area. The central problems now on the agenda are the development of high-level programming languages and the design of algorithms and programs necessary for securing adequate operation of all robot systems under the conditions of complex adaptive control environment.

Robot technology plays a central role in developing the so-called flexible manufacturing systems, highly automated production lines of the—hopefully—not so distant future, in which the human worker will completely, or nearly so, be removed from the shop floor. FMS is a totally revolutionary production engineering concept, implying an extremely high level of computer-based automation not only in the manufacturing process proper but also in such auxiliary services as warehouses and the plant transportation system. FMS offers increased productivity, easier redesign to a new product, improved quality and no, or a minimum of, direct labor.

The ideas and methods of robot technology are being successfully applied to the design of special-purpose transportation systems of wheeled, track-laying or walking varieties with a high cross-country capability. In particular, the modelling method has proved helpful in solving the problem of controlling such systems both in the autonomous regime and under the supervision of a human operator. Other noteworthy achievements in this field include the development of on-board control systems for walking transportation applications and new methods for reducing the power requirement of walking transport propulsive devices (Fig. 3). With these results available R & D work is now underway on large-scale walking vehicles capable of high-speed performance in swamps, lots of sand, mountain slopes, and similar rugged-terrain conditions.

A major area of development for robot technology is also the mining industry where intensive R & D work is being conducted and the first encouraging results have been obtained in the automation of both productive and auxiliary processes.

To summarize, robots and robotic systems are now coming into widespread use in Soviet industry, and some major scientific and technological results have been achieved in this area. Increased efficiency and reliability of robots, reduced power and material requirements, and extention of application field are the main objectives to be pursued in the current five-year plan in the field of robot technology.

Vibration machines and vibration processes. The theory of vibration machines is an area where the leading position of Soviet mechanical science is internationally recognized. Except for a very few special cases, motion is always accompanied by vibration and the control of the latter in favor of the former is the principle underlying the design of vibration machines and the development of vibration-based production processes.

Although vibration technology is a relatively young field, it is rapidly evolving and is showing great promise for the mechanical engineering of the future.

Processing of dispersive systems in a vibrated fluidized bed is one of the earliest and most extensive areas of application for vibration. Conditions that take place in such a bed accelerate mass exchange processes and intensify physical

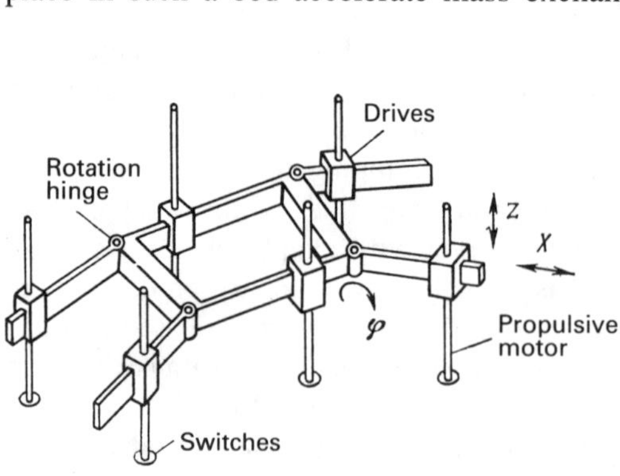

Figure 3 Walking machine, schematic

and chemical reactions such as burning, extraction, dissolving, leaching, as well as catalytical and solid-phase reactions.

Operations such as turning and drilling, abrasive treatment, crushing and grinding, rock and frozen soil breaking are also among the possible applications for vibration technology.

It should be noted that contrary to what their structural simplicity might suggest, the theory of vibration machines is rather complex, and the introduction of vibration technology has required and is requiring much theoretical research, including the extensive use of nonlinear mechanics and both digital and analogue computer technology.

The dynamic schemes with unbalanced vibration excitors (Fig. 4), are usually employed as theoretical models in the study of real vibration machines and, simple as they are, are often capable of providing very useful answers to practical design problems. The minimization of the dynamic load acting on the machine base is, perhaps, the most familiar among the problems facing an engineer in this field. For high-speed machines, a low-amplitude operation regime (low dynamic loads) cannot usually be achieved except in the trans-resonance region. Figure 5 shows the minimum occurring in the circular velocity of a vibrator in the vicinity of the resonance and a rather complicated transition region implying increased dynamic loads on the suspension. The theory of nonlinear vibration dynamics [50] offers effective methods for avoiding dangerous resonance regimes and for vibration technology design in general.

A second mass introduced into the dynamic scheme as shown in Figs. 4, 6 reduces the above-mentioned loads thereby improving vibration isolation and providing a high efficiency of the entire vibration process both in the resonance and the trans-resonance, or operating, regimes. The second mass may be utilized in this case as either a working member or a vibration transport means.

In situations where a certain number of conflicting requirements must be fulfilled, a third mass may sometimes offer a way to a compromise (see Fig. 4c).

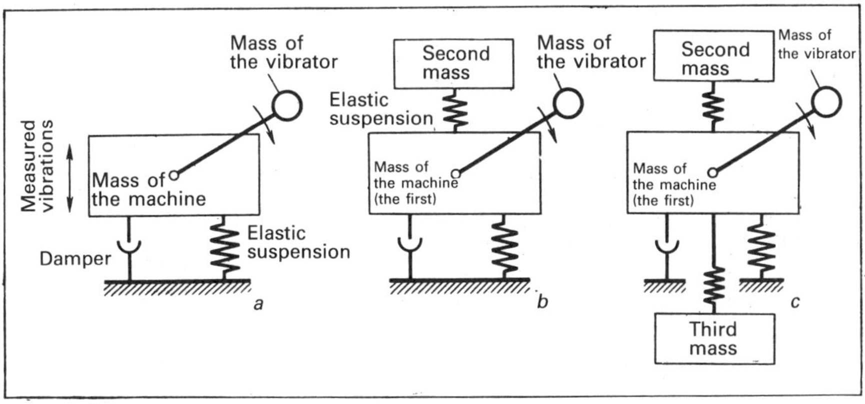

Figure 4 Dynamic models for one-mass, two-mass, and three-mass vibration machines (*a*, *b*, and *c*, respectively)

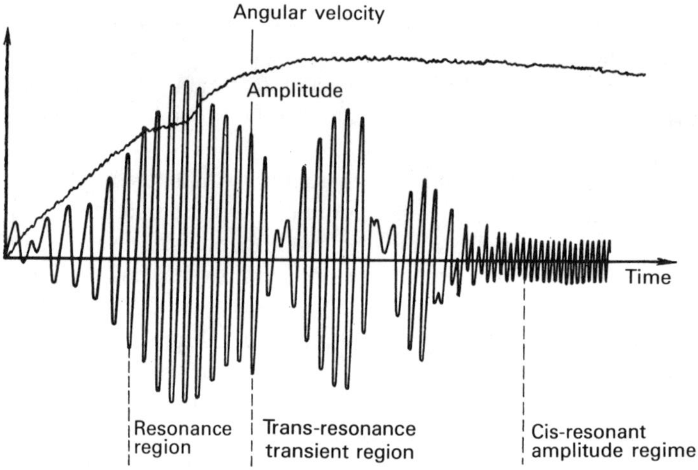

Figure 5 Amplitude of vibration as function of angular velocity for a one-mass vibration machine

The dynamics of vibration machines in general, and the operation stability in particular, depend very heavily on the properties of the specific source of power that generates the perturbing loads. This has been an area of extensive investigations, both theoretically and experimentally, and the results obtained contributed greatly to the development of the vibration technology.

The vibration transport processes most widely used in industry are motion in a horizontal or a slightly declined plane (Fig. 6*a*); ascend in a helical load-carrying surface (Fig. 6*b*); and loading or unloading of volumes. In a vertical vibration conveyor there is no rotation present, and the only type of motion performed by the helical surface is a vibratory one, the upward motion of the material being due to the vibration alone.

The vibratory conveyor is particularly well suited for in-shop transportation and feeding applications and it is anticipated that it will be widely used in un- or little attended future manufacturing processes. The reason for this is not as much its high productivity as its part orientation capability. It turns out that the dynamics of parts being transferred by a conveyor differ from part to part owing to different mass distributions, and if some simple mechanisms are added, it proves possible to use this effect for retaining good while rejecting bad parts. This possibility is especially valuable if the conveyor is combined with a robot because the latter can only (or at least prefers to) deal with strictly oriented objects.

An important application for the vibratory conveyor is the transport of granular (or particulate) materials. One advantage of this technique is that in addition to being transferred, the material may simultaneously be subjected to some kind of treatment, such as drying, dewatering, dust removal, granulation, etc.

Vibratory mills and crushers (Fig. 7) are not only capable of performing their "regular" task of crushing solid materials, but can also be adapted for activating chemical interaction between powders, for surface treatment, and for other ap-

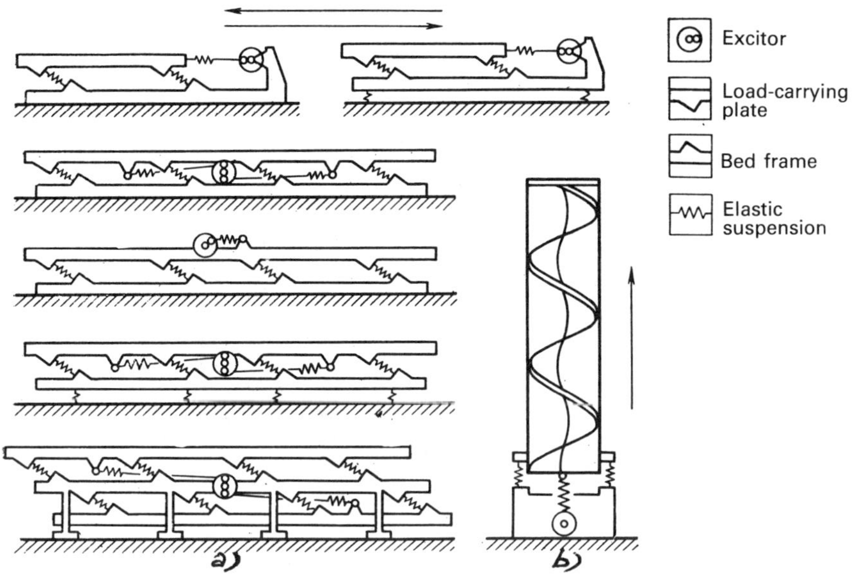

Figure 6 Types of vibratory conveyors: *a*) horizontal transport; *b*) helical surface upward transport

plications. The vibratory motion in these machines is due to the rotation of a decentralized shaft. Apart from the material being treated, special objects like balls, cylinders, or bars may be charged into the chamber(s) to perform the required treatment. It also proves possible to use these machines for combining different kinds of treatment in one process, such as mechanical and thermal treatments (crushing and drying), or mechanical and electrical treatments (vibroelectromechanical treatment).

Vibration can also be used for surface finishing. A part, free or fixed, is placed into a chamber in which a filling agent and a working fluid are then intensively mixed by vibration. The jobs performed in this way include, among others, deburring, removal of corrosion, volume buffing and polishing, surface hardening, and preparation for galvanic or lacquer coating.

An important technological problem to which vibration has been increasingly addressed in the past two decades is the intensification of phsyical and chemical processes in single- and multiphase systems. What makes vibration technology especially valuable in this context is that it allows more efficient use of interacting materials than do other techniques. The possibility of combining several technological processes, as exemplified by the above mentioned mills and conveyors, is another advantage offered by this technology. The two most popular methods used for this purpose employ either a vibrated fluid bed involving a granular substance or the vibratory stirring of suspensions, emulsions, and of gas bubble flows in liquids.

Use of the vibrated fluid bed method not only speeds up production processes but also improves the quality of the product and, rather an important factor, re-

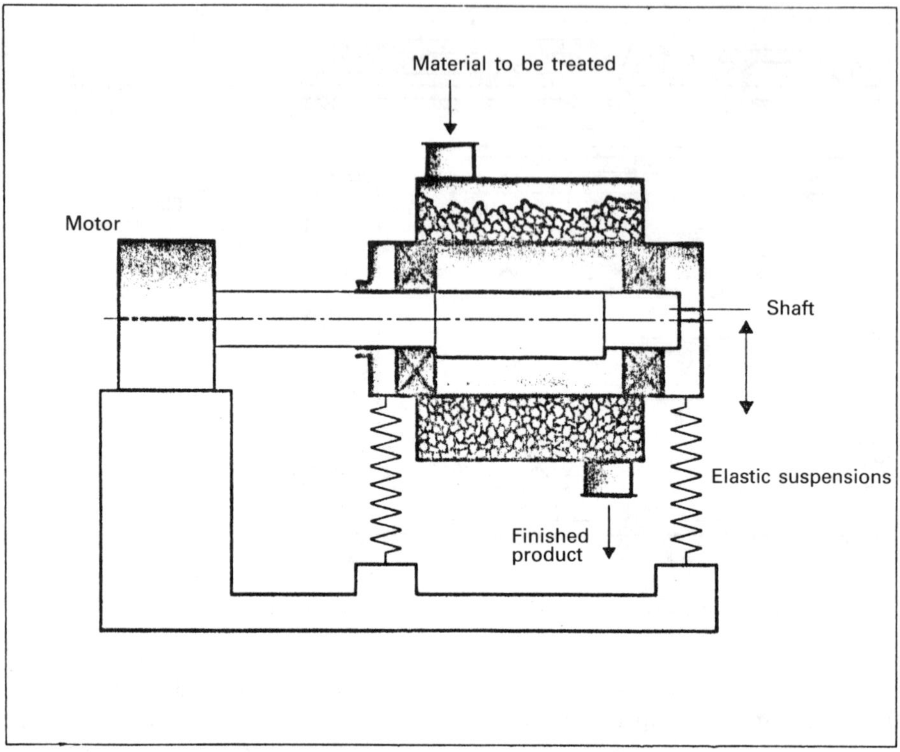

Figure 7 Vibratory mill with a partly decentralized shaft

duces the size of the equipment. Most frequently, a fluidized bed of a granular material in a gaseous medium is employed for promoting heat and mass exchange processes. A typical device shown in Fig. 8 uses convective heat transfer and may be used for cooling, heating, and drying a material.

A vibrated fluidized bed of a special-purpose finely ground material like corund or sand may be used very effectively for the heat treatment of metal workpieces because of the high intensity of the process and added possibility of controlling the heat exchange. When carried out in a protective gaseous medium, this treatment prevents or at least minimizes the oxidation and decarburization of the part's surface [52]. From laboratory break tests performed on medium-carbon steels in an air-and-water medium it is found that when cooled in parallel with intensive vibration stirring, the strength of a metal increases by 10 to 15 percent as compared to the no-stirring case.

In liquid media, an efficient way for intensifying a technological process is, for example, by vibrationally exciting submersed turbulent jets as shown in Fig. 9. The physical phenomenon underlying the operation of the device is that tapered holes pass a liquid much more easily in the direction of their narrowing than the other way around. The result is that the vibrations of the disk in a plane parallel

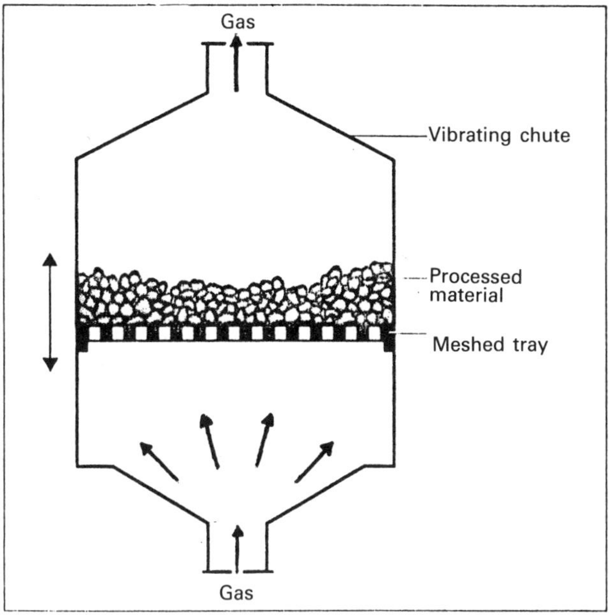

Figure 8 Vibratory tray arrangement for cooling, heating, or drying

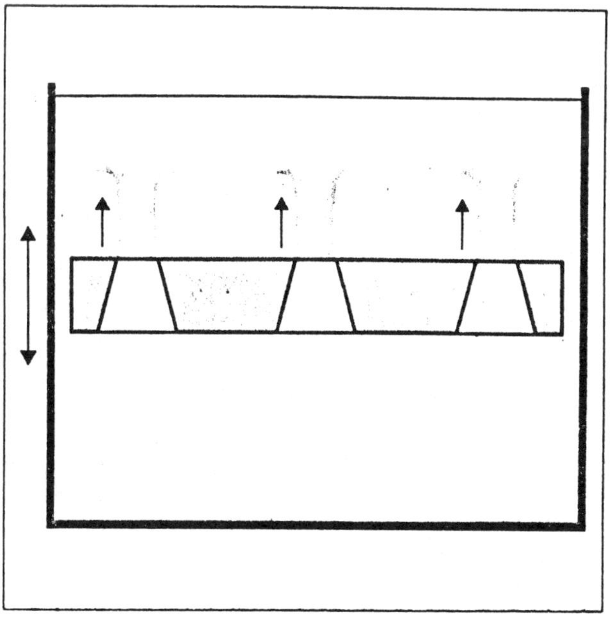

Figure 9 Vibration-aided liquid stirring using submerged turbulent jets

to walls give rise to the so-called vibrated jet effect, in which a submersed jet shoots out through narrow openings.

A comparatively recent innovation now popular in metallurgy is a vibratory mould, which greatly reduces the resistance to the motion of the ingot in the process of continuous casting and also facilitates heat and mass exchange thus increasing the productivity of the process and improving the mechanical properties of the metal being produced [53].

The effectiveness of the above processes depends very heavily on the specific vibratory regime of the machine, which explains that a whole gamut of vibration regimes are found in practice, such as harmonic and polyharmonic, rectilinear, three-dimensional, and two-component ones, with frequencies ranging from the infra- to ultrasound. The trajectory of the working agent of a vibration machine is generally a combined result of a translational and a torsional vibration.

Thus far we have been discussing the benefits that can be and are derived from vibration. Vibrations may have negative effects, however. They reduce the reliability of equipment, violate the operation regime of a machine, cause breakdowns and, last but not least, may exert harmful effects on a human operator. It is fair to say that vibration control is one of the major S & T problems.

The vibratory motion of any real machine involves a host of factors whose combination escapes a precise mathematical analysis. Consequently, mathematical simulation is an insufficient, even though necessary, tool for the study of the problem, and in most cases of interest we depend on experiment for the answer.

The advent of microelectronics has produced many new opportunities in the field of local measurements. Piezoelectric elements, microresistors, strain gauges, optrons, and other highly sensitive detectors are being used as a matter of routine in the analysis of complex vibratory processes in real machines. There are problems, however, the researchers come constantly to grips with in using these devices. On the one hand, the sensors must be highly sensitive and must possess a high mechanical strength for adequately performing; on the other, they must be protected from thermal and electromagnetic fields, random and uncontrollable vibrations, and other possible harmful influences. Much R & D work also has to be done on data-processing methods by which the (oscillatory) electrical signals produced by the sensor system are automatically transformed into intelligible numerical or graphic information in the form of, say, amplitude-frequency characteristics. (Note parenthetically that seismic science must also benefit from this research).

Some of the problems encountered in the conventional vibration-measuring technique may be overcome by replacing sensors with contactless vibration meters. Use of a laser beam, for example, may greatly increase the accuracy of measurement because in this case no links or connections are needed.

The simplest way to combat vibration is, of course, by eliminating the source of vibration which, most typically, is the imbalance of a rotating part, or rotor, of the machine due to errors in assembly or in the machining of the parts, to their deformation or wear during operation, or to some other factors. The laser-beam balancing technique shown in Fig. 10 is one of the most promising innovations in the field of vibration control.

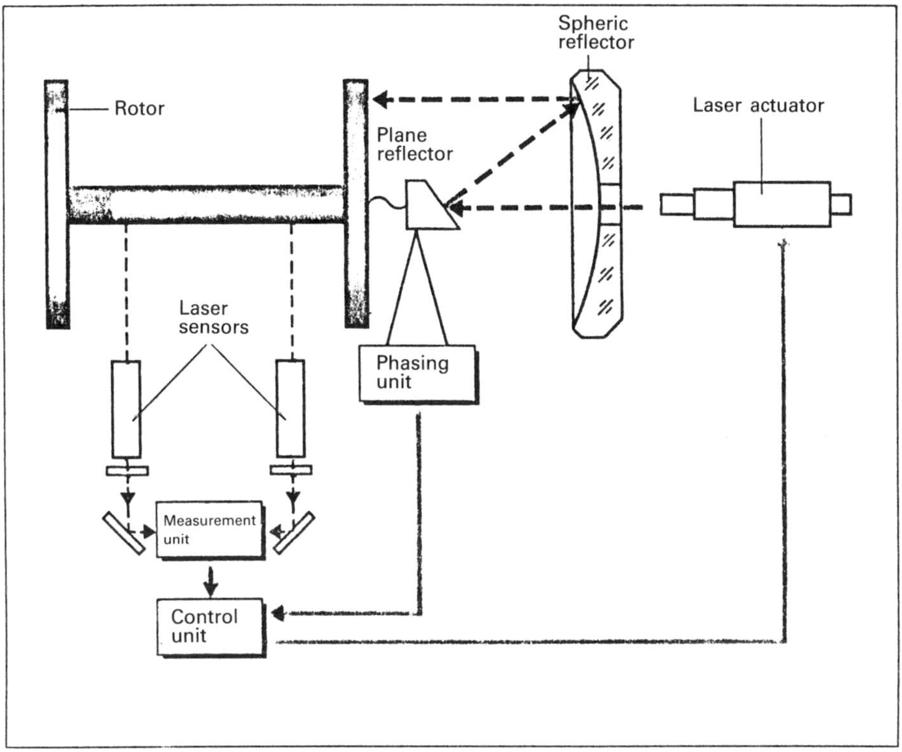

Figure 10 Laser balancing using pulsed metal evaporation in the disbalance region

While the securing and maintenance of balance and the tuning away from resonance frequencies remain the most widely used approaches to vibration isolation, increasingly, more attention is being given to the use of vibration-absorbing materials such as rubber or plastics, and to the incorporation of vibration control systems into the design of machines. Such systems imply of course a need for high-precision vibration meters and noise-proof sensors capable of performing their task when placed within an intensely vibrating structure. The development of such devices is the more important because a knowledge of the vibration spectrum of a machine may help to predict and consequently to prevent its breakdown, in much the same way as a good cardiogram helps to diagnose and avert a heart attack.

Vibration diagnostics is a subject of immense current importance in the aerospace industry, atomic power stations, and other fields where, as experience teaches us, costly damage and even a loss of life may result from a failure.

Reliability and Service Life in Mechanical Engineering

It has already been mentioned earlier that the strength and lifetime on the one hand, and power and material savings on the other are major considerations in machine design. Viewed from a nation-wide perspective, this implies a necessity for development, general acceptance and introduction of unified approaches to major engineering issues, such as the determination of the main service characteristics of structural materials; the calculation of strength and lifetime at the design stage; testing and adjustment; and the evaluation of the current state of an operating machine.

For most critical and heavy-duty machinery, the problem of strength and lifetime is a major concern throughout the whole history of a machine, that is, at the design, adjustment, and operation stages.

At the very earliest stage of design the future machine is usually conceived as a more or less ideal one, with its components all having the same lifetime and functioning at their best. This picture is, however, too happy to be true in most cases, so some compromise must usually be made between the conception of the machine and the real possibilities that the current state of technology may offer. This results, of necessity, in some structural members having an excessive and in others relatively short lifetime. Clearly, the final design will be the more successful, the more technological possibilities are available for the engineer.

The introduction of new advanced materials instead of traditional metals and alloys is probably the most characteristic trend in today's R & D work on the strength and weight properties of critical structural members in many industries.

These *composite* materials offer a number of impressive advantages over their traditional predecessors. Figure 11 demonstrates, for example, the superiority of composites over steel, aluminum and titan alloys in specific strength (by a factor

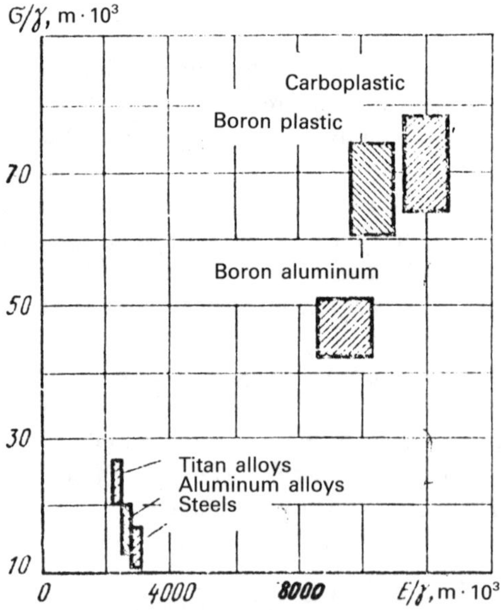

Figure 11 Composite materials versus metals: specific strength and specific stiffness

of 2 to 8) and specific stiffness (5 to 6) for composite material parameters measured along the fibers. That a part made of such a material may be given prescribed properties in a chosen direction is an important advantage allowing more effective material utilization. The ease of manufacture is one further advantage of composite (or fiber reinforced) materials. To produce a part, reinforcing fibers are laid in chosen directions, impregnated with special compounds and then are made into a particular shape by hot forging, none of the operations requiring any complex equipment or great power expenditures. Fibers made of graphite, boron, supphire, light metal, and carbides and nitrides have strength-to-weight ratios 5 to 10 times greater than steel, the factor increasing with temperature, and offer a particularly high potential for future composite materials.

Of composite materials currently in use, carbon fiber reinforced epoxy plastic (CFREP) seems to be the most popular, particularly in the car industry where it is used for manufacturing bodies, frames, doors and other automobile parts. Use of CFREP makes a part lighter by 30% or more, increases its strength, reliability, and impact and corrosion resistance, which means reduced fuel consumption, less direct labor in assembly, longer lifetime, and improved performance of a car (reduced noise and vibration, etc.). With large-sized products, use of CFREP may lead to particularly significant economy. CFREP antennas, for example, can combine huge working areas with perfect shape- and dimension-stability even at considerable temperature fluctuations.

All this does not mean of course, that composite materials will everywhere replace time-honored metals and alloys. On the contrary, a mechanical engineer of today is more and more frequently faced with problems which traditional materials alone may help to solve. Sea-bed studies, for example, require materials capable of withstanding enormous pressures. In the nuclear power industry, particularly in future fusion reactors, the structural materials will be subject to huge alternating loadings and superhigh temperatures.

It is becoming increasingly evident that the solution of major mechanical engineering problems, including reliability, requires new advances and much interdisciplinary effort in both fundamental and applied science as shown in Fig. 12 (see also [54]). The basic directions from which the problem of reliability must be attacked can be identified as follows:

- R & D work on evaluation of stress-strain and ultimate states;
- model and large-scale tests in various kinds of environment (high or criogenic temperatures, magnetic fields, radiation);
- the determination of the residual service life of an individual machine by using on-board computer systems for monitoring the loading history and the current state of the machine;
- the development of machining and reinforcement methods using ultrasonic, electrophysical, laser-beam, explosive, and other advanced techniques.

Fig. 13 shows schematically the main steps to be taken for implementing the strength, reliability, and service life requirements of a specific heavy duty ma-

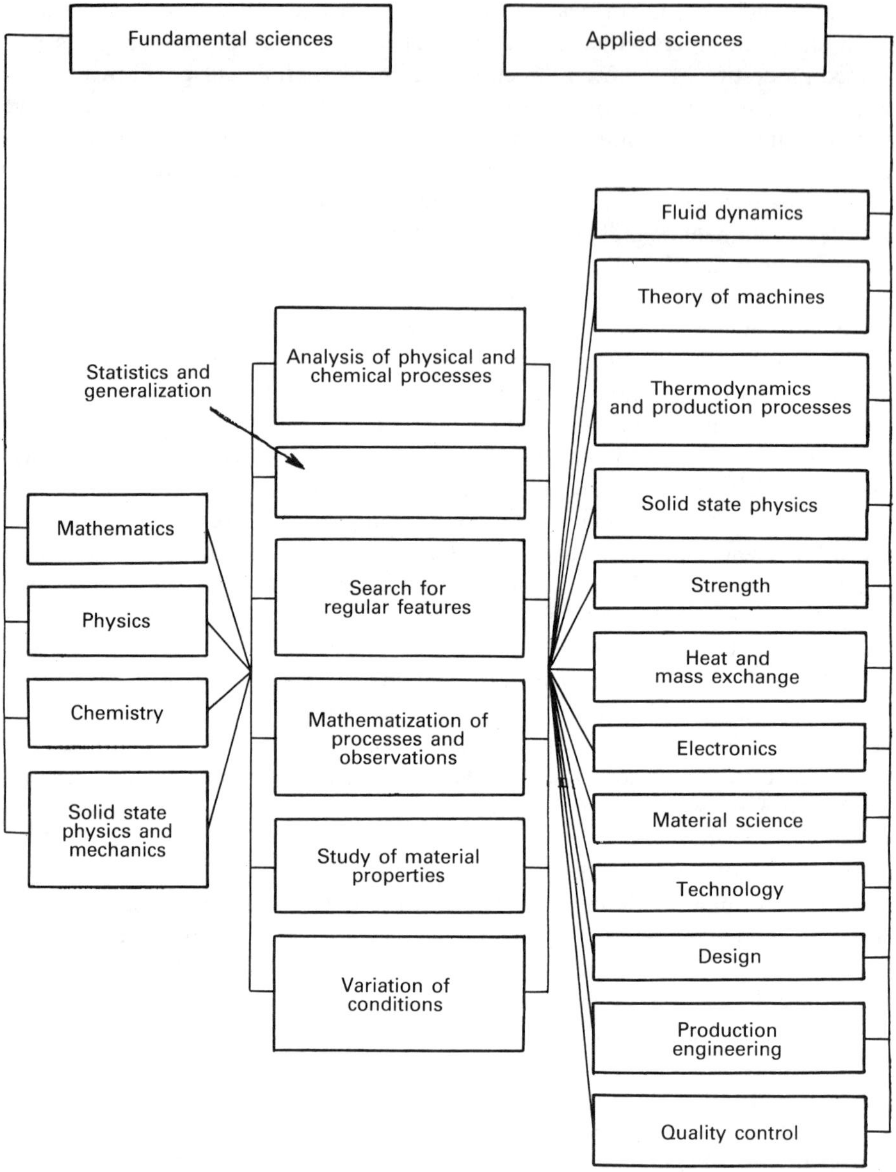

Figure 12 Fundamental and applied research in mechanical engineering

chine. The judicious specification of the safety factor and, consequently, the determination of the initial lifetime of a machine have to be made very early at the design stage using the main operating parameters of the machine as a starting point. The calculation of forces, temperatures, stresses, and strains due to develop in various parts of a machine under operating conditions is usually performed on

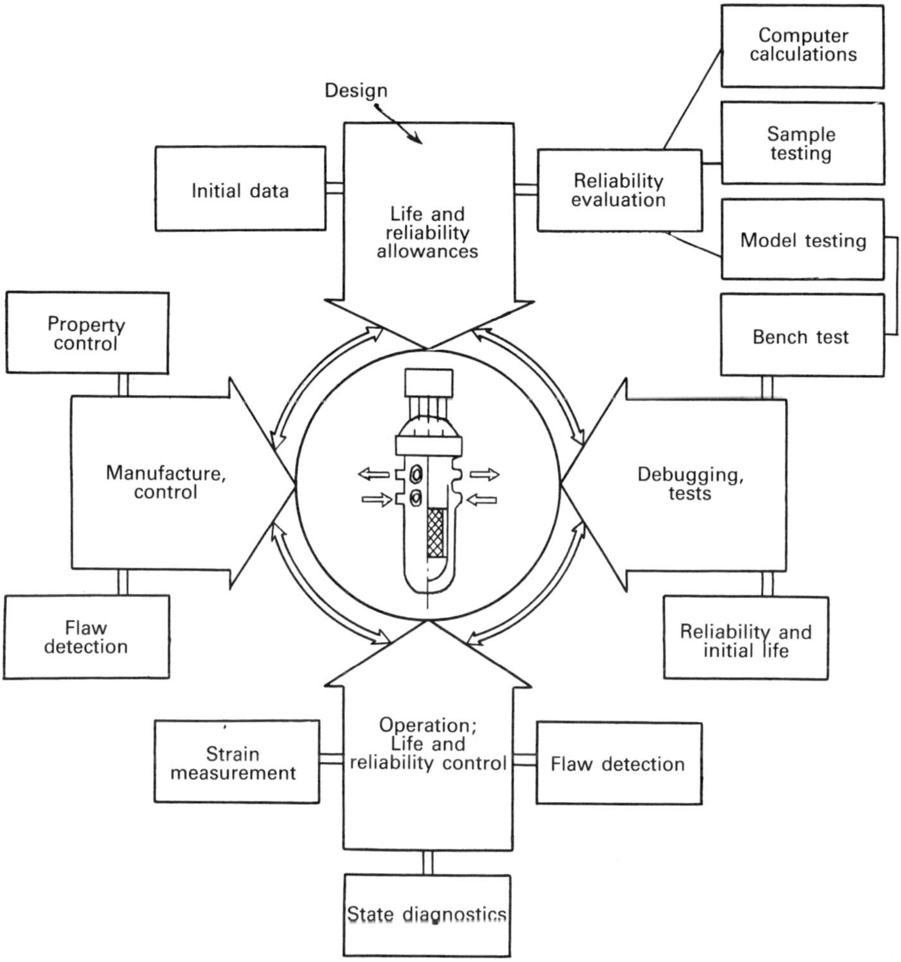

Figure 13 Flow chart of the strength and reliability design process in mechanical engineering

a computer; additional information concerning the mechanical properties of the proposed material is obtained from standard testing techniques. The tests in question are often carried out in a wide temperature range and may involve forces as high as 100 to 500 kN. For the most complicated and critical parts, a model test may be needed. It is estimated that strength and lifetime issues may take as much as 15 to 25 percent of the total amount of design work. This is followed by an optimization procedure (see Fig. 14) in which the methods we have outlined earlier are employed. At the manufacture stage, great attention must be paid to the initial check and subsequent periodic monitoring of the mechanical properties of the structural materials and in particular of the welds involved in the design, the main criterion being the correspondence to specifications (to ensure the fulfillment

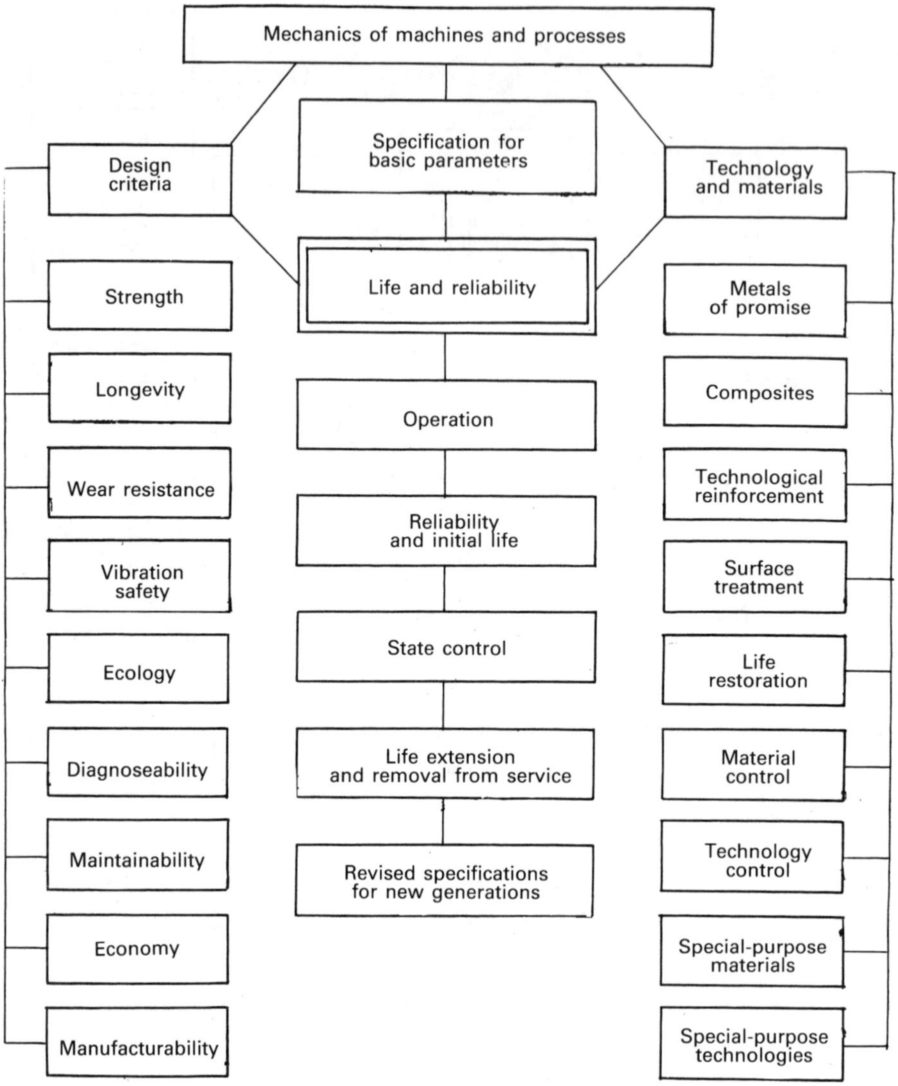

Figure 14 Flow chart for solving life reliability problem

of strength requirements in the manufacturing process). Another major concern at this stage is monitoring the development of defects in bearing structural members. This is necessary for the prediction of lifetime and the determination of the volume and periodicity of the future on-line flaw detection.

At the operation stage, increasingly more attention is being given to the monitoring and surveying of the current state of a machine or a structure, the detection of damages, and the estimation of the residual lifetime of the object.

Two major aspects in the field of strength to which much attention has been directed in recent years are the loading dynamics of and ultimate-state criteria for typical structural elements [55]. In particular, much progress has been made in the study of the strength and the ultimate states of heavy loaded structures under thermal loadings and static and/or dynamic mechanical loadings. Research work is underway on the strength of bearing structural members made of composite materials. New experimental techniques have been developed for measuring material mechanical properties at elevated and varying temperatures. (Note, for example, advanced tensometry methods such as sensitive coatings, holography, replicas etc.). Speaking of theoretical results, mention should be made of the analysis of high- and low-cycle fatigue of structural elements (based on the use of probability theory) and of the crack resistance studies based on the application of (linear or nonlinear) fracture mechanics to simple or repeated loading programs.

It is known from the mechanics of deformed media as it stands today that the lifetime (R), reliability (P), and specific material requirement (G) may each be represented as a function of many variables

$$\{R, P, G\} = F[\sigma_{us}, \sigma_{0.2}, \sigma_{lt}, \sigma_{-1}, \psi, H, t, N, \alpha_\sigma, l_0\, \sigma_d, I_i, K_I, m, \tau] \tag{3}$$

where σ_{us} is the ultimate strength; $\sigma_{0.2}$ is the yield point; σ_{lt} the ultimate long-term strength; σ_{-1} the endurance limit; ψ the plasticity; H the hardness; t the temperature; N the number of cycles; τ the operation time; α_σ the theoretical notch-sensitivity index; l_0 the (characteristic) initial defect dimension; σ_d the design stress; I_i the wear rate; K_I the stress intensity factor; and m the metal reinforcement parameter. By using modern design methods and state of the art technology, the above variables may be varied over wide limits thereby changing R and G by tens percent and P by as much as orders of magnitude [56].

Clearly neither methods nor criteria can be developed for reliability and lifetime evaluation unless the cyclic, dynamic, and corrosion crack resistances of the proposed material are known.

It is well known, for example, that the reliability of equipment and machinery is of paramount importance in such industries as chemistry and power engineering (especially in nuclear power plants). The longevity of structural members in such applications is determined from (static or cyclic) corrosion crack resistance plots, which are curves showing the crack propagation velocity (v) as a function of the stress intensity factor (K_I) in the vicinity of a propagating crack in a given material. Given such a plot, the longevity N of a part may be represented as

$$N = N_1 + N_2, \quad N_2 = \int_{l_0}^{l_*} v^{-1}(K_I)\, dl \tag{4}$$

where l is the length of the crack; l_* is the critical crack length as calculated from static crack resistance data by the Griffith-Irvin method; N_1 the macrocrack formation period in the given material (it is for the evaluation of N_1 by fracture mechanics methods that a knowledge of the function $v(K_I)$ is required); N_2 is the survivability of the structural member having a crack of length l_0 within it (the evaluation of N_2 also depends on information about the cyclic crack resistance of the material).

It thus follows from the above that the study of the crack resistance of structural materials is of tremendous industrial importance and that, consequently, some standards and regulations must be developed for measuring this characteristic under a diversity of conditions. As of this writing, testing methods have been established for obtaining $v(K_I)$ dependences in neutral atmospheres like a vacuum, dry air, or inert gases [57, 58], but clearly there is still much more to be done in this area.

The technologies which appear to be the most promising today for lifetime and reliability improvement include the use of composite and other new materials, heat treatment, and various kinds of machining techniques such as thermomechanical, thermochemical, explosive, laser-beam machining, and diamond burnishing. These methods differ in frequencies, power requirement and in resultant chemical effects and offer the possibility of selectivity controlling strengthening, crack arrest, cumulative damage, and crack growth processes.

It must be admitted, however, that the lack of adequate technological equipment and of state of the art measuring and diagnosing techniques in this country slow down the study of the above methods both on models and under full-scale operating conditions and so prevent full advantage being taken from their application.

Speaking of composite materials alone, it is impossible to recite more than a small fraction of the uses to which a manufacturer can put them. These are, for example, large-size vessels for the storage and transportation of liquid and gaseous products; robot elements and car parts; high-pressure gas-holders; chemical reactor housings; aircraft; the bodies of underwater vehicles for use in sea-bed studies and exploitation; the bodies and parts for internal combustion engines. All these possibilities will remain just that, however, until the design engineer acquires an adequate knowledge of the nature of defects in composite materials and becomes able to predict and, consequently, prevent their development, thus securing the required level of reliability. In more specific terms, the problem is to understand the mechanisms that govern: (1) the formation of the micro- and macrostructure of the composite material's matrix; (2) the evolution of the adhesive bonding between the matrix and the reinforcing agent; and (3) the state-to-state transitions that take place during these processes and often cause drastic changes in the thermoelastic properties of the material as well as deformations and stresses of the order of their ultimate values. Clearly, the mechanisms in question not only depend on the properties of the components but are also determined by the structure itself through the mass distribution of the chemically active polymeric or metallic matrix, the kinetics of the reactions involved, the characteristics of the heat flow in a cooled or heated part, etc. Speaking in mathematical terms, the system of equations necessary to describe the technology mechanics in this situation must generally include:

- the kinetics of structural and state-to-state transformations of solidifying systems, with the frontal character of conversion taken into consideration;
- heat conduction;

- heat release due to the reactions involved;
- the dependence of matrix thermophysical properties on the intensity of the structural transformations;
- equations of motion (of equilibrium);
- the geometry of deformations;
- constitutive equations for a non-uniform, non-thermostable, thermoviscoelastic medium for all of its transformation stages;
- initial and boundary conditions.

Obviously it is a formidable task to write down, let alone to solve, such a system for even the simplest situations. What makes things still more difficult is that in carrying out this program we must be able to express mathematically the formation and motion of ultimate-state surfaces for all times, for various temperatures, and for various depths of structural and state-to-state transformations of the matrix. It may be argued, by analogy with (3), that

$$\{R, P, G\} = F[\dot{\varkappa}, \varkappa_k, \ni_k, Q, c_\rho, \rho, \lambda, T_\sigma, T_\epsilon, \Pi, P(t)] \tag{5}$$

where $\dot{\varkappa}$ is the time rate of structural transformations at the k-th technological state; $\varkappa_k$ the thermodynamic probability of the state of the matrix at this stage; $\ni_k$ the internal energy of the system; Q the specific heat for the transition from the k-th to the $(k + 1)$-th state; c_p, ρ, λ are, respectively, the specific heat, density, and thermal conductivity in the state $\varkappa$; T_σ and T_ε the stress and strain tensors; Π the creep function, and $P(T)$ the operational record.

If the dynamics of the relevant ultimate characteristic is reliably known, then given the solutions for the stress and strain fields occurring in the formation of the structure, it proves possible—at least in principle—to formulate criteria for the formation and prevention of defects in the macrostructure of composite materials [59].

Although there has been much progress in recent years in both fundamental and applied research in composite materials, much work remains to be done in developing computation methods and computer software suitable for solving such practical problems as the calculation of stress and strain fields in a given structure; optimization of major technological parameters; determination of allowable defect densities in typical structures; fatigue failure; and development of composite materials for use under physical loadings (fields) of various kinds.

The achievement of these goals and the introduction of composite materials into practice requires adequate organization of research work both in industry and in the Academy, where the activities of the councils for synthetic materials and composite structures should be properly coordinated. Another obvious requirement is the elimination of the outdated practice of assigning to different research centers such functions as the development of new materials (including new machining and manufacturing techniques) and the development of new design and computation methods. So-called research-and-production associations (Russ. abbr. NPO) appear to offer the most efficient way of speeding up progress in this area.

Improving reliability and increasing trouble-free service life depend very heavily on and underscore the significance of the technical diagnostics of operating machinery. For a present-day heavy-duty machine, diagnostic equipment may account for up to 15 percent of the total cost of the machine. These are the commonly used diagnostic methods:

- measurement of vibrations, temperatures, and pressure pulsations by means of numerous (100 to 1000 per machine) electro-mechanical and electronic sensors;
- multi-point strain measurement at high and cryogenic temperatures, in the flows of heat-transfer agents, in magnetic fields, and under radiation;
- flow detection systems employing ultrasonic, magnetic, electric, acoustic, and radiation effects;
- multi-criterion computer systems for measuring acoustic emission;
- holographic systems of continuous or pulse operation.

The development of testing techniques is of course another important aspect in the field of reliability. New methods are being continuously developed which may be used for testing materials, machines, or structures on strength, wear resistance, durability or longevity, and employ electromechanical, electromagnetic or electrohydraulic equipment and computer-controlled testing stands.

Tribology and Wear Resistance

Reliability, service life, and precision requirements and the general trend toward reducing repair cost and material and power requirements are attracting increasing attention to the subject of tribology in mechanical engineering and other fields. Friction-related problems may sometimes be of vital importance in aerospace and nuclear power industries, in oceanography and mining, agriculture machinery and automated manufacturing technologies.

There are many physical, chemical, and mechanical factors which should be considered when mathematically modeling the destruction of solid surfaces by friction. These include, among others, the specific nature of contact between the bodies involved, the kinematics of motion, the structure and composition of surface and near-surface layers, the state of the lubricant, the formation of chemical compounds at the surface, and absorption and diffusion processes. To incorporate all these factors into a general picture, it is necessary to employ the elastohydrodynamic and hydrodynamic theories of lubrication, the equations of chemical kinetics, the solutions of elasticity theory contact problems, the kinetic theory of strength, the heat dynamics of friction and wear processes; and, of course, a great deal of experimental data. Work on this program is being conducted in a number of research centers in this country, and its realization is expected to accelerate the solution of a whole range of important practical problems as well as to suggest the most promising research directions in the tribology field.

The area in which most intensive and detailed study should be carried out in this connection, both theoretically and experimentally, is the thermophysics of high-speed friction processes that take sometimes no more than a few hundreths of a second for their completion and proceed at speeds as high as tens of thousand meters per second with correspondingly large translational accelerations. In studies of this kind an account should be taken of the viscoelastoplastic deformation of the near-surface layers of the material, the time variation of the contact area, the depth of the layers participating in deformation processes, the nonstationary distribution of heat flows, drastic changes in the thermophysical properties of the rubbing bodies, and the heat and mass transfer taking place during the friction process. Possible macroscopical changes such as buckling due to wear and temperature deformations should also be considered in a general mechanical formulation.

An understanding of the mechanics of contact interactions and of the chemical and physical dissipative processes occurring in the surface and near-surface layers of rubbing bodies is essential for the choice of optimal reinforcing technologies and effective antifriction coatings. Of obvious importance is of course the development of advanced coating methods, such as vacuum, laser, detonation, electron-beam, and other techniques.

It may well be expected that progress in the above outlined directions will increase by 30 to 50 percent the longevity of friction units of various kinds in the nearest future.

It is customary (see e.g., [60]) that a wear-out process is characterized by the intensity of linear wear, I_h, defined by one of the following equations

$$I_h = \frac{dh}{dL_{fr}} = \frac{dv}{A_{fr}\, dL_{fr}} = \frac{\lambda}{\rho_{fr}} \frac{dG}{dL_{fr}} = \lambda \tau_a \frac{d\nu}{dw} \tag{6}$$

where h, v, G are the linear, bulk, and mass wear factors, respectively; L_{fr} the friction path; A_{fr} the rated friction area; $\tau_a = F_{fr}/A_a$ the rated specific friction force, F_{fr} being the friction force and A_a the rated contact area; $w = L_{fr}F_{fr}$ is the work done by the friction force; ρ_{fr} the density of the material being worn out; $\lambda = A_a/A_{fr} = K_{ov}$ is the mutual overlap factor.

There are cases, however, when a wearing process is more conveniently described by the bulk, mass, or energy wear intensities ($I_{\bar{v}}$, I_G, I_w) defined by

$$I_v = \frac{dv}{dL_{fr}}, \quad I_G = \frac{dG}{dL_{fr}}, \quad I_{\bar{w}} = \frac{dv}{dw} \tag{7}$$

and related through equation (6) to the linear wear intensity factor. Some authors characterize wearing by the rate of wear, $\lambda_h = dh/dt_{sl}$, t_{sl} denoting the sliding time. Clearly $\lambda_h = I_h\, v$, where v is the sliding velocity. A useful characteristic is also wear resistance, a quantity inverse to the rate (or intensity) of wear.

It is becoming increasingly clear today that the analytical relationships obtained in the field of tribophysics may be very helpful in the wear design of machines and mechanisms, and that, consequently, further efforts in this direction are worthwhile both in developing theoretical methods and in accumulating reference and standard data on the mechanical and fatigue properties of rubbing surfaces as well as on their parameters and microgeometry. Needless to say, no progress is possible in this field unless specialized computer-assisted measurement and test equipment is available.

A new R & D area which shows particular promise in mechanical engineering is the development of methods of wear-free operation design. Reducing the friction and wear of rubbing surfaces in the boundary-lubricant regime may double the load-carrying capacity of a friction pair with a resultant 15 to 20 percent reduction in the size and weight of friction joints. Other potential benefits are increased efficiency (say, from 0.7 to 0.85 for a globoidal reduction gear, or from 0.25 to 0.50 for a serew-nut pair) and corresponding power savings, improved reliability, and longer MTTR and service life. This implies, of course, reduced (sometimes halved) direct labor, less power expenditures on maintenance and run-in, and lower lubricant consumption.

There has been considerable interest over the last years in the so-called magnetic liquids, which are stable colloids of finely dispersed magnetic materials in neutral liquid carriers such as water, carbohydrates or silico-organic liquids (see [41]). The fact that magnetic liquids may be controlled by means of an external magnetic field for their frictional properties has made them attractive for a great many design and engineering applications in various fields. A significant achievement of Soviet applied science is the development of ethyl-organic-siloxane magnetic liquids, superior in many important ways over their foreign counterparts, particularly under extreme operating conditions. The development of methods for preplanned synthesis and for comprehensive study of such liquids has resulted in a number of highly effective innovations in the field of tribotechnology. One of them, the so-called ML seal, is based on the well known phenomenon of confining a magnetic liquid by a non-uniform magnetic field. ML seals are simple in design, suitable for various types of motion, and have low friction losses, increased service live, and extremely high leak-proofness. Many types of ML seals are being successfully used in various fields in a deep vacuum, in gaseous and vapor-laden atmospheres, in biologically active media and in other hostile environments [61].

Another application, the ML bearing, uses the fact that a magnetized body floats in a magnetic liquid whereas a nonmagnetic one is pushed out by the non-uniform magnetic field. The friction coefficient of an ML bearing is independent of the load and its stiffness may be varied within wide limits by a magnetic field. Note also that ML bearings are virtually noiseless and pose no speed restrictions.

Magnetic liquids, controllable by a magnetic field, have also produced new opportunities in the field of lubrication. A magnetic lubricant synthesized on the basis of a lubricating oil or liquid radically solves the problem of even and reliable lubrication, being as it is automatically pulled into the friction zone by a non-uniform magnetic field arising in the region of the rub contact. This results in a lower friction coefficient and decreases by an order or two the wear of rubbing surfaces as compared with the lubricants currently in use. An important possibility offered by ML lubricants is also the use of thermomagnetic convection for achieving automatic circulation of the lubricating medium. ML lubricants provided a basis for a great number of antifriction machine units developed in our country.

The ability of some ML compositions to drastically change their effective viscosity in a magnetic field has been fruitfully used in the design of magnetically controlled vibration and impact dampers as well as in highly effective magnetorheological devices for the high precision abrasive processing of nonmagnetic materials.

A major direction in which important steps have been taken in engineering tribology in recent years is the use of antifriction materials based on heat-resistant structural materials like polyamides, polyimides, polycarbonates, polyphenilene oxide, and polytetrafluorethylene (teflon).

Gas lubrication is one further area of growing industrial importance. The introduction of gas lubricated bearings into industry poses a number of closely interrelated problems before those involved in the field, such as the development of engineering methods for design, manufacturing, assembly quality control, and for stand and production testing; experimental studies of static and dynamic characteristics of machine units equipped with gas lubricated bearings; and the prototype development of necessary instruments, test stands, and of various units of metal-cutting machine tools.

The low viscosity of lubricating gases drastically reduces frictional losses in friction units thereby allowing operating frequencies as high as about half a million revolutions per minute. On the other hand, the relative motions of structural members with a gas layer between them involve no sharp changes in frictional forces, which allows lowering sliding velocities down to hundredths of a millimeter per minute.

It is a fact of great practical importance for gas lubrication technology that the viscosity of a gas varies only very slowly with pressure and temperature. The viscosity of air, for example, no more than doubles on heating from room temperature to 1000 C at normal pressure or on compressing from 0.1 MPa to 30 MPa at room temperature. On the other hand, the high pressure at the clearance outlet prevents the contamination of working surfaces, which makes such systems an excellent choice in, say, nuclear or laser technologies, where energy precision considerations call for at least some degree of vacuum; or in aerospace, where the possibility of depressurization can never be totally excluded. Note also that the advantage of there being no need for seals or collectors greatly simplifies design work on gas lubricated machines.

Much research has been done [62] on the performance of gas lubricated bearings in transient on-off regimes, including the important practical situations of a rarified medium, when a gaseous lubricating layer is particularly difficult to form. It is transient regime properties which to a large degree control the serviceability of a bearing and determine the material properties of its rubbing surfaces.

The development of metallofluoroplastic bearings is one further major advance in combatting friction. These bearings run dry or are barely greasy and may be relied upon at temperatures from −200 to +300 degrees Celsius, or even up to 350 C for short periods of time. The material is manufactured in the form of ribbon on a continuous operation line, virtually without waste, and is then shaped by stamping into bushes, thrust washers or spherical bearings, again with no waste and with a minimum of machining work. An MFP bearing may be half as large in diameter and 10 to 15 times lighter than an ordinary rolling bearing, which leads of course to further material savings through the size and mass reduction of the machine housing. There is ample evidence that MFP bearings perform quite reliably both when lubricated and when operating under no-lubrication conditions in such liquids as water, gasoline, kerosene, or alcohol.

It is essential that the intensive frictional heat produced at high sliding velocities be carried away as rapidly as possible from the rub contact and that the frictional force be reduced to a minimum both under normal and heavy-duty operating conditions. Since the specific heat of water is two and a half times higher and viscosity about 50 times lower than in a mineral oil, use of water as a lubricant appears to offer the most effective answer to these problems, the more so that its vaporization heat is also very high. It should be remarked, however, that with viscosity that low no given film thickness can guarantee that the lubricant will perform equally well at all possible operating conditions, which of course implies new problems for both the designer and the manufacturer. Whether the structural materials involved will be able to withstand the corrosive effect of water is yet another obvious complication. When these problems are overcome, the service life of a water lubricated bearing may extend to 50,000 hours or so, as is the case in hot-well feed pumps currently in use in heat-electric generation plants.

A subject of immense current importance in the field of tribology is lubrication under extreme conditions, which may include corrosive media, various kinds of radiation, extreme and/or highly variable temperatures and pressures, and possibly other harsh factors acting separately or in some combinations. The most efficient way to attack this problem is to use solid-state lubricants such as salts, minerals, metal sulfides, iodides, or fluorides, polymers, fatty acids, and soft metals. These are usually fed into the contact region by means of the so-called rotation lubrication technique, gradually and in small amounts, the feed of the lubricant being possible at even minimal loads. In gear trains, the role of a lubricator is assigned to a special load-free wheel meshed with the drive wheel, lubrication effect being due to the wear particles being removed from the lubricating wheel and transferred to the gear friction surfaces.

The efficiency of solid lubrication has been greatly improved by development of the so-called magnetic-power method [64], in which

- rubbing parts are manufactured from ferromagnetic materials;
- the mechanism is placed in a steady magnetic field;
- the components of the friction unit are arranged so as to form a common magnetic circuit;
- the powdered lubricating material is provided with appropriate magnetic properties, and most of the material is placed in the vicinity but outside of the friction unit.

All of these requirements are relatively easy to satisfy, and since high-strength structural materials are mostly, or at least frequently, ferromagnetics, the choice of a steel grade is usually an easy problem for a designer.

Ecology and Vibroacoustic Dynamics

Acoustic strength and acoustic dynamics have long been a major preoccupation of mechanical engineers interested in improving the service life and reliability of machinery. The investigation of noise sources, actual and potential, and research work on total or at least partial localization of certain types of industrial noise has the objective of, and has already resulted in, quite a number of noiseless or low-noise design methods. These latter include vibration damping, vibration absorption, and balancing, and presuppose, of course, high-quality manufacturing and assembly technologies. The problems of service life and reliability cannot thus be successfully attacked without proper attention being given to ecological and human engineering issues.

The high speeds and heavy loads characteristic of many advanced technologies, and the widespread introduction of vibration and shock methods into industrial processes have increased the intensities and widened the frequency range of vibration and vibroacoustic fields occurring in machinery and structural elements. It is a well-known fact of mechanical engineering that failures occur only very rarely under static conditions but most frequently are due to time-varying stresses and dynamic loads produced by vibrations and shocks. Vibration reduces the load-carrying capacity and is responsible for the fatigue fracture of machine parts, gives rise to micro- and macroscopical cracks, and aggrevates wear and friction problems. Since vibration, with its attendant noise, affects most directly the human worker, his protection against vibration is a major factor in securing the reliability of integrated man-machine-environment systems.

It is important, in this connection, to investigate the performance and physiological responses of a human operator subjected to various types of vibration and to develop some standards and standard methods to be used in vibration protection design. Much research work in this direction has been conducted at the Institute of Mechanical Engineering of the USSR Academy of Sciences.

An idea that has long been in circulation in the fields of industrial ecology and human engineering is that by modifying vibrations in a proper way it might be possible in some cases not only to eliminate their adverse effects on living organisms but even to render the vibrations beneficial to a human worker (vibrostimulation). The latest results obtained at the IME and other centers concerning man as an integral part of an industrial process have provided useful insight into the problem and suggested possible lines of activity in this direction.

To risk repetition, it is mandatory that high-reliability machines meet all the requirements posed by ecology and human engineering. New types of pneumatic seats now on the market serve to illustrate this principle. The seats are easy in design, are comprised only of standard parts, and protect a worker perfectly from vibrations. Due to the nonlinearities in the dynamic characteristics of the device, significant damping effect and, accordingly, a high vibration transfer factor are achieved both at high- and low-amplitude floor vibrations, the latter being already important at frequencies as low as 4 to 5 Hz.

Detailed experiments on the seats were conducted using a test complex designed for the purpose and pictured in Fig. 15. The experiments showed that the resonant properties of the body of a human operator depend very strongly on the manner in which the operator works when sitting on the pneumatic seat (see Fig. 16). The results obtained provide a starting point for developing a statistically significant model of a human body in this context, in which all the major anthropometric parameters will be taken into account. The study of the resonant properties of such an "average" human body is likely to result in some recommendations as to what protection measures might be taken at dangerous operating frequencies. Based on this type of investigation, permissible vibration levels will be established and state standards on vibration testing facilities introduced.

6. CONCLUSIONS

A few remarks concerning organizational issues now appear to be in order. The State Committee for Theoretical and Applied Mechanics includes many prominent scientists and researchers from across the country and hardly anyone can be found who would question the importance and usefulness of their activity in this capacity. Still, the Committee has not yet grown into the all-Union center for developing research strategy in the field of mechanics and for monitoring the comparative level of domestic and foreign research in order to keep abreast of important developments in science and technology. Nor is the Committee efficient in selecting scientific achievements deserving promotion and subsequent introduction

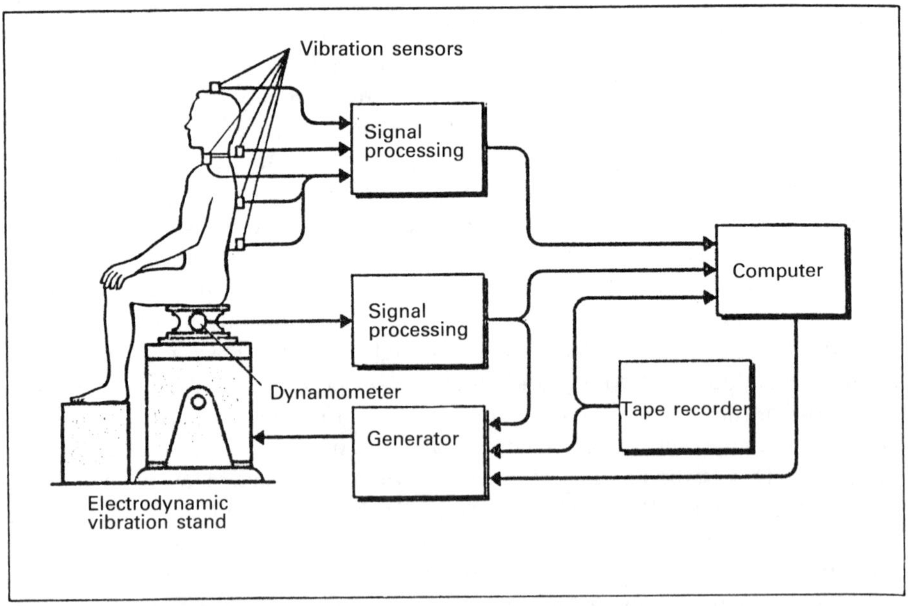

Figure 15 Information-measurement complex for studying human dynamic response to vibration

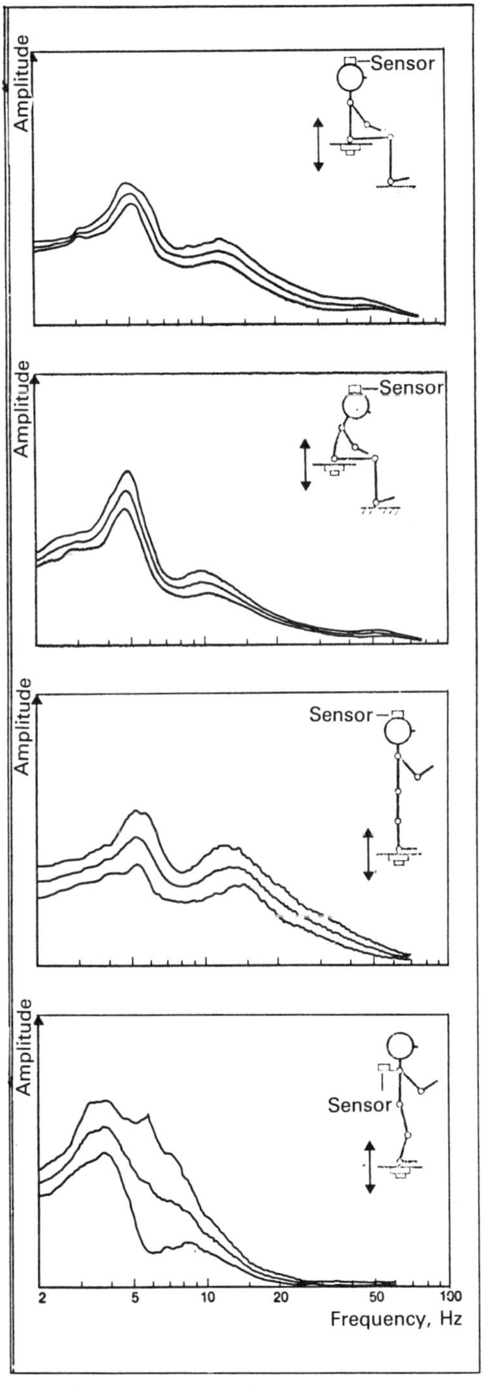

Figure 16 Experimental amplitude vs frequency characteristics for various postures of a human operator

into industry. An analysis of a great number of recent publications on mechanical problems has shown that insignificant or even erroneous papers find their way rather easily onto the pages of scientific journals thus revealing a definite and deplorable lack of attention on the part of our Scientific Councils as to the quality of the submitted material. The decision concerning scientific publications recently adopted by the Presidium of the Academy will probably be instrumental in improving the situation in this area. The quality of publications should be a top-priority problem not only for the State Committee but also for special-problem councils and for journal editorial boards who should, among other things, pay more attention to the quality of paper reviewing. In the Academy, the quality of publications should be also be a concern of the Division for Problems in Mechanics, Control, and Mechanical Engineering.

It is perhaps worthwhile mentioning in this connection that, paradoxically, the proceedings of the previous five Congresses have not been published in any better form than a collection of brief annotations, which are incapable of course of giving more than brief outlines of the problems being considered. From now on it is planned that the materials presented to Congresses will be published on a more solid basis. It is expected, in particular, that the proceedings of this present Congress will be covered in four volumes.

A major area of concern in the field of mechanical science is the poor performance of mechanical problems councils. Regrettably, these have little or no linkage with higher schools and universities, are inefficient in coordinating research and development activities, and issue recommendations which often have little or no effect on either the level or the quality of industrial design. The quality of the councils' reports and documents, it should be admitted, leaves much to be desired. To enlist the aid of scientific journals is one possible step toward improving this situation. The journals should highlight the activities of the councils and devote more space on their pages to research work being conducted by higher school and university research groups.

One further factor bearing on, indeed weakening, the USSR's competitiveness in the field of mechanics is, of course, the state of the teaching of the subject. The irony is that while the science itself is somewhere on the frontier of present-day knowledge, the book market is flooded with texts that stubbornly reiterate the fundamentals of the classical 19th century mechanics, leaving the impression of their authors' full unawareness of the advances in solid state theory, analytical mechanics, fracture mechanics, biomechanics, computer science, microelectronics—you name it. This, of course, is a totally intolerable, if not critical, situation.

The manner in which research itself is being carried out is also open to criticism. An analysis conducted by the USSR Higher Certifying Commission has shown that in the vast majority of candidate and doctorate theses in the field of applied mechanics, too much attention is being paid to refining the classical models such as beams, plates, shell, or rods, often in artificial far-fetched formulations having little or nothing to do with actual loading regimes and operating conditions. On the other hand, far too slow is the progress in adopting and developing the method of boundary integral equations, the final-element computation techniques, and other advanced methods. Clearly, the traditional boast of the high economic effect of such works has no more than a symbolic significance, and symbols can hardly be expected to produce any tangible effect on the level of engineering or on anything else for that matter. It is on the order of the day that the whole texture of our mechanical science should be thoroughly revised and proper attention be given to micro- and macrofracture mechanics, to the technological mechanics of composite materials, to biomechanics, to mechanics of agricultural machinery (in particular, of cotton pickers), and to other problems of similar importance for technological progress [65].

Our scientific and technological cooperation with CMEA member countries is one further area in need of revision as far as mechanical science and engineering are concerned. It is in the framework of the CMEA that the concerted efforts of academia, higher schools, and industry may result in effective research programs in such top-priority fields as new materials and technologies, all-round automation in mechanical engineering, etc. All-Union mechanics Congresses should commit themselves to developing unified strategies in these areas and play a greater role in working out appropriate educational programs both for domestic use and for our CMEA partners.

The special-problem councils, the State Committee, and this Congress' Organizing Committee should take it upon themselves to select the most significant results from those reported here in order to suggest them to the State Planning Committee (Russ. Abbr. Gosplan) for the direct introduction into industrial use.

Finally, it is highly imperative today that academic scientific research should become more oriented toward technology and technological applications. This requirement is accentuated by establishing in the framework of the Academy of the Division for Problems in Mechanics, Control, and Mechanical Engineering intended to coordinate the activities of a number of research institutions located in highly industrial regions around Leningrad, Sverdlovsk, Gorkii, and Ufa. Another development to be mentioned in this connection is the Five Year Program for mechanical-engineering oriented basic research, a united endeavor of the Academy and the Mechanical-Engineering Bureau of the USSR Council of Ministers. Investigations in the field of mechanics are, to be sure, of primary importance in this context.

Mechanics is a scientific discipline of paramount importance in human activities; indeed it is the corner stone of technological progress. No natural phenomenon can be explained without laws of mechanics and no engineering construction can be realized unless its mechanical parameters have been determined. Let me express here my hope—indeed, my confidence—that Soviet mechanical science will remain an important factor in the technological development of our country.

REFERENCES*

1. Materials of the 27th CPSU Congress. Moscow, Politizdat, 1986.
2. Pravda. March 15, 1986.

*All references to Frolov's paper are in Russian.

3. Frolov, K. V., Markov, M. A., and Stroganov, G. B. Science and technology progress and intensified social production. Moscow, Znaniye, 1986.
4. Alomov, O. D., Mamasaidov, M. T., and Putintseva, I. M. Cored flexible screw mechanisms. Frunze, Ilim, 1983.
5. Frolov, K. V. ed. Nonlinear vibrations and transient processes in machinery. Moscow, Nauka, 1972.
6. Frolov, K. V. and Furman, F. A. Applied theory of vibration protection systems. Moscow, Mashinostroyeniye, 1980.
7. Nakoryakov, V. E., ed. Hydrodynamics of one-phase and biphase systems. Institute of Thermal Physics, Siberian Branch of the USSR Academy of Sciences, Novosibirsk, 1982.
8. Rashidov, T. R. Development of mechanics in uzbekistan. Tashkent, Fan, 1986.
9. Ovsyannikov, L. V. Lectures on the principles of gas dynamics. Moscow, Nauka, 1981.
10. Khristianovich, S. A., ed. Gas-vapor intracycle gasification and ecology problems in power industry. Moscow, Nauka, 1983.
11. Davidov, Yu. M., Kruglov, I. G., Mednov, A. A., et. al. Numerical flow studies in internal combustion engines: the large particle method. Academy of Sciences, Computer Center, Moscow, 1983.
12. Svishchev, G. P. et al. Aerodynamics and acoustics of helical fans. Proceedings of ZAGI, Moscow, No. 2189, 1982.
13. Mikeladze, V. G. and Titov, V. M. Geometry and airodynamics of aircraft and rockets. Moscow, Mashinostroyeniye, 1982.
14. Davidov, Yu. M. Transonic gas flows: computational tables. Academy of Sciences Computer Center, Moscow, 1983.
15. Dulov, V. G. and Luk'yanov, G. A. Gas dynamics of outflow processes. Novosibirsk, Nauka, 1984.
16. Kachnov, Yu. S., Kozlov, V. V., and Levchenko, V. Ya. Turbulent processes in a boundary layer. Novosibirsk, Nauka, 1982.
17. Belotserkovskii, O. M. and Davidov, Yu. M. Large particle method in gas dynamics: computation experiment. Moscow, Nauka, 1982.
18. Nepomnyashchii, A. A., ed. Bifurcation transitions in hydrodynamic stability problems. Preprint, Sverdlovsk, 1982.
19. Soloukhin, R. I. and Fomin, N. A. Fluid dynamical displacement lasers. Minsk, Nauka and Tekhn., 1984.
20. Goldstein, M. A. Eddy flows. Novosibirsk, Nauka, 1981.
21. Frolov, K. V., ed. Combustion and explosion: theory. Moscow, Nauka, 1981.
22. Babushkin, V. I. and Goldstein, V. M. Self-ignition range. transient reaction regimes. Math. Inst., Siberian Branch of the USSR Acad. Sci., Preprint, Novosibirsk, 1958.
23. Frolov, K. V., ed. Metal-Saving Technology. Moscow, Znaniye, 1984.
24. Frolov, K. V., ed. New structural materials. Znaniye, 1985.
25. Alimov, O. D., Manzhosov, V. K., and Eremyants, V. E. Wave deformation distributions in impact systems. Frunze, Ilim, 1978.
26. Nonlinear models and mechanics of deformable solids. Moscow, Nauka, 1984.
27. Makeev, V. P., Grigorenko, N. I., and Pavliuk, Yu. S. Statistical problems in the dynamics of elastic constructions. Moscow, Nauka, 1984.
28. Mechanical testing of metals: static-loading crack resistance (fracture toughness) evaluation. All-Union State Standard 25.506–805, Moscow, Izd. Stand., 1985.
29. Strength design and strength testing in mechanical engineering: mechanical testing of metals. dynamic fracture toughness (crack resistance) evaluation. Moscow, 1980.
30. Frolov, K. V. and Gusenkov, A. P., eds. Reliability and service life in mechanical engineering. Moscow, Nauka, 1984.
31. Kravchuk, A. S., Maiboroda, V. P., and Urzhumtsev, Yu. S. Mechanics of polymers and composite materials. Nauka, Moscow, 1985.
32. Frolov, K. V., ed. Biomechanics of the man-machine system. Moscow, Nauka, 1981.

33. Okhotsimskii, D. E. and Golubev, Yu. F. Automated walking machine: mechanics and motion control. Nauka, 1984.
34. Goncharevich, I. F. and Frolov, K. V. Theory of vibration technology. Moscow, Nauka, 1981.
35. Titov, V. M., ed. Mechanics of rapid processes. Institute of Hydrodynamics, Siberian Branch of the USSR Academy of Sciences, Novosibirsk, 1984.
36. Ilyinskii, N. B. and Potashov, A. V. Boundary value problems in the theory of explosion. Kazan, Kazan University Press, 1986.
37. Davidov, Yu. M. Numerical simulation in continuum mechanics: large particle method In: Current problems in continuum mechanics. Moscow, Moscow Physical-Technical Institute, 1985.
38. Nakoryakov, V. E. Hydrodynamics of biphase flows. Institute of Thermal Physics of the Siberian Branch of the USSR Academy of Sciences, Preprint No. 75-78, Novosibirsk, 1981.
39. Dynamic properties of linear vibration protection systems. Moscow, Nauka, 1982.
40. Alifov, A. A. and Frolov, K. V. Nonlinear vibration systems interacting with power sources. Moscow, Nauka, 1985.
41. Frolov, K. V. Mechanical engineering: current problems and machine improvement methods. Moscow, Mashinostroyenie, 1984.
42. Frolov, K. V., ed. Man-machine: vibration problems. Moscow, Nauka, 1973.
43. Fratkin, Yu. B. and Yakubovich, I. M. Automated data acquisition systems. Information Retrieval Complex, Ministry of the Means of Communication Industry of the USSR, Moscow, 1985.
44. Berestova, V. I., Zabrodina, T. V., Kozin, A. V. et al. Workstation: hardware and basic software. Moscow, Moscow Institute of Physics and Engineering, 1984.
45. CAD in technology: principles of development. All-union Research Institute of Information and Technical-and-Economic Studies in Robotics and Machine Science, 1986.
46. Averchenkov, V. I. and Kamaev, V. A. CAD principles. Volgograd, Volgograd Polytechnical Institute, 1984.
47. Zaslavskii, A. I., Zubich, I. A., and Tsipenyuk, M. V. A guide to adaptive CAD data banks based on the automated workstation concept. Information Retrieval Complex, Ministry of the Means of Communication Industry of the USSR, Moscow, 1984.
48. Trends in CAD. Compiled by G. N. Svirin. Moscow, Research Institute of the Automotive Industry, 1985.
49. CAD technology and CNC equipment in mechanical engineering. Kurgan, Kurgan Institute of Mechanical Engineering, 1984.
50. Kononenko, V. O. Nonlinear vibrations in mechanical systems. Kiev, Naukova Dumka, 1980.
51. Frolov, K. V. Multifaced world of vibrations: science and mankind. Moscow, Znanie, 1985.
52. Ganiev, R. F. Vibrations in multi-phase media: technological applications. Kiev, Naukova Dumka, 1980.
53. Frolov, K. V. and Goncharevich, I. F. Theory of vibration technology. Moscow, Mashinostroyenie, 1981.
54. Kuznetsov, P. D. Aeroengine reliability In: Service life and reliability in mechanical engineering. Moscow, Nauka, 1986.
55. Pisarenko, G. S. et al. Strength under extreme conditions. Kiev, Naukova Dumka, Vol. 12, 1980.
56. Kogaev, V. P., Makhutov, N. A., and Gusenkov, A. P. Strength and durability design. Moscow, Mashinostroyenie, 1985.
57. Panasiuk, V. V., Andreikin, A. E., and Kovchik, S. E. Crack resistance of structural materials. Kiev, Naukova Dumka, 1977.
58. Parton, V. Z. and Morozov, E. M. Mechanics of elastoplastic fracture. Moscow, Nauka, 1985.
59. Tomashevskii, V. T. et al. Ship building mechanics: theory and methods. Leningrad, Naval Academy, 1985.
60. Kragelskii, I. V. and Mikhin, N. M. Reference book on friction units. Moscow, Mashinostroyeniye, 1984.

61. Drozdov, Yu. N. and Frolov, K. V. Friction-related surface deterioration: invariant method calculations In: Physics, Chemistry, Mechanics, No. 5, 1985.
62. Pinegin, S. V., Orlov, A. V., and Tabachnikov, Yu. B. Precision antifriction bearings and gas lubrication bearings. Moscow, Mashinostroyenie, 1984.
63. Semenov, A. I. and Istomin, N. I. Antifriction properties of fluoropolymeric composite materials. Moscow, Nauka, 1981.
64. Drosdov, Yu. N., Pavlov, V. G., and Puchkov, V. N. Friction and wear under extreme conditions. Moscow, Mashinostroyeniye, 1986.
65. Frolov, K. V. Mechanical engineering, science, progress. Kommunist, No. 6, pp. 36–47, 1986.

ASYMPTOTIC METHODS IN NONLINEAR MECHANICS

Yu. A. Mitropolskii

The behavior of vibrational systems is usually analyzed in terms of ordinary differential equations, and since most vibrational processes are nonlinear in character, so too are the differential equations. In many cases of interest, however, the nonlinearities involved are small in magnitude, vibrations are almost linear, and the governing differential equations are very nearly linear equations. Such equations usually contain a small parameter ϵ, and in the limit of vanishing ϵ reduce to linear differential equations which describe a harmonic vibrational process. Speaking in geometric terms, the shape of slightly nonlinear vibrations may be very close to, and indeed hardly distinguishable from, a harmonic vibration if viewed over a period or two; in the long run, however, even the slightest departures from linearity may become significant. The important point is that the presence of nonlinearity violates the superposition principle so fruitful in the solution of differential equations. Nonlinear energy sources (absorbers), if present in a system, may be of no consequence over one vibration cycle, but their cumulative effect may be quite considerable over a large number of periods in terms of the attenuation, build-up, or stability of the entire vibrational process. In the same vein, the continuous action of a nonlinear quasi-elastic force may affect the phase and possibly other characteristics of the vibrational process.

Vibrational systems with a small nonlinearity are of course rather easy objects to deal with theoretically, but problems with large nonlinearities are often very difficult to solve and require an individual approach in each particular case. There are several techniques that have been successfully applied to such problems (the Poincaré method, perturbation theory, averaging method, etc.) but Krylov and Bogolyubov's asymptotic method appears to be the most efficient and widely used approach now in nonlinear mechanics.

The idea of the method is best illustrated with a specific example. Following the authors' original argument we start by analyzing the physics of the simple oscillatory process described by the equation of the form

$$d^2x/dt^2 + \omega^2 x = \epsilon f(x, dx/dt) \qquad (1)$$

where ϵ is a small positive parameter.

In the case of no perturbation ($\epsilon = 0$) it is clear that we shall have purely harmonic vibrations of the form

$$x = a \cos \psi, \quad da/dt = 0, \quad d\psi/dt = \omega(\psi = \omega t + \theta)$$

where the amplitude a and the phase θ are constant in time and are determined by the initial conditions.

Introduction of a nonlinear perturbation (nonzero ϵ) gives rise to: (1) the appearance of higher harmonics in the solution of (1), (2) the dependence of the instantaneous frequency on the amplitude, and (3) the systematic increase or decrease of the amplitude, depending on whether the perturbation increases or decreases the total energy of the system. All these effects are clearly absent in the limiting case of zero ϵ.

With the above in mind, the approximate solution of the differential equation (1) may be represented in the form of the asymptotic expansion

$$x = a \cos \psi + \epsilon U_1(a, \psi) + \epsilon^2 U_2(a, \psi) + \ldots \qquad (2)$$

with the assumption that the functions $U_1(a, \psi)$, $U_2(a, \psi)$, etc., are 2π-periodic in ψ, and the time dependence of a and ψ is determined by the differential equations

$$\begin{aligned} \frac{da}{dt} &= \epsilon A_1(a) + \epsilon^2 A_2(a) + \ldots \\ \frac{d\psi}{dt} &= \omega + \epsilon B_1(a) + \epsilon^2 B_2(a) + \ldots \end{aligned} \qquad (3)$$

The problem thus is reduced to that of choosing the functions $U_1(a, \psi)$, $U_2(a, \psi)$, ..., $A_1(a)$, $A_2(a)$, ..., $B_1(a)$, $B_2(a)$, ... in such a way that, on substituting a and ψ as found from (3), the asymptotic expansions (2) will solve the original equation (1). Once such functions are found and the expansion coefficients in the right-hand sides of (2) and (3) are explicitly known, the solution of (1) is reduced to the integration of the system of equations (3), which can be solved by separating the variables in the usual manner.

The above scheme has been successfully extended to non-stationary processes in nonlinear vibrational systems with slowly varying parameters. If the motion is in one dimension, the systems of this type are governed by equations of the form

$$\frac{d}{dt}\left[m(\tau)\frac{dx}{dt}\right] + c(\tau)x = \epsilon F\left(\tau, \theta, x, \frac{dx}{dt}\right) \tag{4}$$

where ϵ is a small parameter; $\tau = \epsilon t$ is the slow time; $m(\tau)$ may be thought of as the slowly varying mass of the system; $c(\tau)$ is the rigidity; $d\theta/dt = \nu(\tau)$ is the slowly varying instantaneous frequency of the perturbing force. The same argument as above allows us to represent the asymptotic solution of (4) in the form

$$x = a\cos\psi + \epsilon U_1(\tau, a, \psi) + \epsilon^2 U_2(\tau, a, \psi) + \dots \tag{5}$$

where $U_1(\tau, a, \psi)$, $U_2(\tau, a, \psi)$, ... are 2 π-periodic functions of the angle ψ, and the time dependence of a and ψ is determined by the system of differential equations

$$\frac{da}{dt} = \epsilon A_1(\tau, a) + \epsilon^2 A_2(\tau, a) + \dots \tag{6}$$

$$\frac{d\psi}{dt} = \omega(\tau) + \epsilon B_1(\tau, a) + \epsilon^2 B_2(\tau, a) + \dots \tag{7}$$

where $\omega(\tau) = \sqrt{c(\tau)/m(\tau)}$ is the "eigenfrequency" of the vibrational system under study.

The above approach has also been applied to systems with many degrees of freedom and to systems with resonances; to distributed parameter systems; to almost integrable systems, etc. In each of these cases effective computational formulas have been developed, and a detailed analysis carried out of stationary and non-stationary regimes possible in a particular system.

It seems worthwhile discussing here one major aspect of the behavior of systems with slowly varying external parameters, namely, the problem of adiabatic invariants in nonlinear vibrational systems.

The term "adiabatic invariant" has long been known in classical mechanics and usually refers to the quantities emerging in the study of situations in which forces or constants involved in the Hamiltonian vary with time very slowly on the frequency scale of the system. The evolution of this concept from the works of Boltzman and Ehrenfest is retraced in [1] where the reader will also find a very useful survey of the physical (or rather mathematical) properties of adiabatic invariants. It is shown, in particular, that the conservation of these quantities is related to the fact of conservation of invariant integral varieties during the slow variation of physical parameters. The authors also examine the behavior of adiabatic invariants at singular points, where the system is no longer an adiabatic one, and apply the method to systems with many degrees of freedom.

It is instructive to consider some aspects of the actual construction of adiabatic invariants for dynamic systems with one or more degrees of freedom.

For a low-amplitude mathematical pendulum it has been shown by M. Born that if ω is the frequency of oscillations and W the total energy of the pendulum, then the quantity W/ω remains constant. M. Born obtained this result by analyzing the averaged motion of the pendulum. We shall demonstrate now (see [4]) that the same result can be derived rigorously.

Writing the equation of motion in the form

$$\frac{d}{dt}\left[ml^2(\tau)\frac{dx}{dt}\right] + mgl(\tau)x = 0 \tag{8}$$

and applying the asymptotic method, we find in the first approximation that

$$x = a\cos\psi, \quad \dot{x} = -a\omega(\tau)\sin\psi \tag{9}$$

where a and ψ must be determined from the system which is obtained by averaging the right-hand sides of the exact equations,

$$\begin{aligned}\frac{da}{dt} &= -\frac{a}{\omega(\tau)}\left[\frac{2l(\tau)}{l(\tau)}\omega(\tau) + \dot{\omega}(\tau)\right]\sin^2\psi \\ \frac{d\psi}{dt} &= \omega(\tau) - \frac{1}{\omega(\tau)}\left[\frac{2\dot{l}(\tau)}{l(\tau)}\omega(\tau) + \dot{\omega}(\tau)\right]\sin\psi\cos\psi\end{aligned} \tag{10}$$

and has the form

$$\frac{da}{dt} = -\epsilon\frac{3l'_\tau(\tau)}{4l(\tau)}a, \quad \frac{d\psi}{dt} = \omega(\tau), \quad \omega(\tau) = \sqrt{\frac{g}{l(\tau)}} \tag{11}$$

(in the first approximation). Integrating, we get

$$a = a_0(l_0/l(\tau))^{3/4}$$

the suffix zero denoting initial values.

The total energy of the pendulum is

$$W = T + V = ml^2(\tau)\dot{x}^2 + \frac{1}{2}mgl(\tau)x^2 \tag{12}$$

in the first approximation or, using [9]

$$W = \frac{1}{2}ml^2(\tau)a^2\,\omega^2(\tau) \tag{13}$$

$$\frac{W}{\omega(\tau)} = \frac{mga_0^2 l_0^{3/2}}{2\sqrt{g}} = \text{const} \tag{14}$$

showing that a pendulum with a slowly varying length, equation (8), does possess an adiabatic invariant (again to the first approximation, that is, for those values of a that result from the system (11)). It is to be understood that the quantity $W/\omega(\tau)$, equation (14), is an exact invariant for the system (11) accurate to the first order.

The quantity $W/\omega(\tau)$ is an adiabatic invariant for equations (10), and its time derivative, in view of (10), differs from zero and is of the first order in ϵ

$$\frac{d}{dt}\left[\frac{W}{\omega(\tau)}\right] = \frac{\epsilon}{2} mg \cdot \frac{3}{2} l'(\tau) \cdot l^{1/2}(\tau) a^2 \cos 2\psi \tag{15}$$

It is an important point that the knowledge of an adiabatic invariant makes it possible to gain some useful insight into the behavior of a system without actually integrating the differential equation(s). Let us consider, for example, a one-dimensional oscillating system with a slowly varying rigidity

$$\frac{d^2x}{dt^2} + c(\tau)x = 0 \tag{16}$$

Given the adiabatic invariant of the system

$$J(\tau, a) = \frac{1}{2}\sqrt{c(\tau)} \cdot a^2 = \text{const} \tag{17}$$

it is a simple matter to conclude that the amplitude of vibrations is

$$a = \frac{\text{const}}{\sqrt[4]{c(\tau)}} \tag{18}$$

A convenient and relatively general method for deriving an adiabatic invariant has been suggested by Djukic [2]. The idea of the method is as follows.

Consider the equation

$$m(\tau)\ddot{q} + m(\tau)\,\omega^2(\tau)q = \epsilon f(\tau, q, \dot{q}) \tag{19}$$

with slowly varying parameters, where $\tau = \epsilon t$, t is the time, q—the generalized coordinate, ϵ—a small positive parameter, and $\omega(\ )$, $m(\tau)$—arbitrary functions of the slowly varying time.

An adiabatic invariant of the dynamical system (19) is, as already noted, a certain function $D_\epsilon(t, \tau, q, \dot{q})$ whose time derivative contains only terms proportional to ϵ in view of the differential equation (19). But if this derivative does equal zero, then $D_\epsilon(t, \tau, q, \dot{q})$ is an exact invariant, i.e., the first integral of equation (19).

Let the time, the generalized coordinate, and the generalized momentum be changed by the equations

$$\begin{aligned}
\bar{t} &\cong t + \mu\psi(t, \tau, q, p) \\
\bar{q} &\cong q + \mu\phi(t, \tau, q, p) \\
\frac{d\bar{q}}{dt} &\cong \frac{dq}{dt} + \mu[\dot{\phi} - \dot{q}\dot{\psi} + \Phi(t, \tau, q, p)] \\
\bar{p} &\cong p + \mu\nu(t, \tau, q, p)
\end{aligned} \tag{20}$$

where μ is a small parameter, and the functions ψ, φ, Φ and ν have the time, slow time, and generalized coordinates and momenta as their arguments. Assuming that

$$\nabla = (p\dot{q} - H)\, dt \tag{21}$$

is invariant in the sense

$$\Delta\nabla \equiv \left[\bar{p}\frac{d\bar{q}}{dt} - \bar{H}(\tilde{t}, \bar{p}, \bar{q})\right] d\tilde{t} - \left[p\frac{dq}{dt} - H(t, p, q)\right] dt = 0 \tag{22}$$

$$H = \frac{1}{2}\left[\frac{1}{m(\tau)}p^2 - m(\tau)\omega^2(\tau)q^2\right] \tag{23}$$

we arrive after some manipulations at equations for determining the functions ψ, φ, Φ, ν and the adiabatic invariant D_ϵ, which the latter can be written

$$D_\epsilon = H - \epsilon\left\{\frac{\omega'}{2\omega^2}H\cdot(2\varphi + \sin 2\varphi) + \frac{m'}{2\sqrt{m}}\sqrt{2H}\cdot\varphi\right\}$$

$$+\,\epsilon\frac{\sqrt{2H}}{\sqrt{m\omega}}\int_0^\varphi f\sin\varphi\, d\varphi \tag{24}$$

under certain not too restrictive assumptions, the leading term H representing the total energy of the unperturbed system (that is, for $f(\tau, q, \dot{q}) = 0$).

To illustrate the use of the above method we consider a double pendulum with slowly varying lengths, $l_1(\tau)$ and $l_2(\tau)$, and point masses m_1 and m_2, as shown in Fig. 1 (see [5]). It is assumed that both the lengths oscillate in the vertical plane and are slow functions of the time $\tau = \epsilon t$, ϵ being a small parameter. The kinetic and potential energies of the system will then be given respectively by the equations

$$T = \frac{1}{2}m_1 l_1^2(\tau)\dot{\varphi}_1^2 + \frac{1}{2}m_2[l_1^2(\tau)\dot{\varphi}_1^2 + 2l_1(\tau)l_2(\tau)\dot{\varphi}_1\dot{\varphi}_2\cos(\varphi_1 - \varphi_2) + l_2^2(\tau)\dot{\varphi}_2^2] \tag{25}$$

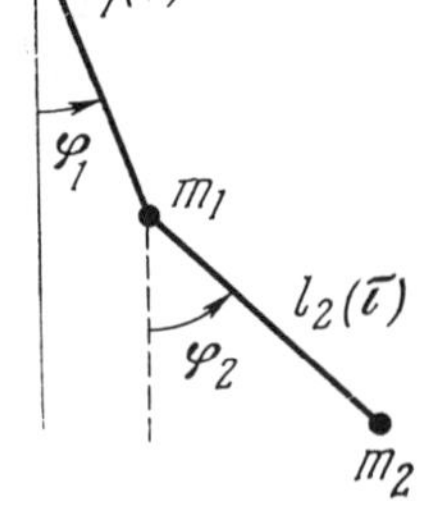

Figure 1

and

$$V = m_1 g l_1(\tau)(1 - \cos\varphi_1) + m_2 g[l_1(\tau)(1 - \cos\varphi_1) + l_2(\tau)(1 - \cos\varphi_2)] \tag{26}$$

Assuming $m_1 = m_2 = 1$ for the sake of simplicity and dropping terms of an order higher than the third in the expansions of trigonometric functions we obtain the system of equations

$$\begin{aligned} &2l_1^2(\tau)\ddot{\varphi}_1 + l_1(\tau)l_2(\tau)\ddot{\varphi}_2 + 2gl_1(\tau)\varphi_1 = \epsilon f_1(\tau, \dot{\varphi}_1, \dot{\varphi}_2) \\ &l_1(\tau)l_2(\tau)\ddot{\varphi}_1 + l_2^2(\tau)\ddot{\varphi}_2 + gl_2(\tau)\varphi_2 = \epsilon f_2(\tau, \dot{\varphi}_1, \dot{\varphi}_2) \end{aligned} \tag{27}$$

where

$$\begin{aligned} &\epsilon f_1(\tau, \dot{\varphi}_1, \dot{\varphi}_2) = -4l_1(\tau)\dot{l}_1(\tau)\dot{\varphi}_1 - (\dot{l}_1(\tau)l_2(\tau) + l_1(\tau)\dot{l}_2(\tau))\dot{\varphi}_2 \\ &\epsilon f_2(\tau, \dot{\varphi}_1, \dot{\varphi}_2) = -(\dot{l}_1(\tau)l_2(\tau) + l_1(\tau)\dot{l}_2(\tau))\dot{\varphi}_1 - 2l_2(\tau)\dot{l}_2(\tau)\dot{\varphi}_2 \end{aligned} \tag{28}$$

Putting $l_1(\tau) = kl_2(\tau)$ and making a number of other assumptions, the adiabatic invariant is found to be

$$D = \bar{H}\sqrt{l(\tau)} = \text{const} \tag{29}$$

with

$$\bar{H} = \frac{g l_0^{3/2}}{2\sqrt{l(\tau)}}[K\,\varphi_{20}^2 + 2\,\varphi_{10}^2] \tag{30}$$

where $l_0 = l(0)$, and φ_{10}, φ_{20} are the initial deviations of the pendulum's components. In the general case, the double-pendulum adiabatic invariant is given by

$$D = \frac{\bar{H}_1}{\nu_1(\tau)} + \frac{\bar{H}_2}{\nu_2(\tau)} = \text{const} \tag{31}$$

where

$$\begin{aligned} &\bar{H}_1 = \frac{1}{2}[x_0^2\Delta_{10}\nu_{10}\nu_1(\tau)] \\ &\bar{H}_2 = \frac{1}{2}[y_0^2\Delta_{20}\nu_{20}\nu_2(\tau)] \end{aligned} \tag{32}$$

$\bar{H}_i$ is the averaged Hamiltonian in the averaged motion regime; a and y—the normal coordinates, and $\nu_1(\tau)$, $\nu_2(\tau)$—the normal frequencies.

The recent results obtained by A. K. Lopatin and the present author [6, 7] in decomposing differential equations with a small parameter appear to be interesting in themselves and are also indicative of the progress in the field of the asymptotic methods of nonlinear mechanics. The work is based on using transformations in the form of Lie expansions in powers of a small parameter. The perturbation theory using Lie series and Lie transformations offers several advantages over other existing methods, not the least of which is the simplicity of the algorithm, and has already been successfully used by a number of investigators.

One of them is A. Ya. Povzner, who, based on the well-known Campbell-Hausdorff formula, has succeeded in deriving a number of delicate theorems bearing on the properties of transformed systems [8]. Although we also employ the CH formula, the cornerstone of our decomposition method is in fact the analysis of the structure of the Lie groups generated by the original and transformed systems, which allowed us to derive a number of major results by using the powerful formalism of the classical theory of continuous Lie groups.

We will illustrate the general algorithm of asymptotic decomposition by considering the system of ordinary differential equations

$$dx/dt = \omega(x), \quad x(0) = x_0 \tag{33}$$

where x = colon $\|x_1, x_2, \ldots x_n\|$; $\omega(x)$ = colon $\|\omega_1(x), \ldots \omega_n(x)\|$, $\omega_i(x) \in 0(G)$, $0(G)$ denoting the manifold of analytical functions defined on G; $i = \overline{1, n}$; $G_0 = I \times G \times R^{n+1}(G \in R^n,\ t \in I)$ is the existence and uniqueness domain of the solution of the Cauchy problem for the system (33). In order to investigate the system (33), let us consider its structural properties in terms of its invariance group. The system (33) is said to be invariant under a local one-parameter Lie group

$$x = \exp\,(\mu Z(\bar{x}))\bar{x} \tag{34}$$

where μ is a characteristic parameter of the group and Z a certain operator, if this transformation converts the system into

$$d\bar{x}/dt = \omega(\bar{x}) \tag{35}$$

which occurs if, and only if, the operator Z solves

$$[U, S] = 0 \tag{36}$$

where $U = \omega_1(x)\ \delta/\delta\ x_1 + \ldots + \omega_n(x)\ \delta/\delta\ x_n$ is the linear differential operator associated with the system (33), and [,] is the Poisson bracket. The set of all the solutions of the equation (33) forms a Lie algebra L, which specifies completely the initial system (33).

We now subject the zero-approximation system (33) to small perturbations $\epsilon\tilde{\omega}(x)$ to obtain the perturbed system

$$\frac{dx'}{dt} = \omega(x') + \epsilon\ \tilde{\omega}(x'), \quad x'(t_0) = x_0 \tag{37}$$

where $\tilde{\omega}(x')$ = colon $\|\tilde{\omega}_1(x'), \ldots, \tilde{\omega}_n(x')\|$.

It is appropriate at this stage to associate a reference system with the system (33) by using the substitution of variables in the form of the Lie series

$$x_j' = \exp\,(\epsilon S(x))x_j, \quad j = \overline{1, n} \tag{38}$$

where $S = S_1 + \epsilon S_2 + \ldots$; $S_i(x) = \gamma_{i1}(x)\ \partial/\partial x_1 + \ldots + \gamma$ in $(x)\ \partial/\partial x_n$.

Using the relationship between the system (33) and its associated differential operator

$$U_0' = U' + \epsilon\tilde{U}' \tag{39}$$

where $\tilde{U}' = \tilde{\omega}_1(x') \, \partial/\partial x' + \ldots \tilde{\omega}_n(x') \, \partial/\partial x'_n$, we subject U'_0 to the transformation (38) and apply the Campbell-Hausdorff formula to obtain, after a little manipulation

$$U'_0 \rightarrow U_0 = U + \epsilon(-[U, S_1] + F_1) + \ldots + \epsilon^\nu(-[U, S_\nu] + F_\nu) + \ldots \tag{40}$$

where $F_1 = U$, $F_2 = -[\tilde{U}, S_1] + \frac{1}{2}[U, [U, S_1]]$, $\ldots$, F_ν is a known function of the operators U, $\tilde{U}$, $\tilde{S}_1, \ldots, S_{\nu-1}$.

The transformed operator U_0 and hence the corresponding system of differential equations both depend on the manner in which the operators S_1, $S_2 \ldots$ are chosen from the sequence of recurrent operator equations

$$[U, S_j] = F_j - pr\, F_j, \quad j = 1, 2, \ldots \tag{41}$$

where $pr\, F_j$ denotes the projection of the operator F on the algebra I_0.

The transformed operator U_0 takes the form

$$U_0 = U + \epsilon N_1 + \epsilon^2 N_2 + \ldots + \epsilon^\nu N_\nu + \ldots \tag{42}$$

and yields the transformed system

$$\frac{dx_j}{dt} = \omega_j(x) + \epsilon N(x) x_j, \quad j = \overline{1, n} \tag{43}$$

with $N(x) = N_1 + \epsilon N_2(x) + \epsilon^{\nu-1} N_\nu(x) + \ldots$

The system (43) is usually referred to as the centralized system and has two important properties: (1) its zero-th approximation is identical to its zero-th approximation system (33), and (2) it remains invariant under the one-parameter transformation group

$$x = \exp(\mu U(\bar{x}))\bar{x} \tag{44}$$

where the operator $U(x)$ is associated with the zero-th approximation system (33). We will use the term "asymptotic decomposition algorithm" to denote the above transition from the perturbed system (37) to the centralized system (43).

The centralized system (43), which is thus a reference system for the perturbed system (37), is exactly invariant with respect to the one-parameter transformation group (44), whereas the system (37) itself is invariant only to the zero-th approximation. The centralized system (43) is simpler to integrate than is the initial perturbed system (37). Note also that all the properties of the centralized system are completely determined by the algebra of the centralizer.

A very important question associated with the averaging method and the asymptotic methods of nonlinear mechanics is whether the system under study allows separation of variables (decomposition) into slow and rapid ones. We shall say that the centralized system (43) does allow such a separation if the substitution

$$\begin{aligned} y_1 &= f_1(x), \quad y_2 = f_2(x), \ldots, y_k = f_k(x) \\ Z_1 &= \varphi_1(x), \quad Z_2 = \varphi_2(x), \ldots, Z_r = \varphi_r(x) \end{aligned} \tag{45}$$

(see [9]) converts it into a pair of subsystems, one of which, of the order $r = n$

$-\ k$, is to be integrated after the other, of the order k. This fundamental statement may be expressed as **Theorem** [9]. Suppose a one-parameter group generated by an operator U associated with a zero-th approximation system, satisfies the following conditions:

1) there exists a certain area of the variables x and functions $\omega(x)$ in which a system of invariants may be found with a basic set $P_k = \{f_1(x), \ldots, f_k(x)\}$ defined as a solution to the equation $Uf = 0$;
2) in the same area there exists an imprimitivity system with a basis set $T_r = \{\varphi_1(x), \ldots, \varphi_r(x)\}$ defined as a solution to the system $Mf = 0$, where the M's are the generatrices of a certain ideal contained in the defining algebra, and
3) the functions of the defining sets P_k and T_r form a compatible system of $k + r = n$ independent functions in their definition area.

Then the change of variables

$$\begin{aligned} y_1 &= f_1(x), \quad y_2 = f_2(x), \ldots, y_k = f_k(x) \\ Z_1 &= \varphi_1(x), \quad Z_2 = \varphi_2(x), \ldots, Z_r = \varphi_r(x) \end{aligned} \tag{46}$$

transforms the centralized system

$$dx_j/dt = \omega_j(x) + \epsilon N x_j, \quad j = \overline{1, n}$$

where $N = N_1 + \epsilon N_2 + \ldots + \epsilon^{\nu-1} N + \ldots, [U, N] \equiv 0, \nu = 1, 2, \ldots$ into a pair of subsystems, one of which, to be integrated first

$$\frac{dy_j}{dt} = \sum_{\nu=1}^{\infty} \epsilon^{\nu}\, \Phi_{\nu j}(y_1, \ldots, y_k), \quad j = \overline{1, k} \tag{47}$$

determines the k slow variables and contains the slow time $\tau = \epsilon t$, and the other

$$\frac{dz_i}{dt} = F_i(z) + \sum_{\nu=1}^{\infty} \epsilon^{\nu} \sum_{\varkappa=1}^{m\nu} (\Phi_{0\nu}(y) F_i(z) + \Psi_{\varkappa\nu}(y) \Phi^{(1)}_{\varkappa\nu}(z)), \quad i = \overline{1, r} \tag{48}$$

determines the $r = n - k$ rapid variables and is integrated second.

It will be noted that conditions 1 to 3 above are both sufficient and necessary for separating the variables into slow and rapid ones in the centralized system.

As an illustration, the decomposition method just discussed will be applied to the well-known Van der Pol equation

$$\frac{d^2U}{dt^2} + \epsilon(U^2 - 1)\frac{du}{dt} + U = 0 \tag{49}$$

(see [10]). Changing from U, du/dt to new slow variables b_1, b_2 defined by

$$\begin{aligned} U &= b_1 \cos t + b_2 \sin t \\ \frac{dU}{dt} &= -b_1 \sin t + b_2 \cos t \end{aligned} \tag{50}$$

averaging the resulting system of equations for b_1, b_2, and transforming again to the variables x_1, x_2, τ ($x_1 = r^2 = b_1^2 + b_2^2$, $x_2 = b_2^2/b_1^2$, $\tau = \epsilon t$), we obtain the system

$$\frac{dx_1}{dt} = \left(1 - \frac{x_1}{4}\right)x_1$$
$$\frac{dx_2}{dt} = -\frac{\epsilon}{8}\left(1 - \frac{3}{2}x_1 + \frac{11}{32}x_1^2\right)(1 + x_2^2) \tag{51}$$

Applying our method we find the substitution

$$x_1 = z_1 + \epsilon^2 \ldots$$
$$x_2 = z_2 + \epsilon(1 + z_2^2)\int^{z_2} \frac{\left(\frac{3}{2}z_1 - \frac{11}{32}z_1^2\right)}{2(4 - z_1)z_1}\,dz_1 + \epsilon^2 \ldots \tag{52}$$

after some manipulations, which convert the system (51) to

$$\frac{dz_1}{dt} = \left(1 - \frac{z_1}{4}\right)z_1$$
$$\frac{dz_2}{dt} = -\frac{\epsilon}{8}(1 + z_2^2) \tag{53}$$

where the variables are completely separated, and either equation may be integrated independently.

If there is a small periodic force term in the Van der Pol equation

$$\frac{d^2U}{dt^2} - \epsilon(1 - U_2)\frac{dU}{dt} + U = \epsilon E \cos \nu t \tag{54}$$

substitutions similar to those above lead to the system

$$\frac{dx_1}{dt} = \epsilon\left\{\left[1 - \frac{1}{4}\left(1 + \frac{\Delta}{\nu}\right)x_1\right]x_1 + \frac{E}{\nu}\left(1 + \frac{\Delta}{2\nu}\right)\sqrt{\frac{x_1}{1 + x_2^2}}\right\}$$
$$- \frac{\epsilon^2 E}{16\nu^2}(x_1 + 4)x_2\sqrt{\frac{x_1}{1 + x_2^2}}$$
$$\frac{dx_2}{dt} = -\Delta(1 + x_2^2) - \frac{\epsilon}{2\nu}E\left(1 + \frac{\Delta}{2\nu}\right)\sqrt{\frac{1 - x_2^2}{x_1}} - \frac{\epsilon^2}{8\nu}\left\{(1 + 4\nu\Delta^2)(1 + x_2^2)\right. \tag{55}$$
$$\left. + \frac{11}{32}x_1^2(1 + x_2^2) - \frac{3}{2}x_1(1 + x_2^2) + \frac{E}{4\nu}\left[\frac{4 - 5x_1}{\sqrt{x_1}\sqrt{1 + x_2^2}} + \frac{4x_2^2 - 7x_2^2x_1}{\sqrt{x_1}\sqrt{1 + x_2^2}}\right]\right\}$$
$$\Delta = (\nu^2 - 1)/2\nu$$

after an appropriate averaging procedure. The substitution (46) now takes the form

$$x_1 = z_1 - \epsilon \frac{E}{\Delta\nu}\left(1 + \frac{\Delta}{2\nu}\right)\sqrt{z_1}\int^{z_2} \frac{-\frac{z_2^2}{2} + \frac{3}{8}z_2^4}{1 + z_2^2}\,dz_2 \tag{56}$$

$$x_2 = z_2$$

and the system (55) becomes

$$\frac{dz_1}{dt} = \epsilon\left\{\left[1 - \frac{1}{4}\left(1 + \frac{\Delta}{\nu}z_1\right)\right\} + \frac{E}{\nu}\left(1 + \frac{\Delta}{2\nu}\right)\sqrt{z_1}\right. \tag{57}$$

$$\frac{dz_2}{dt} = -\Delta(1 + z_2^2) - \frac{\epsilon}{2}\cdot\frac{E}{\nu}\left(1 + \frac{\Delta}{2\nu}\right)z_2\sqrt{\frac{1 + z_2^2}{z_1}}$$

to the first order in ϵ, showing that the variables are not completely separable in this particular case. The first equation in the transformed system (57) depends on z_1 alone and once this equation has been integrated we may proceed to the second equation, whose right-hand side depends on both z_1 and z_2.

A further and more complex example which deserves attention is the asymptotic decomposition of the perturbed motion of flying vehicles. Following [3] we consider an airplane-type vehicle with a vertical plane of symmetry $x0y_1$, see Fig. 2, and denote respectively as $0x_3y_3z_3$, $0x_1y_1z_1$, and $0xy_* \, z_*$—the coordinate systems fixed in space and fixed in the vehicles, and the half-speed system.

The equations of motion relative the half-speed system contain the following 13 variables.

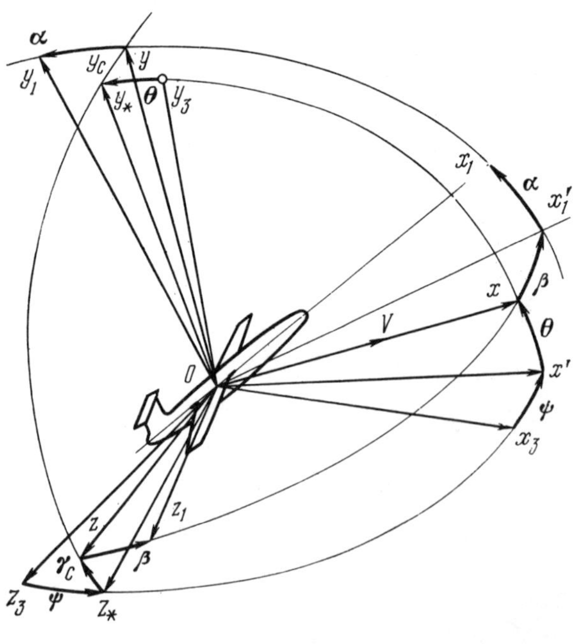

Figure 2

1. Longitudinal variables: V—the velocity of the center of gravity; θ—the angle the flight path makes with the horizontal plane; ω_z—the z-component of the angular velocity vector; ϑ—the pitch angle; x—the x_1 coordinate of the center of gravity; H—the flight altitude; m—the mass of the vehicle;
2. Lateral variables: Ψ—the turning angle of the flight path; ω_x and ω_y—respectively the x_1 and y_1 components of the angular velocity vector; ψ—the hunting angle; γ—the angle of bank.

The longitudinal motion of the flying vehicle includes the translation of the center of gravity in the symmetry plane $0x_1y_1$ and rotation around the $0z_1$ axis.

The lateral motion includes the translation of the center of gravity along the $0z_1$ axis and rotation with respect to the $0x_1$ and $0y_1$ axes.

The longitudinal and lateral components of the motion generally affect each other, and it is their combination that forms the general motion of the object.

The equations of motion of a flying vehicle include seven equations for the longitudinal motion

$$
\begin{aligned}
&\frac{dV}{dt} = F_v(V, \theta, \vartheta, H, m, \Psi, \psi, \gamma) \\
&\frac{d\theta}{dt} = F_\theta(V, \theta, \vartheta, H, m, \Psi, \psi, \gamma) \\
&\frac{d\omega_z}{dt} = F_{\omega_z}(V, \theta, \vartheta, H, m, \Psi, \omega_z, \omega_x, \omega_y, \psi, \gamma) \\
&\frac{\partial\vartheta}{dt} = F_\vartheta(\omega_z, \omega_y, \gamma), \quad \frac{dx}{dt} - F_x(V, \theta, \Psi) \\
&\frac{dH}{dt} - F_H(V, \theta), \quad \frac{dm}{dt} = F_m(V, H, t)
\end{aligned}
\tag{58}
$$

and six equations for the lateral motion,

$$
\begin{aligned}
&\frac{d\Psi}{dt} = F_\Psi(V, \theta, H, \vartheta, m, \Psi, \psi, \gamma) \\
&\frac{d\omega_x}{dt} = F_{\omega_x}(V, \theta, \omega_z, \vartheta, H, \Psi, \omega_x, \omega_y, \psi, \gamma) \\
&\frac{d\omega_y}{dt} = F_{\omega_y}(V, \theta, \omega_z, \vartheta, H, \Psi, \omega_x, \omega_y, \psi, \gamma) \\
&\frac{d\psi}{dt} = F_\psi(\omega_z, \vartheta, \omega_y, \gamma), \quad \frac{d\gamma}{dt} = F_\gamma(\omega_z, \vartheta, \omega_x, \omega_y, \gamma) \\
&\frac{dz}{dt} = F_z(V, \theta, \Psi)
\end{aligned}
\tag{59}
$$

Explicit expressions for the right-hand sides of these equations may be found in [3].

It is advantageous to rewrite the system (58), (59) in the vector form

$$\frac{d\eta}{dt} = F(t, \eta) \tag{60}$$

where the vectors η = colon $\|\eta_1, \eta_2, \ldots, \eta_{13}\|$ and F = colon $\|F_1, F_2, \ldots, F_{13}\|$ have as their component the variables and functions in the right-hand sides of the equations of motions, the ordering of the components being evident from the way the equations are written down.

Supposing now that $\eta = \eta_*$ is a certain program motion of the flying vehicle, that is,

$$\frac{d\eta_*}{dt} \equiv F(t, \eta_*) \tag{61}$$

we wish to investigate the behavior of the object in the close vicinity of this path and, accordingly, we put

$$\eta = \eta_* + \epsilon\Delta\eta \tag{62}$$

with ϵ as a small parameter.

Assuming that there is no slip in the main program motion $\eta_* = \eta_*(t)$, that is $\beta = 0$, and taking into account the symmetry of the vehicle, the system of equations for the perturbed motion is

$$\begin{aligned}\frac{d\Delta\eta_i}{dt} &= \left(\Sigma_\eta \Delta\eta \frac{\partial}{\partial\eta}\right)F_i + \frac{\epsilon}{2}\left(\Sigma\Delta\eta\frac{\partial}{\partial\eta}\right)^2 F_i + \epsilon^2 \ldots \\ \frac{d\Delta\eta_j}{dt} &= \left(\Sigma_\delta \Delta\eta \frac{\partial}{\partial\eta}\right)F_j + \frac{\epsilon}{2}\left(\Sigma\Delta\eta\frac{\partial}{\partial\eta}\right)^2 F_j + \epsilon^2 \ldots \quad (i = \overline{1, 7}, \quad j = \overline{8, 13})\end{aligned} \tag{63}$$

where

$$\left(\Sigma\Delta\eta\frac{\partial}{\partial\eta}\right)F_i = \sum_{j=1}^{13} \Delta\eta_i \frac{\partial F_i(t, \eta_*)}{\partial\eta_j} \tag{64}$$

and the superscripts n and δ denote that in sums of the form (64) only the longitudinal or, respectively, the lateral parameters of motion are present.

If ϵ is put to zero, the system (63) of 13 perturbed equations resolves into two independent systems, one of the sixth and the other of the seventh order.

Applying the decomposition technique discussed above and calculating to the first order in ϵ, we find the substitution

$$\eta_i = \left(1 - \epsilon\sum_{\varkappa=1}^{13}\gamma\varkappa\frac{\partial}{\partial\eta_\varkappa}\right)\bar{\eta}_i \quad (i = \overline{1, 13}) \tag{65}$$

(see [11]) which resolves the system (63) for the perturbed motion into the two independent systems

$$\frac{d\Delta\bar{\eta}_i}{dt} = \left(\Sigma_\eta \Delta\bar{\eta}\frac{\partial}{\partial\bar{\eta}}\right)F_i + \epsilon\left(\Sigma_\eta \Delta\bar{\eta}\frac{\partial}{\partial\bar{\eta}}\right)^2 F_i \quad (i = \overline{1, 7}) \tag{66}$$

of the seventh order, and

$$\frac{d\Delta\bar{\eta}_j}{dt} = \left(\Sigma_\delta \Delta\bar{\eta}\frac{\partial}{\partial\bar{\eta}}\right)F_j + \epsilon\left(\Sigma_\delta \Delta\bar{\eta}\frac{\partial}{\partial\bar{\eta}}\right)^2 F_j \quad (j = \overline{8, 13}) \tag{67}$$

of the sixth.

In conclusion, it seems worthwhile to mention some recent developments in the field of algorithms and in the study of the differential equations governing nonlinear vibrational systems.

A new method suggested by the present author and coworkers [12] for treating evolution equations combines analytical and numerical features and has proved useful in synthetizing an algorithm of construction—and in the study of the properties—of periodic and quasiperiodic solutions to various classes of differential equations, including those with deviating arguments, integro-differential and finite-difference equations, etc. Among other problems relevant to these equations we have studied the construction of invariant toroidal manifolds; the manner in which trajectories behave on manifolds and in the vicinity of an exponentially stable toroidal manifold; the conditions under which a linear system of finite-difference equations with quaziperiodic coefficients may be reduced to a similar system with constant coefficients. We have also succeeded in proving the existence theorem for a continuous torus in a discrete dynamic system.

Interesting results have been obtained by applying the averaging method to systems of differential equations with pulsed excitation. For one class of such systems a detailed study has been carried out concerning their reducibility to the canonical form in the vicinity of a compact integral set filled with equilibrium positions [13]. Two important practical problems to which applications have been made are

- two-frequency beam vibrations with a longitudinal periodic pulsed excitation and nonautonomous and nonlinear fixture conditions;
- nonlinear hydroelastic effects arising in an elastic pipe with a viscous noncompressible liquid flowing under the action of instantaneous pressure pulses.

A number of important achievements in the field of the asymptotic methods of nonlinear mechanics are associated with the application of the averaging method to random vibration problems in nonlinear mechanical systems. One problem of this kind deals with the special case of viscoelastic vibrations governed by the stochastic integro-differential equations of the form

$$\ddot{x} + \omega^2 x = \epsilon F(x, \dot{x}, t) + \epsilon \int_0^t [K(t-s)x(s) + R(t-s)\dot{x}(s)]\, ds + \sqrt{\epsilon}\, h(x, \dot{x})q(t) \tag{68}$$

(see [14, 15]), where $q(t)$ is a centered, stationary, random process, and F and h are periodic functions in t and polynomials in x, $\dot{x}$. Assuming the existence of a slowly varying solution and introducing some restrictions on the relaxation kernels $K(t - s)$ and $R(t - s)$, it proves possible, by means of the statistical linearization method, to approximate (68) by an (also stochastic) ordinary differential equation which can then be treated by asymptotic methods of nonlinear mechanics and by standard techniques of statistical dynamics.

REFERENCES

1. Bakai, A. S. and Stepankovskii, Yu. P. Adiabatic invariants. Kiev, Naukova Dumka, 1981 (in Russian).
2. Djukic, Dj. S. Adiabatic invariants for dynamical systems with one degree of freedom. Intern. J. Non-Linear Mech., 1981, Vol. 16, No. 5–6, pp. 489–498.
3. Lebedev, A. A. and Chernobrovkin, L. S. Flight dynamics of unmanned flying vehicles. Moscow, Oborongiz, 1962 (in Russian).
4. Mitropolskii, Yu. A. Averaging method in nonlinear mechanics. Kiev, Naukova Dumka, 1971 (in Russian).
5. Mitropolskii, Yu. A. On adiabatic invariants in nonlinear mechanics problems. Prepr. Inst. of Math. of the Ukr. SSR Academy of Sciences, Kiev, 1983, No. 85.73, (in Russian).
6. Mitropolskii, Yu. A. and Lopatin, A. K. Vector fields, algebras and groups generated by a system of ordinary differential equations. Ibid, Kiev, 1985.
7. Mitropolskii, Yu. A. and Lopatin, A. K. Decomposition of systems of ordinary differential equations. Prepr. Inst. Math. Ukr. SSR Academy of Sciences, Kiev, 1985, No. 85.74, (in Russian).
8. Povzner, A. Linear methods in problems of nonlinear differential equations with small parameter. Intern. J. Nonlinear Mech., 1974, No. 9.
9. Mitropolskii, Yu. A. and Lopatin, A. K. Asymptotic decomposition of differential systems with a small parameter. Ukr. Math. J., 1987, No. 2.
10. Mitropolskii, Yu. A. On the reducibility of the differential equations of nonlinear mechanics. In: Mathematical models and methods in mechanics. Warsaw, PWN-Pol. Sci. Publ., 1985, pp. 409—486.
11. Mitropolskii, Yu. A. Averaging method in development. In: International conference on nonlinear oscillations. Analytical methods of the theory of nonlinear oscillations. Kiev, Naukova Dumka, 1984, Vol. 1, pp. 23–34.
12. Mitropolskii, Yu. A. Samoilenko, A. M., and Martynyuk, D. I. Systems of evolution equations with periodic and conditional periodic coefficients. Kiev, Naukova Dumka, 1985 (in Russian).
13. Mitropolskii, Yu. A., Samoilenko, A. M., and Perestyuk, N. A. Averaging method in pulsed systems. Ukr. Math. J., 1985, 37, No. 1, pp. 56–64.
14. Mitropolskii, Yu. A. and Nguen Dong An'. Random vibrations in certain viscoelastic nonlinear systems. Ibid, No. 4, pp. 468–472.
15. Mitropolskii, Yu. A. and Kolomiyei, V. G. On random forces acting on nonlinear oscillatory systems. Math. Phys. and Nonlinear Mechanics, Kiev, 1986, No. 5, pp. 23–24.

ABSOLUTE STABILITY OF NONLINEAR PERIODIC SYSTEMS

V. A. Yakubovich

Although the absolute stability problem has been explored rather thoroughly for nonlinear systems with stationary linear parts [1–5], this paper appears to be one of the first to investigate the situation (no less important in practice) in which the linear part is governed by differential equations with periodic coefficients. In fact, Schilman's sufficiency condition [6, 7] is, to the writer's knowledge, the only result yet available concerning the absolute stability of such systems. The absolute stability criteria presented here differ in their form from those derived in [5, 6]. Other topics discussed in the paper are dichotomy, absolute instability criteria, and the majorant method as a tool in absolute stability studies. The paper concludes with an example of high-frequency parametric excitation.

§1. ABSOLUTE STABILITY PROBLEM FORMULATION

We assume in this paper that the linear part of a nonlinear system[1] is periodically non-stationary, that is,

$$dx(t)/dt = A(t)x(t) + b(t)\xi(t) \tag{1.1}$$

where $A(t)$ and $B(t)$ are $n \times n$ and $n \times m$ matrix-functions that are real, T-periodic ($A(t + T) = A(t)$, $b(t + T) = b(t)$), and have measurable and (almost everywhere)

1) Note that any nonlinear system may be resolved into linear and nonlinear parts.

bounded elements; while $x(t)$ and $\xi(t)$ are real vector-functions with dimensions n and m, respectively.

The nonlinear part of the system will be written

$$\xi = \varphi(x, t) \tag{1.2}$$

without a T-periodicity assumption. Were it not for the T-periodicity in (1.1), these two equations would represent an arbitrary nonlinear system of differential equations, with nonlinearity separated out. Many absolute stability studies have shown, however, that if Lyapunov functions from certain standard classes are used, the mathematical entities actually of importance are $x(t)$ and the output $\xi(t)$ of the nonlinearity (1.2). Accordingly, (1.2) may be replaced, for example, by a local quadratic constraint

$$\mathfrak{G}[t, x(t), \xi(t)] > 0 \tag{1.3}$$

Here and hereafter

$$\mathfrak{G}(t, x, u) = \tfrac{1}{2}[x^*G(t)x + 2x^*g(t)\xi + \xi^*\Gamma(t)\xi] \tag{1.4}$$

is a quadratic form with T-periodic coefficients, i.e., $\mathfrak{G}(t + T, x, u) = \mathfrak{G}(t, x, u)$. Two more complex options are, in the order of increasing generality, a differential quadratic constraint

$$\mathfrak{G}(t, x(t)), \xi(t)) \geqslant d\Phi[x(t)]/dt \tag{1.5}$$

where $\Phi(x)$ is differentiable, and an integral quadratic constraint

$$\exists t_k \to t_* \leqslant +\infty, \gamma \geqslant 0: \int_{t_0}^{t_k} \mathfrak{G}[t, x(t), \xi(t)]dt \geqslant -\gamma \tag{1.6}$$

The above replacements enable the analysis of many important nonlinearities of the form $\xi(t) = \varphi[x(t)|_0^t, t]$, examples being hysteresis nonlinearities and various types of pulse modulators [1–5].

The subsequent discussion deals with the system (1.1), (1.6). It is understood that the functions $x(\cdot)$ and $\xi(\cdot)$ are defined over the existence interval $[t_0, t_*]$, $t_* \leqslant +\infty$, and satisfy (1.1) and (1.6) with t_k and γ generally dependent on the functions themselves. As is natural for ordinary differential equations, the existence interval satisfies the condition that t_* is finite only if $\operatorname{Sup}|x(t)| = +\infty$ for $t \in [t_0, t_*]$.

Example 1.1. Consider the equation $\ddot{y} + a(t)\dot{y} + b(t)y + c(t)y^3 = 0$ with the coefficients a, b, c being T-periodic. Setting $\xi = y^3$ and $x = \operatorname{col}[y, \dot{y}]$ converts this into the system (1.1), (1.6)

$$\frac{dx}{dt} = \begin{bmatrix} 0, & 1 \\ -b(t), & -a(t) \end{bmatrix} x + \begin{bmatrix} 0 \\ -c(t) \end{bmatrix} \xi, \quad \xi = y^3$$

showing that the constraints (1.3) and (1.6) are fulfilled with $\mathfrak{G} = \xi y$ and $\mathfrak{G} =$

$\xi y + \xi\vartheta\dot{y}$, respectively (in the latter case, $\vartheta \geqslant 0$, t_k may assume any value, and $\gamma = \vartheta y(0)^4/4$).

The reader is referred to [1–4] for other examples of quadratic constraints.

Definition. The system (1.1), (1.6) is said to be absolutely stable if for any of its solutions: (1) $t_* = \infty$ and (2) there exists a positive constant C such that the L_2 norm

$$|x(\cdot)|^2 + |\xi(\cdot)|^2 = \int_0^\infty (|x(t)|^2 + |\xi(t)|^2)dt \tag{1.7}$$

is finite in the sense

$$|x(\cdot)|^2 + |\xi(\cdot)|^2 \leqslant C(\gamma + |x(0)|^2)$$

which means that $|x(t)| \to 0$ for $t \to \infty$.

A similar definition holds for the class $\mathfrak{M}$ of all nonlinearities satisfying the constraint (1.6).

In subsequent analysis two implicit assumptions are made. First, it is assumed that $\Gamma(t) < \Gamma_0 < 0$ in (1.4). Although the condition $\Gamma(t) \leqslant 0$ can be shown to be necessary for absolute stability, the strict inequality is advantageous in greatly simplifying the presentation. Second, it is assumed that the pair $(A(\cdot), b(\cdot))$ in (1.1) is stabilizable. This means that there exists a matrix-function $r(t)$ (in our case, $r(t + T) = r(t)$ without any loss of generality) such that the matrix-function $B(\cdot) = A(\cdot) + b(\cdot)r(\cdot)^*$ is Hurwitzian; that is, given $dw/dt = B(t)w$, we have

$$\exists\epsilon > 0, \exists K > 0, \forall t \geqslant s: |w(t)| \leqslant Ke^{-\epsilon(t-s)}|w(s)|$$

The function $r(\cdot)$ in this definition is often given the additional property $G(t, x, r(t)^*x) \geqslant 0$ $(\forall x, t)$. The system (1.1), (1.6) is then called strongly minimally stable, and the Hurwitzian property $B(\cdot) = A(\cdot) + b(\cdot)^*r(\cdot)$ is obviously necessary to assume absolute stability; in this sense, absolute stability necessarily implies strong absolute stability.

The system (1.1), (1.6) is called minimally stable if for any $t_0 \epsilon [0, T]$, $a \in \boldsymbol{R}^n$ it has a solution $x^M(\cdot)$, $\xi^M(\cdot)$ (with $t_* = \infty$, $t_k = t_k^M$, $\gamma = \gamma^M(t_0, a)$) such that $x^M(t_0) = a$, $|x^M(t)| \to 0$ for $t \to +\infty$, and $\operatorname{Inf}[\lambda^{-2}\gamma^M(t_0, \lambda a)|\lambda > 0] < 0$. It is obvious that strong minimal stability implies minimal stability (with $\gamma(t_0, a) \equiv 0$).

§2. ADJOINT CANONICAL SYSTEMS. QUADRATIC CRITERION FOR ABSOLUTE STABILITY

If

$$H = \psi^*(A(t)x + b(t)\xi + \mathfrak{G}(t, x, \xi) \tag{2.1}$$

the adjoint canonical system is defined by

$$\frac{dx}{dt} = \left(\frac{\partial H}{\partial \psi}\right)^*, \quad \frac{d\psi}{dt} = -\left(\frac{\partial H}{\partial x}\right)^*, \quad \frac{\partial H}{\partial \xi} = 0 \tag{2.2}$$

Introducing

$$z = \begin{bmatrix} x \\ \psi \end{bmatrix}, \quad J = \begin{bmatrix} 0, & -I_n \\ I_n & 0 \end{bmatrix}$$
$$H(t) = \begin{bmatrix} -g\Gamma^{-1}g^* + G, & (A - b\Gamma^{-1}g^*)^* \\ (A - b\Gamma^{-1}g^*), & -b\Gamma^{-1}b \end{bmatrix} \tag{2.3}$$

where I_n is a unitary $n \times m$ matrix, the system (2.2) is converted to the vector canonical form

$$J\, dz/dt = H(t)z \tag{2.4}$$

We assume an arbitrary T-periodic matrix Hamiltonian $H(t) = H(t)^*$ and use $Z(t)$ to denote the evolution matrix of the system (that is, the $2n \times 2n$ matrix of its fundamental solutions, $Z(0) = I_{2n}$). We will use the frequency condition

$$\det\,[Z(T) - e^{i\omega}I_{2n}] \neq 0\ (\forall \omega \in [0, 2\pi]) \tag{2.5}$$

implying (see, e.g., [8]) that (2.4) has no solutions bounded on $(-\infty, +\infty)$ and is therefore completely unstable. When (2.5) holds, (2.4) has $2n$ solutions $z_1, \ldots,$ z_{2n} such that (for $j = 1, 2, \ldots n$) the solutions $z_j(t)$ and $z_{j+n}(t)$ behave, respectively, as e^{-t} and e^{+t} as $t \to +\infty$. The set of decreasing solutions define the $n \times n$ matrices $X(t)$ and $\Psi(t)$ thus

$$\|z_1(t), \ldots, z_n(t)\| = \begin{bmatrix} X(t) \\ Y(t) \end{bmatrix} \tag{2.6}$$

A perfectly unstable equation (2.4) is called nonoscillatory if

$$\det X(t) \neq 0 (\forall t \in [0, T]) \tag{2.7}$$

Theorem 2.1. (Absolute stability criterion). Let system (1.1), (1.6) be minimally stable. It is absolutely stable if and only if its adjoint system (2.2) is perfectly unstable and nonoscillatory.

Thus (2.5) and (2.7) are, respectively, the necessary and sufficient conditions for absolute stability, and the stability problem for all nonlinear systems (1.1), (1.6) is reduced to that for a linear system (2.2) on $[0, T]$. It must be investigated, that is, whether or not (2.5) and (2.7) are satisfied for this system.

While this definition implies the perfect instability of (2.4), it also proves possible to define this property for an arbitrary equation (2.4) [9]. Recall that $z_1(t), \ldots, z_n(t)$ define a Lagrangian system of solutions to (2.4) if they are real, linearly independent, and $(Jz_j(t), z_h(t)) = 0$, $j, h = 1, 2, \ldots, n$. (Since $(Jz_j(t),$ $z_h(t)) \equiv \text{const}$, it suffices to pick linearly independent $z_j(0), \ldots, z_n(0)$ such that $(Jz_j(0), z_h(0)) = 0$). The above system of equations is obviously Lagrangian because $z_j(t) \to 0$ and $(Jz_j(t), z_h(t)) \to 0$ as $t \to \infty$, and hence $(Jz_j(t), z_h(t)) \equiv 0$. In the general case, (2.6) is again used to define $n \times n$ matrices $X(t)$ and $\Psi(t)$, and (2.4) is called nonoscillatory [9] if $|\Delta \operatorname{Arg} \det [X(t) + i\Psi(t)]| \leqslant \text{const}$ as $t \to +\infty$, and oscillatory otherwise. Equation (2.4) is called strongly nonoscillatory if it is nonoscillatory and remains so for any small deformations of the Hamiltonian $H(\cdot)$ (in terms of the L_1 metric $\|H(\cdot)\|_{L_1}$). In particular, (2.4) is strongly nonoscillatory if the $n \times n$ matrix $X(t)$ is defined in terms of a Lagrangian system of solutions (cf. (2.6) and $\det X(t) \neq 0$ for $t \geqslant t_*$).

Given the system (1.1), (1.6), we are in a position to formulate:

Theorem 2.2. The following statements are equivalent: 1) The adjoint system is nonoscillatory and perfectly instable; 2) The adjoint system is strongly nonoscillatory; 3) There exists a "Lyapunov function" $V(t, x) = x^*R(t)x$ such that $R(t) = R(t + T) = R(t)^*$ is absolutely continuous and

$$\dot{V} \leqslant -2G - \delta(|x|^2 + |\xi|^2)(\exists \delta > 0, \forall t, x, \xi) \tag{2.8}$$

4) For any complex vector-functions $\tilde{x}(\cdot) \in L_2^n(0, \infty)$, $\tilde{\xi}(\cdot) \in L_2^m(0, \infty)$ satisfying (1.1) and $\tilde{x}(0) = 0$

$$J = \int_0^\infty G(t, \tilde{x}, \tilde{u})\, dt \leqslant -\delta(\|\tilde{x}(\cdot)\|^2 + |\tilde{\xi}(\cdot)|^2) \tag{2.9}$$

5) For any T-periodic complex vector-functions $\tilde{x}(\cdot) \in L_2^n(0, T)$, $\tilde{\xi}(\cdot) \in L_2^m(0, T)$ satisfying $d\tilde{x}/dt + i\omega\tilde{x} = A(t)\tilde{x} + b(t)\xi$ with some $\omega \in [0, 2\pi]$,

$$J_T = \int_0^T G(t, \tilde{x}, \tilde{u})dt \leqslant -\delta \int_0^T (|\tilde{x}(t)|^2 + |\tilde{\xi}(t)|^2)dt \tag{2.10}$$

for a certain $\delta > 0$.

Note that the last statement is close in form to Schilman's sufficient conditions [6, 7].[2]

Theorems 2.1 and 2.2 are proved in [10–12]. The proof of Theorem 2.1 [2] parallels the stationary case derivation and depend heavily on the equivalence of conditions (1) and (5) of Theorem 2.2 (that is, on the "frequency theorem" of the periodic system problem [10]).

For stationary systems ($A(t) = A =$ const, $b(t) = b =$ const, $G(t, x, \xi) = G(x, \xi)$) it follows from (2.3) that $H =$ const. The frequency condition (2.5) is equivalent to

$$\det [i\omega J - H] \neq 0 \tag{2.11}$$

which in turn, is equivalent to

$$\begin{aligned} &\mathfrak{G}(\tilde{x}, \tilde{\xi}) \leqslant -\delta(|\tilde{x}|^2 + |\tilde{\xi}|^2), \\ &(\forall \tilde{x}, \xi, \omega\colon i\omega\tilde{x} = A\tilde{x} + b\xi) \end{aligned} \tag{2.12}$$

in view of the equivalence of statements (1) and (5) of Theorem 2.2.

It can be shown that a stabilizable system is nonoscillatory in the stationary case and hence Theorem 2.1 leads to a quadratic criterion for absolute stability

2) Indeed, it turns out to be equivalent to these conditions for the problem under study (Schilman, private communication).

[1.2]. As is the case with the stationary case [2.5], Theorem 2.1 yields nonstationary analogues of several absolute stability criteria, including the circular and Popov criteria [3, 4].

§3. DICHOTOMY AND ABSOLUTE INSTABILITY

In this section the nonlinear blocks are subjected to the additional condition

$$\text{if } |x(t)| \leqslant \text{const, then } |\xi(t)| \leqslant \text{const} \tag{3.1}$$

which is obviously weakly restrictive and is satisfied, for example, by blocks $\xi(t) = \varphi[x(t)]$ with φ continuous, or by ordinary hysteresis nonlinearities.

The system (1.1), (1.6) is called dichotomic if for any of its solutions: (1) either $|x(t)| \leqslant$ const or $|x(t)| \to 0$, as $t \to \infty$; (2) $\|x(\cdot)\|^2 + \|\xi(\cdot)\|^2 < \infty$.

Theorem 3.1. Each of the (equivalent) conditions (1)–(5) of Theorem 2.2 is also equivalent to the dichotomy of the system (1.1), (1.6).

It should be noted that theorem 3.1 cannot establish the instability of a given nonlinear system (with $\xi = \varphi(x, t)$, for example) fulfilling (1.6), and the absolute instability criterion must be applied.

Let $\mathfrak{M}$ be subset of pairs $[x(\cdot), \xi(\cdot)]$ that satisfy the system (1.1), (1.6).

Definition. The system (1.1), (1.6) is called absolutely unstable in $\mathfrak{M}$ if for any $\delta > 0$ there exist $a_0 \in \boldsymbol{R}^n$, $t_0 \in \boldsymbol{R}^1$ such that for any solution $[x(\cdot), \xi(\cdot)] \in \mathfrak{M}$ satisfying $|x(t_0) - a_0| < \delta$, $|x(t)|$ is unbounded on the existence interval.

The system (1.1), (1.6) is called minimally unstable in $\mathfrak{M}$ if there exists a solution $[x_*(\cdot), \xi_*(\cdot)] \in \mathfrak{M}$, with an existence interval $(0, \infty)$, such that $\|x_*(\cdot)\|^2 + \|\xi_*(\cdot)\|^2 = +\infty$.

As was remarked earlier, a matrix-function $r(t)$ may often be found for which $G(t, x, r(t)^*x) \geqslant 0(\forall t, x)$. The system (1.1), (1.6) is then minimally unstable in $\mathfrak{M}$ if the system $dx/dt = Ax + b\xi$, $\xi = r(t)^*r$ has a solution $[x_*(\cdot), \xi_*(\cdot)] \in \mathfrak{M}$ such that $\|x_*(\cdot)\| = \infty$.

Theorem 3.2. The system (1.1), (1.6) is absolutely unstable in $\mathfrak{M}$ if it is minimally unstable in this class and its adjoint canonical system in nonoscillatory and perfectly unstable.

Theorem 3.1 and 3.2 extend to T-periodic systems the analogous statements known for the stationary case [5] and are proved in a very similar manner except for the use of the specialized frequency theorem [10].

The same approach has been used to generalize some other results to T-periodic systems, e.g., the absolute stability criterion for forced oscillations and the dissipation criterion [2, 3, 5, 8].

§4. MAJORANT METHOD

The stability problem of all nonlinear systems is thus reduced to the analysis of the above class of only one adjoint system, namely, of the linear canonical system (2.2). What is necessary in this connection is to be able to establish whether or not the system satisfies the properties (1) and (2) of Theorem 2. If the initial system depends on some parameters, which is often the case, the question arises, which regions of the parameter space do, and which do not, satisfy the property. Which makes useful the following "comparison theorem," or

Theorem 4.1 [9][3]. Consider two equations (2.4) of positive type, with matrix Hamiltonians $H_1(t)$ and $H_2(t)$ such that $H_1(t) \leqslant H_2(t)$. If the equation with the Hamiltonian $H_2(t)$ is strongly nonoscillatory, the same is true of the equation with $H_1(t)$. If the equation with the Hamiltonian $H_1(t)$ is oscillatory, the same is true of the equation with $H_2(t)$.

Note that a result similar to Theorem 10 in [9] as well as to the statement given at the end of §1 in the same paper has recently—and independently—been derived by Arnold [15] for equations (2.4) of a special type obtained from Lagrange equations. Arnold also analyzes the relationship between the oscillatory properties of these equations with Maslov's indices.

Let us now assume that, given equation (2.4) with a Hamiltonian $H(t)$ of the type (2.3), we have contracted a majorant Hamiltonian

$$H^M(t) \geqslant H(t) \tag{4.1}$$

It is then possible to prove that $H^M(t)$ can always be represented in the form

$$H^M(t) = \begin{bmatrix} g_M\Gamma_M^{-1} g_M^* + G_M; & (A_M - b_M\Gamma_M^{-1} g_M^*)^* \\ A_M - b_M\Gamma_M^{-1} g_M^*; & b_M\Gamma_M^{-1} b_M^* \end{bmatrix} \tag{4.2}$$

similar to (2.3), and hence to construct a form

$$\mathfrak{G}_M = \tfrac{1}{2}\,(x^*Gx + 2g_M^*x\xi + \xi^*\Gamma_M\xi) \tag{4.3}$$

and a *majorant* system

$$\begin{gathered} dx/dt = A_M(t)x + b_M(t)\xi \\ \int_0^{t_k} \mathfrak{G}_M(t, x, \xi)\,dt \geqslant -\gamma \end{gathered} \tag{4.4}$$

The following statement is readily deduced from Theorems 4.1, 2.1, and 3.2, and may be considered as a formulation of the majorant method.

3) Here we use the strong oscillatory property instead of the oscillatory property used in [9]; the proof is obtained immediately by an obvious modification of the proof 10 in [9].

Theorem 4.2. If the initial system (1.1), (1.6) is minimally stable (minimally unstable) and the majorant system (4.4) is absolutely stable (absolutely unstable), then the initial system is absolutely stable (absolutely unstable).

An absolute stability analysis can be carried out analytically if the majorant system is represented by a system with constant coefficients. This yields sufficient conditions for the problem, even though of somewhat questionable accuracy.

§5. HIGH-FREQUENCY PARAMETRIC EXCITATION

Consider the system

$$dx/dt = A(\theta t)x + b(\theta t)\xi,$$
$$\int_0^{t_k} \mathfrak{G}(\theta t, x, \xi)\, dt \geqslant -\gamma, \tag{5.1}$$

where

$$A(s + 2\pi) = A(s),\ b(s + 2\pi) = b(s),\ \mathfrak{G}(s + 2\pi, x, \xi) = \mathfrak{G}(s, x, \xi)$$

The question we are going to address is, how does the system behave at a high frequency? It turns out that, as far as its stability is concerned, the system behaves as one with constant coefficients

$$dx/dt = A_{cp}x + \hat{b}\xi,$$
$$\int_0^{t_k} \hat{\mathfrak{G}}(x, \xi)\, dt \geqslant -\gamma. \tag{5.2}$$

Here and hereafter $(\ldots)av$ denotes a time average of the quantity $(\ldots)$, $\mathfrak{G}(x, \xi) = \frac{1}{2}(x^*G_{av}x + 2x^*\hat{g}\xi + \xi^*\hat{\Gamma}\xi)$, and $\hat{b}$, $\hat{g}$, $\hat{\Gamma} = \hat{\Gamma}^* < 0$ are any constant matrices satisfying

$$(g\Gamma^{-1}g^*)_{av} = \hat{g}\hat{\Gamma}^{-1}\hat{g}^*,\ (b\Gamma^{-1}g^*)_{av} = \hat{b}\hat{\Gamma}^{-1}\hat{g}^* \tag{5.3}$$

The transition from the system (5.1) to (5.2) will be referred to as a nonlinear averaging. Note that if g and Γ are constant, not to mention the case of stationary form $\mathfrak{G}(s, x, \xi) = \mathfrak{G}(x, \xi)$, we may take $\hat{b} = b_{av}$, $\hat{g} = g$, and $\hat{\Gamma} = \Gamma$, and the nonlinear averaging becomes equivalent to the linear one. Note also the evident relationship $\hat{H} = H_{av}$ for the Hamiltonians $H(\theta t)$ and $\hat{H}$ constructed by formulas (2.3) for the systems (5.1) and (5.2).

Theorem 5.1. Assume that the initial system is minimally stable (minimally unstable) for any sufficiently large $\theta > 0$ and the nonlinearly averaged system (5.2) is absolutely stable (absolutely unstable with respect to the frequency criterion (2.12)). Then there exists $\theta_0 > 0$ such that the initial system (5.1) is absolutely stable (unstable) for any $\theta > \theta_0$.

It can be shown that the absolute stability of the system (5.2) is not only sufficient but is almost always necessary for the absolute stability of the system (5.1) at large θ, "almost always" meaning that the zero-measure set is excluded from the parameter space of the system (5.2).

The following well-known statement (obtainable, for example, by the Bogolyubov-Mitropolskii method) is useful in analyzing the minimal stability (nonstability) of the initial system at large θ. The linear equation $dy/dt = B(\theta t)y$ with a 2π-periodic matrix $B(s)$ is asymptotically stable (unstable with a characteristic exponent $\text{Re}\beta > 0$) for large $\theta > 0$ if so is the averaged equation $dy/dt = B_{\text{av}}y$, B_{av} being the average value of the matrix $B(s)$. Note that this statement may also be deduced from a discussion given by Yakubovich and Starzhinskii [8] who showed that the characteristic indices of the initial and averaged equations (α_j and λ_i, respectively) are related by $\lambda_j = \theta^{-1}(\alpha_j + 0(1))$ at $\theta \to +\infty$.

For the case of a stationary form $\mathfrak{G}(t, x, \xi) = \mathfrak{G}(x, \xi)$, it follows from the above statement that the strong minimal stability (strong minimal instability) of the averaged system (5.2) implies the minimal stability (minimal instability) of the initial system for any sufficiently large $\theta > 0$.

The idea of the proof of theorem 5.1 is as follows. Equation (2.4) for the system 5.1 is of the form

$$J\,dz/d\tau = \epsilon H(\tau)z \tag{5.4}$$

where $\epsilon = \theta^{-1}$, $\tau = \theta t$, $H(\tau + 2\pi) = H(\tau)$ ($H(\tau)$ is defined by replacing t by τ in (2.3)). We must find out the conditions under which this equation is unstable at small ϵ. It can be shown by applying Bogolyubov and Mitropolskii's averaging method [8] to equation (5.4), that this equation is strongly nonoscillatory; if so is the averaged equation

$$J\,dz/dt - H_{\text{av}}z \tag{5.5}$$

If equation (5.5) is strongly nonoscillatory, which is equivalent to the frequency condition $\det[\iota\omega J - H_{\text{av}}] \neq 0(\forall\omega \in R^1)$ being fulfilled, this means that so too is equation (5.4), and, by means of theorems 2.1 and 3.2, the absolute stability (absolute instability) of the initial equation is deduced.

A useful by-product of the proof is an estimate for the quantity θ_0 involved in theorem 5.1. Let

$$P(\tau) = \int_0^\tau [H(s) - H_{\text{cp}}]ds, \quad |P(\tau)| \leqslant p = p(\tau),$$

$$q_M = \max_\tau \left\{ \frac{e^{\epsilon p} - 1}{\epsilon} [e^{\epsilon p}|H(\tau) - H_{\text{cp}}| + |H_{\text{cp}}| + e^{\epsilon p}|H(\tau)|] \right\}$$

Then $\theta_0 = \epsilon_0^{-1}$, where the number $\epsilon_0 > 0$ is such that, for any $\epsilon \in [0, \epsilon_0]$, the frequency condition

$$\det[iwJ - H_{\text{cp}} + \epsilon q_M I_{2n}] \neq 0 \quad (\forall\omega \in \mathbf{R}^1)$$

is fulfilled. (See [8], pages 238 and 239, for a more detailed discussion).

It has thus far been assumed that $\Gamma(t) < 0$. Theorem 5.1 also holds for $\Gamma(t) \equiv 0$ if the additional condition

$$\lim_{\omega \to \infty} \omega^2 \hat{\mathfrak{G}}(\tilde{x}, \tilde{\xi}) < 0 \quad (\forall \tilde{\xi} \neq 0)$$

is fulfilled, with $\tilde{x} = (i\omega I_n - A_{av})^{-1} \hat{b}\tilde{\xi}$. (It will be noted that $\Gamma(t) \equiv 0$, the frequency condition (2.12) for the nonlinearly averaged system (with $\mathfrak{G} = \hat{\mathfrak{G}}$) is invalidated at $\omega = +\infty$).

§6. EXAMPLES

The results obtained in Sections 2 to 4 reduce the problems we are considering to the analysis of linear nonstationary systems, which cannot generally be carried out without a computer. The examples given below are intended to illustrate the discussion in Section 5, do not require any complex calculations, and may be treated analytically.

Example 1. Consider the scalar equation

$$\ddot{y} + a(\theta t)\dot{y} + b(\theta t)y + \xi = 0 \tag{6.1}$$

where

$$\xi = \varphi(t, y), \ |\varphi(t, y)| \leqslant c|y| \tag{6.2}$$

$$a(s + 2\pi) = a(s); \quad b(s + 2\pi) = b(s) \tag{6.3}$$

It follows from the second of equations (6.2) that the local quadratic constraint

$$\mathfrak{G}(y, \xi) = c^2y^2 - \xi^2 \geqslant 0 \tag{6.4}$$

and hence the corresponding integral constraints (1.6) are satisfied. For $x = \mathrm{col}(y, \dot{y})$, equation (6.1) takes the form of (1.1).

Now let a_{av} and b_{av} be the average values of the functions $a(s)$ and $b(s)$, respectively. We define

$$\alpha = \alpha_{av}^2/2c, \quad \beta = b_{av}/c \tag{6.5}$$

and assume that

$$\beta^2 > 1, \quad \alpha > \beta - \sqrt{\beta^2 - 1} \tag{6.6}$$

We are going to demonstrate that, if (6.6) is valid for sufficiently large $\theta > 0$, then there exists dichotomy or even absolute stability if $a_{av} > 0$, $b_{av} > 0$, and absolute instability if either $a_{av} > 0$, $b_{av} < 0$ or $a_{av} < 0$, $b_{av} > 0$.

The "nonlinearity" $\xi \equiv 0$ satisfies the constraints (6.4); equation (6.1) with $\xi \equiv 0$ is asymptotically stable (unstable) at large $\theta > 0$ if $a_{av} > 0$, $b_{av} > 0$ (if $a_{av}b_{av} < 0$), since so is the averaged equation.

It thus follows that the system (6.1), (1.6) is minimally stable (unstable) at large $\theta > 0$ if $a_{av} > 0$, $b_{av} > 0$ (if $a_{av}b_{av} < 0$). In this case $\mathfrak{G}(\tilde{y}, \tilde{\xi}) = c^2|\tilde{y}|^2 - |\tilde{\xi}|^2$ ($\tilde{y}$ and $\tilde{\xi}$ being complex numbers), nonlinear averaging is equivalent to ordinary one, and the averaging of equation (6.1) yields $\tilde{y} = -(p^2 + a_{av}p +$

$b_{av})^{-1}\tilde{\xi}$. If $p = i\omega$, the frequency condition (2.12) is converted to (6.6). We thus conclude that equations (6.6) coupled with an appropriate condition on a_{av}, b_{av}, represent a criterion for absolute stability (instability) of the averaged system. The above statement follows from theorems 2.1, 3.2, and 5.1. If $a_{av} > 0$, $b_{av} > 0$, then we can use theorem (2.1) and results obtained in [8] to show that at large θ the condition (6.6) is (almost always) both sufficient and necessary for the absolute stability of the system (6.1), (1.6) with the form (6.4).

It is easily verified that if the (6.2) we set $c = c(\theta t) = c(\theta t + 2\pi)$, then the result remains the same if c is replaced by $\sqrt{(c^2)_{av}}$ in (6.5).

Note that if the condition (6.6) is not satisfied, then there may exist oscillatory processes in the system (6.1), (6.2).

Example 2. Consider the nonlinear equation

$$\frac{d^2y}{dt^2} + (\alpha_0 + \alpha_1 y^2)\frac{dy}{dt} + (\beta_0 y + \beta_1 y^2 + \beta_2 y^3) = 0 \tag{6.7}$$

where $\alpha_j = \alpha_j(\theta t)$; $\beta_j = \beta_j(\theta t)$; the function $\alpha_1(s) > 0$ is absolutely continuous; $\alpha_j(s)$ and $\alpha_j(s)$ are measurable bounded 2π-periodic functions and $\alpha_{j,av}$, $\beta_{j,av}$ their average values.

Assuming

$$\xi_1 = y^2, \quad \xi_2 = y^3, \quad G = \xi_2 y - \xi_1^2 + \tau\dot{y}\xi_2 \tag{6.8}$$

where $\tau = \text{const} \geqslant 0$, we find that the integral quadratic constraint (1) is satisfied if $\gamma = \tau y(t_0)^3/3$.

It will be shown below that theorem 5.1 implies the relationships

$$\begin{aligned} &\alpha_{0,av} > 3(\beta_2\alpha_1^{-1}) \\ &4\beta_{0,av}(\beta_2\alpha_1^{-1})_{av} > |\beta_{1,av}|^2(\alpha_1^{-1}) \end{aligned} \tag{6.9}$$

(A relaxation of these restrictions, although possible, would complicate the analysis).

Let us show now that, given $\theta > \theta_0$ and the conditions (6.9), equation (6.7) is stable in whole (the corresponding system (1.1), (1.6) is absolutely stable) if $(\alpha_0)_{av} > 0$, $(\beta_0)_{av} > 0$, and unstable in whole (the system (1.1), (1.6) is absolutely unstable) if $(\alpha_0)_{av}(\beta_0)_{av} < 0$.

Denoting

$$x_1 = y, \quad x_2 = \dot{y} + \alpha_1\xi_2/3 \tag{6.10}$$

equation (6.7) and the form (6.9) may be rewritten in the form of equations (1.1) and (1.4)

$$\begin{aligned} \dot{x}_1 &= x_2 - \alpha_1\xi_2/3, \\ \dot{x}_2 &= -\beta_0 x_1 - \alpha_0 x_2 - \beta_1\xi_1 + \alpha_2\xi_2, \\ \mathfrak{G} &= (x_1 + \tau x_2)\xi_2 - \xi_1^2 - \tau\alpha_1\xi_2^2/3 \end{aligned} \tag{6.11}$$

with $\alpha_2 = \dot{\alpha}_1/3 + \alpha_0\alpha_1/3 - \beta_2$.

The nonlinearly averaged system is found as follows. Since g = const in our case, it follows that the condition $(g\Gamma^{-1}g^*)_{av} = g\hat{\Gamma}^{-1}\hat{g}^*$ is satisfied by putting $\hat{g} = g$, $\hat{\Gamma} = (\Gamma^{-1})_{av}^{-1}$. From $(b\Gamma^{-1}g^*)_{av} = \hat{b}\hat{\Gamma}^{-1}\hat{g}^*$ we obtain the relationship $(b\Gamma^{-1})_{av}g^* = \hat{b}\Gamma^{-1}g^*$ yielding equations for the elements of the matrix b. It is easy to see that these equations are satisfied if in the nonlinearly averaged system and in the form $\hat{\mathfrak{G}} = \mathfrak{G}$ represented as (6.11), we replace the functions $\alpha_j = \alpha_j(\theta t)$ and $\beta_j = \beta_i(\theta t)$ respectively by the constants $\bar{\alpha}_j$ and $\bar{\beta}_j$, where

$$\bar{\alpha}_0 = \alpha_{0,av}, \quad \bar{\alpha}_1 = [(\alpha_1^{-1})_{av}]^{-1}, \quad \bar{\alpha}_2 = (\alpha_2\alpha_1^{-1})_{av}[(\alpha_1^{-1})_{av}]^{-1}$$
$$= [\alpha_{0,av}/3 - (\beta_2\alpha_1^{-1})_{av}][(\alpha_1^{-1})_{av}], \quad \bar{\beta}_0 = \beta_{0,av}, \bar{\beta}_1 = \beta_{1,av} \tag{6.12}$$

Denoting $y = x_1$, the first two equations in (6.11) (in which α_j, β_j are replaced by $\bar{\alpha}_j$, $\bar{\beta}_j$) can be rewritten in the form

$$\ddot{y} + \hat{\alpha}_0\dot{y} + \hat{\alpha}_1\dot{\xi}_2 + \hat{\beta}_0 y + \hat{\beta}_1\xi_1 + \hat{\beta}_2\xi_2 = 0 \tag{6.13}$$

similar to (6.7), with

$$\hat{\alpha}_0 = \bar{\alpha}_0, \quad \hat{\alpha}_1 = \bar{\alpha}_1/3, \quad \hat{\beta}_0 = \bar{\beta}_0, \quad \hat{\beta}_1 = \bar{\beta}_1, \quad \hat{\beta}_2 = \frac{\bar{\alpha}_0\bar{\alpha}_1}{3} - \bar{\alpha}_2 \tag{6.14}$$

whereas the form $\mathfrak{G}$ remains the same as in (6.8). The problem with variable coefficients is thus reduced to a similar problem with constant coefficients by replacing the initial system by a nonlinearly averaged one.

The problem we address next is absolute stability of the nonlinearly averaged system. For complex quantities $\tilde{\xi}_j$, $\tilde{y}$ we have

$$\mathfrak{G} = \mathrm{Re}\,(\tilde{\xi}_2^*\tilde{y}) - |\tilde{\xi}_1|^2 + \mathrm{Re}\,(\tau i\omega\tilde{y}\tilde{\xi}_2^*) \tag{6.15}$$

(see [2] for details), and from (6.13)

$$\tilde{y} = -W_1(p)\tilde{\xi}_1 - W_2(p)\tilde{\xi}_2$$

where

$$W_1(p) = \Delta(p)^{-1}\hat{\beta}_1, W_2(p) = \Delta(p)^{-1}(\hat{\alpha}_1 p + \hat{\beta}_2), \Delta(p) = p^2 + \hat{\alpha}_0 p + \hat{\beta}_0 \tag{6.16}$$

The frequency condition $\mathfrak{G}(\tilde{x}, \tilde{\xi}) \leqslant -\delta\,(|x|^2 + |\tilde{\xi}|^2)$ is in this case equivalent to the condition $\mathfrak{G}(\tilde{x}, \tilde{\xi}) < 0$ for $|\tilde{\xi}| \neq 0$, or, in view of (6.15), to the condition

$$4\mathrm{Re}[(1 + i\omega\tau)W_2(i\omega)] > |(1 + i\omega\tau)W_1(i\omega)|^2 \tag{6.17}$$

for $0 \leqslant \omega \leqslant +\infty$. We consider the case $\tau = 0$ first. Substituting (6.16) into frequency condition $4\mathrm{Re}W_2(i\omega) > |W_1(i\omega)|^2$ we obtain the relationship $4[\hat{\beta}_2(\hat{\beta}_0 - \omega^2) + \hat{\alpha}_0\hat{\alpha}_1\omega^2] > |\beta_1|^2$, $0 \leqslant \omega < \infty$, satisfied by the inequalities $4\hat{\beta}_0\hat{\beta}_2 > |\hat{\beta}_1|^2$, $\hat{\alpha}_0\hat{\alpha}_1 > \hat{\beta}_2$, which can be written in the form (6.9) by making use of (6.14) and (6.12). At $\tau = 0$, then, (6.17) holds for $0 \leqslant \omega < \infty$. For $\omega = \infty$, we have $4\tau\hat{\alpha}_1 > 0$ from (6.16), (6.17), so that in this case the inequality (6.17) holds for $\tau > 0$ and does not hold for $\tau = 0$. Let us assume that $\tau > 0$ is small and $\hat{\alpha}_1 > 0$. Since the functions-coefficients of τ^2 and τ in (.17) are bounded, inequality (6.17) will be satisfied for all $0 \leqslant \omega \leqslant \infty$ at small $\tau > 0$. Note that if all possible τ are considered, condition less restrictive than (6.9) may be derived.

Let us now analyze the conditions of minimal stability (instability).

For $\xi_1 = 0$, $\xi_2 = 0$, we have $\mathfrak{G} \geqslant 0$. The averaged system (6.11) with $\xi_1 = \xi_2 \equiv 0$ is asymptotically stable (unstable) if $\hat{\alpha}_0 = \alpha_{0,av} > 0$ and $\hat{\beta}_0 = \beta_{0,av} > 0$ (if $\hat{\alpha}_0\hat{\beta}_0 < 0$). According to the remark following theorem (5.1), this is also true for the initial equation. At $\xi_1 = \xi_2 \equiv 0$, the nonlinearly averaged system coincides with the averaged system. We thus see that the initial and nonlinearly averaged systems are both minimally stable (unstable) if $\alpha_{0,av} > 0$, $\beta_{0,av} > 0$ (if $\alpha_{0,av}\beta_{0,av} < 0$) which, by theorem (5.1), justifies the above statement on the absolute stability (absolute instability) of the initial system at sufficiently large θ.

Example 3. Consider the third-order system

$$\frac{dx_1}{dt} = x_3 - x_1, \quad \frac{dx_2}{dt} = x_1$$
$$\frac{dx_3}{dt} = -h(\theta t)x_1 - x_2 - b(\theta t)\xi, \quad \xi = \varphi(x_1) \tag{6.18}$$

where $\nu = \text{const} > 0$, $h(s + 2\pi) = h(s)$, $b(s + 2\pi) = b(s) > 0$ and $\mu_1 \leqslant \varphi(\sigma)/ \leqslant \mu_2$. Let $\mu_{av} = (\nu - h_{av})b_{av}^{-1}$. The above nonlinearities satisfy the integral quadratic constraint with the form

$$G = (\mu_2 x_2 - \xi)(\xi - \mu_1 x_1) + \tau(\xi - \mu_1 x_1)\dot{x}_1, \ \tau \geqslant 0$$

Since this form is stationary, it follows that the nonlinearly averaged system is identical to the averaged system. Applying the Popov criterion (or the quadratic criterion [2]) it is easily found that $S(-\infty, \mu_{av})$ and $S(\mu_{av}, +\infty)$ are, respectively, the absolute instability and absolute stability domains of the averaged system. Which means [2] that the averaged system is stable in whole and absolutely stable (unstable in whole and absolutely unstable) in the nonlinearity class $\mu_1 \leqslant \varphi(\sigma)/\sigma \leqslant \mu_2$, where $\mu_{av} < \mu_1 < \mu_2 < +\infty$ $(-\infty < \mu_1 < \mu_2 < \mu_{av})$. It turns out also that this system exhibits strong minimal stability (instability) with $\xi \equiv \mu_0 x_1$, $\mu_1 \leqslant \mu_0 \leqslant \mu_2$. According to the argument of Section 5, the initial system is also minimally stable (unstable) for large $\theta > 0$. By theorem 5.1, the initial system (6.18) at large $\theta > 0$ is stable in whole if $\mu_1 > \mu_{av}$ and unstable in whole if $\mu_2 < \mu_{av}$. We also have absolute stability (absolute instability) in the class of nonlinearities satisfying an integral constraint with the form $\mathfrak{G}$ above, which class includes some hysteresis nonlinearities, for example.

REFERENCES

1. Voronov, A. A. Stability, controllability, observability. Moscow, Nauka, 1979 (in Russian).
2. Yakubovich, V. A. Investigation of nonlinear automatic control systems. Moscow, Nauka, 1975 (in Russian).
3. Gelig, A. Kh., Leonov, G. A., and Yakubovich, V. A. Stability of nonlinear systems. Moscow, Nauka, 1978 (in Russian).
4. Rezvan, V. A. Absolute stability of dead-timed systems. Moscow, Nauka, 1983 (in Russian).

5. Yakubovich, V. A. Absolute instability of nonlinear controlled systems. Autom. Telem., 1970, No. 12, pp. 5–14 and 1971, No. 6, pp. 25–33.
6. Shilman, S. V. Method of generating functions in the theory of dynamic systems. Moscow, Nauka, 1978 (in Russian).
7. Shilman, S. V. Autom. Telem., 1972, No. 2, pp. 164–168.
8. Yakubovich, V. A. and Statzhinskii, V. M. Linear differential equations with periodic coefficients. Theory and applications. Moscow, Nauka, 1972 (in Russian).
9. Yakubovich, V. A. Oscillatory properties of the solutions of canonical equations. Matem. Sbornik, 1962, No. 56(98), pp. 3–42.
10. Yakubovich, V. A. Frequency theorem for periodic systems. Doklady USSR Acad. Sci., 1986, Vol. 287, No. 1, pp. 70–73.
11. Yakubovich, V. A. Absolute stability of nonlinear systems with periodical nonstationary nonlinearity. Doklady USSR Acad. Sci., 1987, Vol. 289, No. 2.
12. Yakubovich, V. A. Linear-quadratic optimization problem and frequency theorem for periodic systems. Sib. Matem. Zhurn., 1986.
13. Gelfand, I. M. and Lidskii, V. B. On the stability domains of the linear canonic differential systems with periodic coefficients. Uspekhi Matem. Nauk, 1955, Vol. 10, No. 1 (63), pp. 3–40.
14. Yakubovich, V. A. Arguments on a symplectic matrix group. Matem. Sborn., 1961, No. 55(97), pp. 255–280.
15. Arnold, V. I. Sturm theorems and symplectic geometry. Functional analysis and its applications. 1985, Vol. 19, No. 14, pp. 1–10.

SOME PROBLEMS OF STEADY-STATE MOTIONS OF A SOLID BODY

A. Yu. Ishlinsky, S. V. Malashenko, V. A. Storozhenko, M. E. Temchenko, P. G. Shishkin

1. A string-suspended solid body can perform a wide variety of steady-state motions and possesses exotic stability properties [1–6], but as far as mechanical stability is concerned, steady-state motions having an even number of the degrees of instability (and hence susceptible of gyroscopic stabilization) seem to arouse the greatest interest and are certainly the most difficult to study [7, 8, 5]. A typical example of this kind is presented by one of the stationary motions of a string-suspended axisymmetric body with point of suspension lying on the axis of symmetry as shown in Fig. 1 (see [1, 2]) and with the special property that the string and the symmetry axis of the body deviate different sides of the fixed vertical. With the angular velocity ω unboundedly increased, the symmetry axis tends to assume a horizontal position, whereas the mass center of the body is virtually on the fixed vertical (Figs. 1, 6). Although the gyroscopic nature of stability [9] has long prevented this motion from being adequately studied, some results have been obtained by investigating the distribution of the roots of a first-order perturbation-theory characteristic equation for the high angular velocity no proper rotation case [10]. It turned out that for $\omega \to \infty$, a pair of complex roots exists, indicative of the instability of this limiting form of the stationary motion in question, and that the roots become real at a certain angular velocity whose value $\omega = \omega_*$ may be approximated by

$$\omega_* = \frac{\sqrt{2gl}}{a} \sqrt{\frac{C + ma^2}{C}} \tag{1}$$

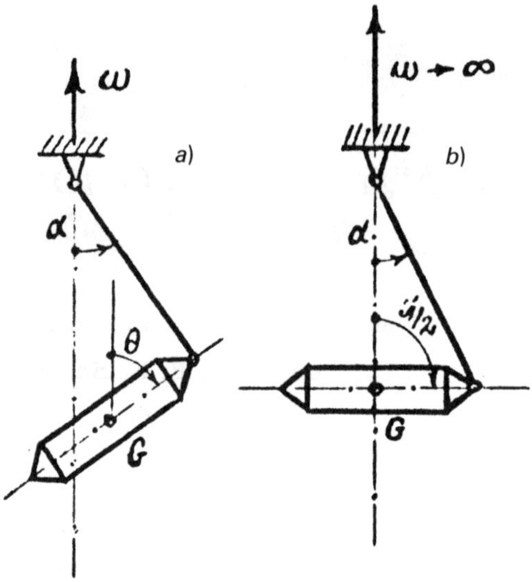

Figure 1

Here l is the length of the string; a—the distance between the center of masses and the fixing point; C—the axial moment of inertia, and g—the free fall acceleration. At $\omega < \omega_*$, it has been shown that this stationary motion is stable to the first approximation, and as ω is increased from ω_*, loss of stability occurs, a rather unusual (even though experimentally observed) fact because many gyroscopic systems, for example, increase their stability with increasing angular velocity (or the proper angular momentum).

A similar high-angular velocity destabilization occurs in one form of the stationary motion of the axisymmetrical body with a fixture point outside the axis, see Fig. 2*a*. A feature of this form of motion is that at $\omega \to \infty$ the symmetry axis tends to coincide with the fixed vertical (Fig. 2*b*). It is this form of stationary motion which enables the practical application of the dynamic balancing method to the string-type drive of rotating bodies [11]. In this case an angular velocity has been shown to exist, $\omega_{**} = \sqrt{2gl/l_1}$ (l_1 being the distance between the fixture point and the symmetry axis), such that the stationary motion becomes unstable whenever $\omega > \omega_{**}$. This high-angular velocity stability loss must necessarily be taken into account in the practical application of the dynamic balancing method to solid bodies rotating on a string-type suspension. A balancing procedure may take several stages to carry out. First, the position of the principal central axis of inertia in the body is found at small angular velocities and then the fixture point is moved closer to the axis, after which the process may be repeated as many times as required.

2. It has been found that some types of steady-state motion have properties which may be used to good advantage in certain practical applications, one example of this kind being the determination of the mass geometry of the solid body itself.

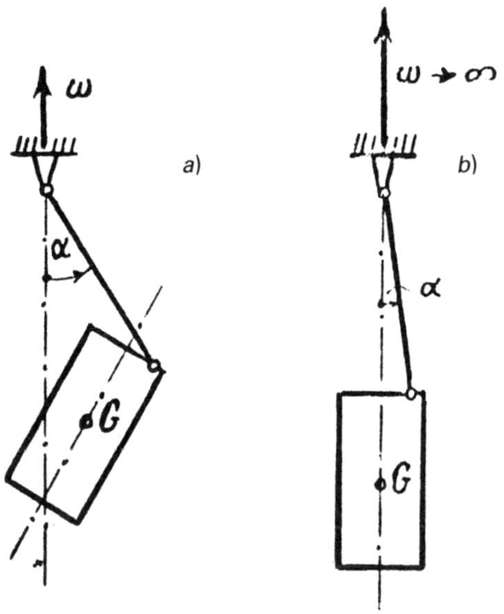

Figure 2

A gimbal suspension shown in Fig. 3 is one of the most elementary illustrations for the steady-state motion of a solid body [12]. The role of the inner ring of the device is played by the solid body being studied, whose center of masses is located exactly in the center of the (outer) ring, the axis of the latter being supported by base bearings. An arbitrary central axis of the body, z, coincides with the ζ axis around which the body can rotate relative to the ring. We fix in the body axes x and y constituting a right-handed orthogonal coordinate system together with the z axis. If the ring and the body inside are now set rotating with a (constant) angular velocity ω then, denoting by φ the angle the body makes with the ring, the equation of motion of the body reads

$$J_{zz}\ddot{\varphi} + \omega^2[(J_{yy} - J_{xx})\sin\varphi\cos\varphi + J_{xy}(\cos^2\varphi - \sin^2\varphi)] = 0 \tag{2}$$

where J_{xx}, J_{yy}, ..., J_{yz}, ... are the components of the central tensor of inertia of the body referred to the xyz coordinate system.

One partial solution of (2) is

$$\varphi = \varphi_1 = \frac{1}{2}\tan^{-1}\frac{2J_{xy}}{J_{xx} - J_{yy}} + k\frac{\pi}{2}\ (k = 0, \pm 1, \ldots) \tag{3}$$

with a simple mechanical meaning that the axis of rotation of the ring coincides with one of the axes of the ellipse formed by cutting the central inertial ellipsoid with the xy plane. As a result, the principal axes of this ellipse may be fixed, and a certain new trihedral, $x_1y_1z_1$ (z_1 coinciding with z) is obtained with the property of minimizing the sum $D = J^2_{x_1y_1} + J^2_{x_1z_1} + J^2_{y_1z_1}$ with respect to all other trihedrals having z as a common axis. In other words, the trihedral $x_1y_1z_1$ is "closer" to the principal central axes of inertia of the body than any other trihedral xyz with the

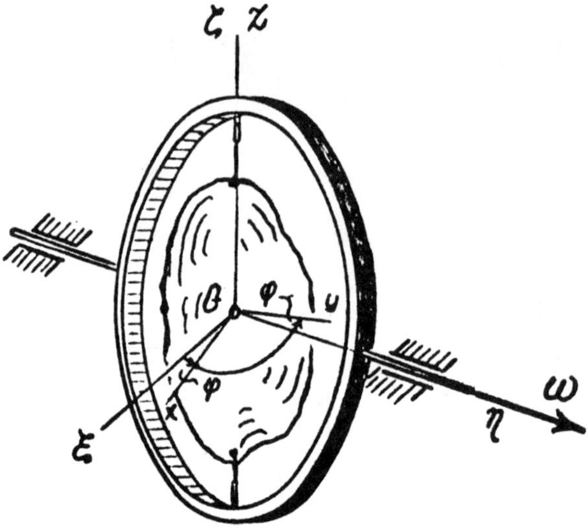

Figure 3

same $z_1(z)$ axis. If we now choose one of the newly determined axes, x_1 or y_1, as a rotation axis, a stationary motion will take place which will allow us to find a new coordinate trihedral $x_2y_2z_2$ located closer to the principal central inertial axes than the trihedral $x_1y_1z_1$ is. Repetition of the same procedure yields a sequence of coordinate trihedrals $x_jy_jz_j$ for

$$\lim_{j\to\infty} (J^2_{y_jz_j} + J^2_{x_jz_j} + J^2_{x_jy_j}) = 0 \tag{4}$$

showing that by using the above technique we can determine as accurately as we please the directions of the principal central inertial axes for an arbitrarily shaped solid body. Numerical studies have shown that the axes arrived at after six repetitions of the procedure for all practical purposes coincide with the true principal axes.

A similar property may be found in some oscillatory steady-state motions of solid bodies. Consider, for example, a pendulum capable of freely rotating about an axis perpendicular to its horizontal axis of oscillation (Fig. 4). If the free axis of rotation of the pendulum is simultaneously its principal central axis, the equations of motion are

$$\begin{aligned} &(A\cos^2\gamma + B\sin^2\gamma)\ddot{\alpha} - (A - B)\dot{\alpha}\dot{\gamma}\sin 2\gamma + mgl\sin\alpha = 0 \\ &C\ddot{\gamma} + (A - B)\dot{\alpha}^2 \sin\gamma\cos\gamma = 0 \end{aligned} \tag{5}$$

where α and γ are, respectively, the angle of oscillation and the angle of rotation, and A, B, C are the principal central moments of inertia of the pendulum.

Equations (5) have partial solutions

$$\gamma = \gamma_0 = 0; \quad \pi/2, \quad \alpha = \alpha_0(t) \tag{6}$$

where the function $\alpha_0(t)$ solves

$$(A\cos^2\gamma_0 + B\sin^2\gamma_0)\ddot{\alpha} + mgl\sin\alpha = 0 \tag{7}$$

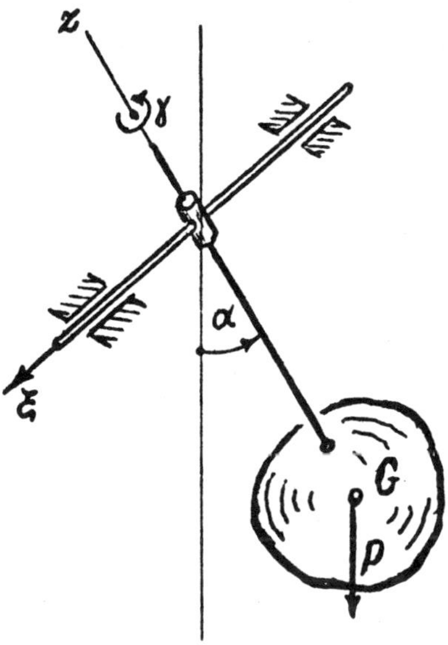

Figure 4

We thus see that a solid body can perform steady-state oscillations in such a way that, apart from the principal central axis z, the plane of oscillations also contains one of the principal axes of inertia, $y(\gamma_0 = 0)$ or $x(\gamma_0 = \pi/2)$. Of such oscillations, it has been shown that the stability property is exhibited by those maximizing the moment of inertia with respect to the axis of oscillation.

For the case when the free axis of rotation does not coincide with the principal axis of inertia, the equations of motion take the form of the system

$$\begin{aligned}
&[J_{xx}\cos^2\gamma + J_{yy}\sin^2\gamma + 2J_{xy}\sin\gamma\cos\gamma]\ddot{\alpha} - 2[(J_{xx} - J_{yy})\sin\gamma\cos\gamma \\
&- J_{xy}(\cos^2\gamma - \sin^2\gamma)]\dot{\alpha}\dot{\gamma} - (J_{xz}\cos\gamma - J_{yz}\sin\gamma)\ddot{\gamma} \\
&+ (J_{xz}\sin\gamma + J_{yz}\cos\gamma)\dot{\gamma}^2 + mgl\sin\alpha = 0 \qquad (8)\\
&J_{zz}\ddot{\gamma} = [(J_{yy} - J_{xx})\sin\gamma\cos\gamma + J_{xy}(\cos^2\gamma - \sin^2\gamma)]\dot{\alpha}^2 + (J_{xz}\cos\gamma - J_{yz}\sin\gamma)\ddot{\alpha}
\end{aligned}$$

which has an approximate partial solution

$$\alpha = \alpha_*(t); \quad \gamma = \gamma_* + \frac{J_{xz}\cos\gamma_* - J_{yz}\sin\gamma_*}{J_{zz}}\alpha_*(t) \qquad (9)$$

where

$$\gamma_* = \frac{1}{2}\tan^{-1}\frac{2J_{xy}}{J_{xx} - J_{yy}} + k\frac{\pi}{2}\ (k = 0, \pm 1, \ldots) \qquad (10)$$

and $\alpha_*(t)$ is the solution of

$$[J_{xx}\cos^2\gamma_* + J_{yy}\sin^2\gamma_* + 2J_{xy}\sin\gamma_*\cos\gamma_*]\ddot{\alpha} + mgl\sin\alpha = 0 \qquad (11)$$

The mechanical interpretation of the solution (9) is as follows. On the one hand, there are oscillations close to those of an ordinary physical pendulum. On the other, there are undamped oscillations relative to the plane of the fundamental oscillations of the pendulum; these undamped oscillations are performed by one of the principal axes of the ellipse formed by passing a section through the central ellipsoid of inertia with a plane perpendicular to the axis of rotation. We may therefore fix the principal axis of the ellipse, which means, in turn, that in this case the above procedure may also be used for finding the principal central axes of inertia of the body.

Mechanical phenomena similar to those we have just described can also be observed in the behavior of a physical pendulum with one fixed point. It is a well-known fact dating back to Mlodzeevskii that a solid body having one fixed point can perform plane-parallel oscillatory motions if its center of masses lies in one of its principal planes of inertia with respect to the fixed point. The analysis of the stability of such motions has stimulated the study of systems of differential equations with periodic coefficients, and as a result the possibility of stable plane-parallel motions has been shown in principle and confirmed by numerous experiments [13, 14]. If the center of masses of a pendulum does not lie in any plane of inertia with respect to the fixed point, it has been shown by an approximate analysis of equations of motion that, as far as finite periods of time are concerned, a body can simultaneously perform undamped oscillations in a virtually fixed plane and low-amplitude oscillations about the axis through the fixed point and center of masses. Now if we look at the ellipse formed by cutting the central ellipsoid of inertia by a plane horizontal in the equilibrium state, it is seen that the average position of one of the principal axes of the ellipse is in the plane of fundamental oscillations of the body. But it is then possible to approximately fix this axis and again to determine the principal axes of inertia by the above step-by-step procedure (which has actually been done on a mock-up installation devised for the purpose).

Some interesting possibilities for determining the principal central axes of inertia arise in the steady of the stationary motion of an arbitrarily shaped solid body suspended on a string in the manner shown in Fig. 5. The equations of such stationary motion can be reduced to the form

$$g \sin \alpha - \omega^2 \cos \alpha (c = l \sin \alpha + \alpha \sin \theta) = 0$$

$$\frac{mga}{\omega^2} \sin(\alpha - \theta) + (J_{xx} - J_{yy}) \sin^2 \varphi \sin \theta \cos \theta - (J_{zz} - J_{yy}) \sin \theta \cos \theta$$

$$-2J_{xy} \sin \varphi \cos \varphi \sin \theta \cos \theta - (J_{xz} \sin \varphi + J_{yz} \cos \varphi)(\cos^2 \theta - \sin^2 \theta) = 0 \qquad (12)$$

$$-(J_{xx} - J_{yy}) \sin \varphi \cos \varphi \sin \theta + J_{xy} \sin \theta (\cos^2 \varphi - \sin^2 \varphi)$$

$$+ (J_{xz} \cos \varphi - J_{yz} \sin \varphi) \cos \theta = 0.$$

The second and third of these may rightfully be regarded as equations for the unknowns $J_{xx}, \ldots, J_{xy}$ because of the remaining parameters involved, we may specify the string length l, the rod length c, and the distance a from the center of mass to the fixed point, and to measure experimentally the angular velocity,

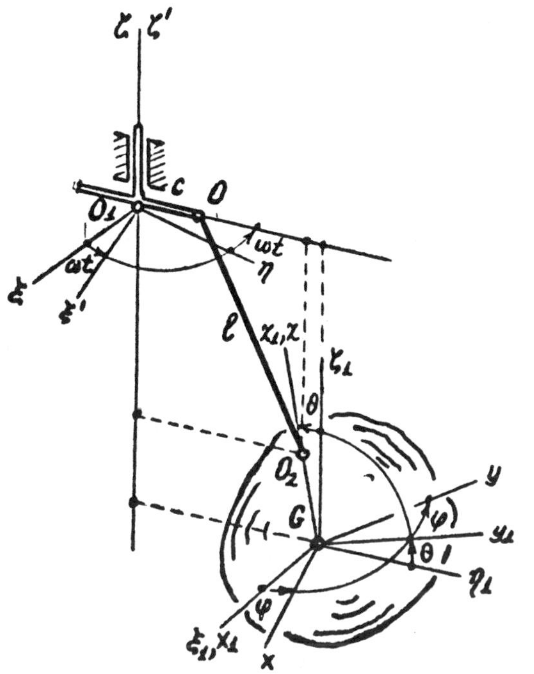

Figure 5

ω, the rotation about the z axis φ, and the angles the string and the axis of the body make with the vertical (respectively, α and θ). Since in the measurement of θ some technical difficulties may arise, it is sometimes more convenient to calculate it by the formula

$$\sin\theta = \frac{g}{\omega^2 a} n\alpha - \frac{c + l\sin\alpha}{a} \tag{13}$$

which follows from the first equations (12).

Now there are six unknown components of the tensor of inertia and only two equations for their determination, which makes it necessary that the quantities α, θ and φ for three essentially different values of ω be measured. We shall then have three sets of values ω_i, α_i, θ_i, φ_i ($i = 1, 2, 3$) and, accordingly, six equations from which the (unknown) components of the tensor of inertia can be determined. Of these six equations, only five should actually be used because in the last two equations in (12), the centroidal moments of inertia are only involved in the form of differences, $J_{xx} - J_{yy}$ and $J_{zz} - J_{yy}$. These equations will yield both the directions of the principal central axes of inertia and the differences A–B, A–C, and B–C, between the principal central moments of inertia of the body being studied.

A second method of finding the principal central axes of inertia employs an interesting property of the stationary motion of a body rotating on a string [15]. It turns out, namely, that if an arbitrary body is in a stationary motion when suspended on a string, one of its principal central axes of inertia is vertical when (and only when) the center of masses lies strictly on the straight line along which the string is aligned in this motion. It turns out that this property may be used in a sufficiently wide range of angular velocities for identifying the angular velocity at which the center of masses lies on the extension of the string. The vertical axis through the center of masses then becomes the principal central axis of inertia of the body.

3. There are a number of idealizations which have been used in the treatment of rigid body steady-state motions and which are no longer acceptable as far as the practical application of such motions is concerned. The weightless and perfectly flexible string, for example, offering no resistance to a twisting force applied to it, is an inadequate model in studying the motion of string-suspended large-sized bodies. A real "string" has a considerable mass of its own and, to add to the problem, usually serves to translate rotational motion from an electric motor to the body suspended. The simplest possible model of a heavy string is a cylindrical rod connected by a hinge to the body being studied. The other end of the rod is connected by a spherical hinge to the rotor axis of a vertical motor rotating with a constant angular velocity ω. The solid body and rod system has six degrees of freedom (see Fig. 6). A detailed analysis of the system has been conducted by Temchenko [16], who used Lagrange's equations of the second kind to derive the equations of motion and dicussed in terms of the direct Lyapunov method the stability of the vertical motion in which the axes of the rod and the body each coincide with the vertical through the fixed point in Fig. 6. This form of motion has been found to be stable at any values of angular velocity.

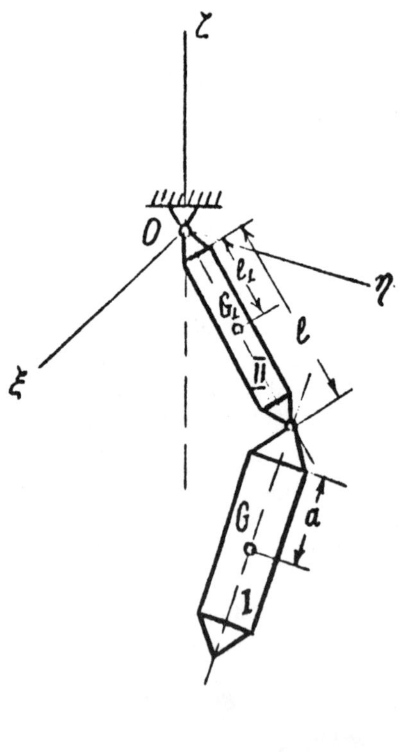

Figure 6

An important result obtained in this study is the equation for determining the bifurcation points of the system, that is, the critical values of angular velocity at which stationary forms other than vertical ramify from this vertical mode. This equation is of the form

$$f_1(\Omega) = [(A^*C_1)(A^0 - C) - m^2a^2l^2]\Omega^2 - g[ma(A^*C_1)$$

$$+ K(A^0 - C)]\Omega + K mag^2 = 0 \tag{14}$$

where $\Omega = \omega^2$; $A = B$, C, and m are the principal central moments of inertia and the mass of the solid; $A_1 = B_1$, C_1, and m_1 are the same quantities for the rod; $A^* = A_1 + m_1 l_1^2 + ml^2$, $A^0 = A + ma^2$; $K = m_1 l_1 + ml$; and g is the free fall acceleration. It was shown by analysis of this equation that there are two critical values of angular velocity in this case: one at which the symmetry axes of the body and the rod deviate the same side of the vertical; the other at which they deviate the opposite sides (see Fig. 6). It is worthwhile noting that these critical values both lie within the interval between the corresponding ideal-string values and differ only marginally from these even for very heavy rods.

An often used real-string suspension model (see Fig. 7) consists of n bodies connected by means of Hooke's joints with each other and with the shaft of a vertically oriented electric motor. Each of these bodies is an upright circular cylinder of height l and radius r. In the first order approximation it has been shown [17] that the problem of stability of the vertical rotation of a body suspended on such a "string" reduces to that of investigation of the roots of the characteristic equation

$$\det \left\| -d_{jk}(1+\lambda)^2 + g_{jk}(1+\lambda) + q_{jk}\frac{g}{\omega^2 l} \right\| = 0 \quad (j, k = 1, 2, \ldots, n+1) \tag{15}$$

$$d_{jk} = 1 + \frac{m_1}{m}\left(n + \frac{1}{2} - \max(j, k)\right) - \delta_{jk}\left(-\frac{A_1}{ml^2} + \frac{m_1}{4m}\right)$$

$$g_{jk} = \delta_{jk}\frac{C_1}{ml^2}; \quad q_{jk} = \delta_{jk}\left[1 + \frac{m_1}{m}\left(n + \frac{1}{2} - j\right)\right]; \quad d_{j,n+1} = d_{n+1,j} = 1 \tag{16}$$

$$g_{j,n+1} = g_{n+1,j} = q_{j,n+1} = q_{n+1,j} = 0; \quad d_{n+1,n+1} = 1 + \frac{A}{ma^2}$$

$$q_{n+1,n+1} = \frac{C}{ma^2}, \; q_{n+1,n+1} = \frac{l}{a} \quad (j, k = 1, 2, \ldots, n)$$

where m, A, and C are the mass and the principal central moments of inertia of the body, and m_1, A_1, and C_1 are the same quantities for each of the cylinders comprising the "string."

The equation has all its roots real if

$$(\overset{*}{r}Gr)^2 + 4(\overset{*}{r}Qr)(\overset{*}{r}Dr)\frac{g}{\omega^2 l} > 0 \tag{17}$$

Here $(\overset{*}{r}Dr)$, $(\overset{*}{r}Gr)$, and $(\overset{*}{r}Qr)$ are hermitian forms which are constructed on the

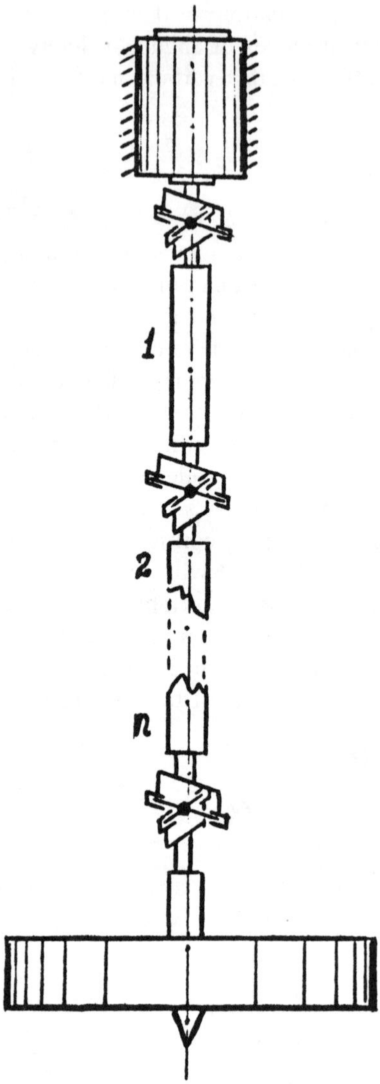

Figure 7

matrices $D = \|d_{jk}\|$, $G = \|g_{jk}\|$, $Q = \|q_{jk}\|$, and turn out to be all positive, thus indicating that the motion we are studying is stable in this approximation. Concerning other possible forms of bifurcation from the vertical stationary motion, it has been shown that there exist $n + 1$ such forms if the body is an elongated one, and n forms if it is oblate.

Fig. 8 shows possible forms of stationary motion for the case $n = 4$. In one of them, occurring at low angular velocities, the system has all its rods virtually aligned in one straight line, so that the center of gravity of the body is outside the fixed vertical (Fig. 8*a*). Other forms of stationary motion depend on suffi-

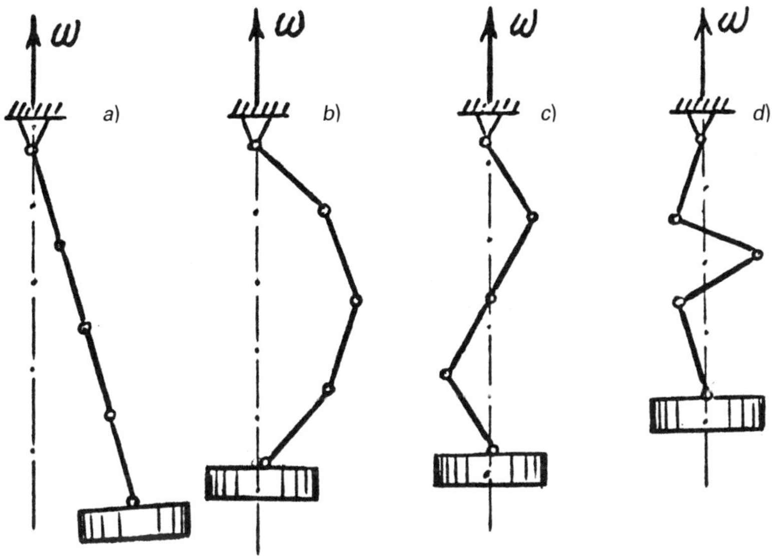

Figure 8

ciently rapid rotation for their existence. In these cases the center of gravity and the symmetry axis of the body deviate slightly from the fixed vertical, and the constituting rods may have different mutual arrangements (consistent, of course, with the geometric conditions of the system).

4. The method of pendular oscillations proves hardly, if at all, suitable for determining the principal central axes of inertia of large arbitrarily shaped bodies because these cannot generally be suspended in the manner suggested by the method. An alternative approach is to make the support of the pendulum in the form of a ball and to allow the ball to move on the surface of a fixed spherical base (Fig. 9). The behavior of such a system is conveniently studied by applying the theorem of the change of the momentum and the angular momentum of a mechanic system. Assuming some amount of friction between the base and the spherical support on the pendulum, the equations of motion turn out to be

$$m\frac{d^2\mathbf{r}_G}{dt^2} = \mathbf{P} + \mathbf{N} + \mathbf{T}$$
$$\frac{dG}{dt} = \mathbf{a}x\left(p - m\frac{d^2\boldsymbol{\rho}}{dt^2}\right) + \mathbf{r} \times \mathbf{T} + \mathbf{M} \tag{18}$$

where a and $\mathbf{r}_G$ are the radius vectors of the center of masses of the pendulum relative to the pole A and the fixed center 0, respectively; $\boldsymbol{\rho}$ is the radius vector of the point A (the center of the support of the ball) with respect to the fixed center 0; $\mathbf{r}$ is the radius of the spherical support; $\mathbf{P}$ is the pendulum weight force, applied at the center of masses of the pendulum and pointing in the negative direction; $\mathbf{M}$ is the principal moment of the forces on the pendulum relative the

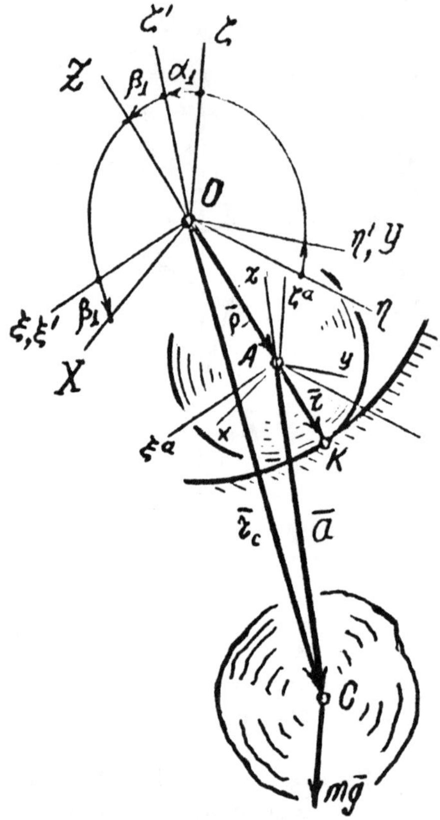

Figure 9

contact point K. The quantities **N** and **T** in equation (18) are the components of the principal vector **F** of the forces acting on the pendulum: **N**, the normal reaction of the support, is directed along the normal to the surface of the contact between the spheric support and the spheric base; the component **T** lies in the plane of the contact.

Employing cartesian coordinates X, Y, Z with origin at the fixed point 0 (see Fig. 9), the first of the vector equations (18) takes the form

$$m\left[\frac{d^2\xi_G}{dt^2}\cos\xi X + \frac{d^2\eta_G}{dt^2}\cos\eta X + \left(\frac{d^2\zeta_G}{dt^2} + g\right)\cos\zeta X\right] = T_X$$

$$m\left[\frac{d^2\xi_G}{dt^2}\cos\xi\, y + \frac{d^2\eta}{dt^2}\cos\eta y + \left(\frac{d^2\zeta_0}{dt^2} + g\right)\cos\zeta y\right] = T_Y \qquad (19)$$

$$m\left[\frac{d^2\xi_G}{dt^2}\cos\xi Z + \frac{d^2\eta_G}{dt^2}\cos\eta Z + \left(\frac{d^2\zeta_G}{dt^2} + g\right)\cos\zeta Z\right] = N$$

Note that the coordinate system XYZ is obtained from the fixed system $\xi\eta\zeta$ by two successive rotations through Euler angles α_1 and β_1. It is seen from Fig.

9 that the Z axis lies along the negative direction of the vector r. The quantities ξ_G, η_G, and ζ_G in (19) are the coordinates of the center of masses G of the pendulum referred to the $\xi\eta\zeta$ system.

To expand of equations (18) we use axes xyz fixed in the body to obtain, after some manipulations

$$\frac{dG_x}{dt} + \omega_y G_z - \omega_z G_y = -ma\left[\frac{d^2\xi_A}{dt^2}\cos\xi y + \frac{d^2\eta_A}{dt^2}\cos\eta y + \left(\frac{d^2\zeta_A}{dt^2} + g\right)\cos\zeta y\right]$$
$$+ r(T_Y \cos Xx - T_X \cos Y_x) + M \cos Zx$$
$$\frac{dG_Y}{dt} + \omega_z G_x - \omega_x G_z = ma\left[\frac{d^2\xi_A}{dt^2}\cos\xi x + \frac{d^2\eta_A}{dt^2}\cos\eta x + \left(\frac{d^2\zeta_A}{dt^2} + g\right)\cos\zeta x\right] \tag{20}$$
$$+ r(T_Y \cos Xy - T_X \cos Yy) + M \cos Zy$$
$$\frac{dG_z}{dt} + \omega_x G_y - \omega_y G_x = r(T_Y \cos Xz - T_X \cos Yz) + M \cos Zz$$

where ω_x, ω_y, ω_z are the x-, y-, and z-components of the angular velocity of the xyz system itself; G_x, G_y, G_z are the x-, y-, and z-components of the vector of the angular momentum of the pendulum relative of the pole A; ξ_A, η_A, ζ_A are the coordinates of the pole A in the coordinate system $\xi\eta\zeta$.

An assumption made in [13] is that the force $\mathbf{T}$ and the moment $\mathbf{M}$ are proportional, respectively, to the linear velocity v_c of the pendulum at the point of contact and to the z-component of the angular velocity of the pendulum.

Performing some manipulations on equations (19) and (20) and using the table of the cosines of the angles the XYZ axes make with xyz axes, a system of equations is obtained which completely describes the behavior of a spherical support pendulum moving with some amount slip in a fixed spherical surface [13]. It proves possible to find the plane-parallel motion of this pendulum if we assume that the pendulum deviates only slightly from its stable equilibrium (at which the center of masses is in the lowest position possible) and assume also the simplest possible friction law for the contact between the spheric support and the spheric base. It has been found under these assumptions that these are two kinds of oscillations which the principal axis of inertia of the body can perform, namely the fundamental, slowly diminishing, pendulum-like, oscillations, and additional, rapidly decaying, high-frequency and low-amplitude motions. It has also been found that after a sufficiently long period of time the center of masses of the pendulum begins to oscillate in a certain fixed plane. It turned out possible, further, to prove the non-asymptotic stability of these motions under the assumption of the fixed point being sufficiently distant from the center of mass [13].

5. Equations (19), (20) may also be applied to the stability study of a spinning top moving on a spheric support in the surface of a fixed spherical base. This will be done if, the above manipulations having been performed, we change the sign of the quantity a (see Fig. 9) and consider the axis of z as the principal central

axis of inertia of the top and x and y axes as its principal axes of inertia at point A. We also take advantage of the symmetry of the top by equating the components J_{xx} and J_{yy} of the tensor of inertia in the xyz coordinate system.

It should be noted that the motion of the top on the fixed spherical surface gives rise both to sliding friction and to pivoting friction about the axis normal to the surface, and although the latter has only a slight direct effect on the motion, the rotation of the top about the vertical does affect the quantitative characteristics of the sliding friction. While the free sliding frictional force remains constant in magnitude whatever the velocity of sliding (Coulomb's law), the inclusion of pivoting gives rise to the free averaging effect and results in the actual resistance being approximately proportional to the absolute magnitude of the slippage velocity [18, 19]. This being so, let us put to zero the moment M arising from the pivoting of the top about the axis normal to the surface of the base. This additional simplification makes it possible to use equations (19) and (20) for investigating the stability of the vertical motion in the framework of the linear theory of differential equations. It turns out that, for any radius R of the fixed spherical support, the stability condition of the motion is the positivity of the imaginary parts of all of the roots of the characteristic equation

$$g(p) + ikh(p) = 0 \tag{21}$$

in which $i = \sqrt{-1}$ and

$$g(p) = \lambda^2 p^4 - \omega\rho^2 p^3 + g[a - \epsilon(a^2 + \lambda^2)]p^2 + g\epsilon\omega\rho^2 p - ag^2\epsilon$$

$$h(p) = -(\lambda^2 + h^2)p^3 + \omega[\rho^2 + rh - r\epsilon(\lambda^2 + ah)]p^2$$

$$-[-ag + r\epsilon(rg + \rho^2\omega^2)]p - g\epsilon\omega rh = 0 \tag{22}$$

where k is the friction coefficient: $\lambda^2 = A^*/m$; $\rho^2 = C^*/m$; $h = r + a$; A^*, C^* are the principal central moments of inertia of the top; $\epsilon = 1/R$.[1]

The roots of (21) have been investigated for all non-negative values of the friction coefficient. In the cases $k = 0$ and $k \to \infty$ it has been shown that the condition of stability for top motion is equivalent to that of reality of all the roots of the equations $g(p) = 0$ and $h(p) = 0$. These roots have been found by Ishlinsky and Temchenko [20, 21] by the grapho-analytical method usually employed for determining the roots of an algebraic equation and based on the variation of the characteristic parameters of the equation. The parameters dealt with in our particular case were ω and ϵ, both ranging from zero to infinity, and the solution procedure was as follows. First the equation $g(p; \omega, \epsilon) = 0$ was solved for the parameter ϵ and the curve $\epsilon_g(p, \omega)$ versus ω was plotted. The plot was then subdivided into a number of ω-intervals, for each of which the derivative of $\epsilon_y(p, \omega)$ and the number of roots of the numerator and denominator of this function were known. For each of these intervals the function $\epsilon_g(p)$ in the (p, ϵ) plane was

1) A characteristic equation totally identical to (21) has been derived by Contensou [18], who analyzed it for fast and slow motions but restricted himself to limiting case of plane motion, $R \to \infty$.

plotted. This done, lines parallel to the abscissa (or p) axis were drawn and for every fixed ϵ-value the number of real roots of the function $\epsilon_g(p)$, and hence the number of roots of the equation $\epsilon_g(p) = 0$, was determined. The same procedure was then applied to the equation $h(p) = 0$.

In the case of a finite nonzero k, we have mentioned earlier that a condition must be found which would secure that the imaginary parts of all the roots of the equation (21) be possible (note that the coefficients of this equation are complex in this particular case). The search for this condition has been carried out by combining the well-known Hermite-Biehler formula with the Neumark method [22]. It can be shown by analysis of equations (21) and (22) that the polynomials $g(p)$ and $h(p)$ differ by unity in their degree and that their respective highest-degree coefficients form a negative product. Hence, by the Neumark method, equation (21) will necessarily have one of its roots located in the upper half-plane.

For all the roots of (21) to lie on one side of the real axis, it follows from the Hermite-Biehler theorem that it is necessary and sufficient that the roots of the polynomials $g(p)$ and $h(p)$ be real and alternating. In our particular case these roots will lie in the upper half-plane.

The grapho-analytical method described above was also used to find the conditions of reality for the roots of the polynomials $g(p)$ and $h(p)$. This enabled us to plot the functions $\epsilon_g(p)$ and $\epsilon_h(p)$ (on the same scale) for various values of the parameter $\Omega = \omega^2$ of the spinning top, and to find those regions of ϵ values in which all the roots of $\epsilon_g(p)$ and $\epsilon_h(p)$ are real and alternating.

A possible behavior of the functions $\epsilon_g(p)$ (dashed line) and $\epsilon_h(p)$ (solid line) over an Ω interval is indicated schematically in Fig. 10. It will be seen that for all $0 < \epsilon < \epsilon_*$ the roots of the two polynomials alternate showing that the motion of the top is a stable one.

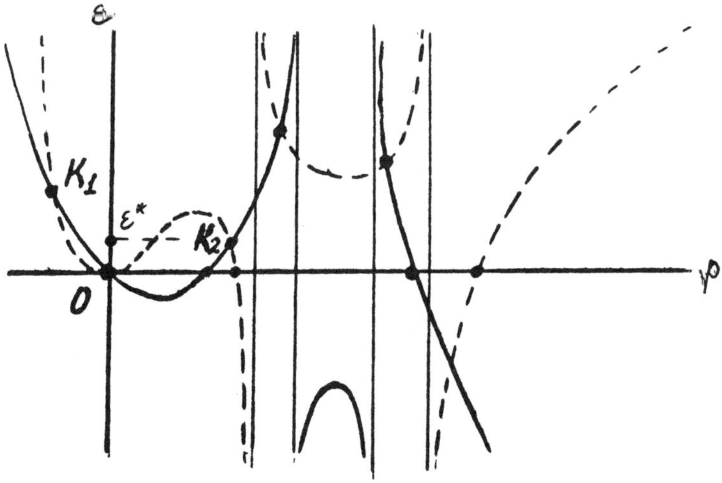

Figure 10

REFERENCES

1. Ishlinsky, A. Yu. On bifurcations preserving the stability of stationary motions. Dokl. USSR Acad. Sci., 1957, Vol. 117, No. 1, pp. 47–49.
2. Ishlinsky, A. Yu., Malashenko, S. V., and Temchenko, M. E. Dynamic equilibrium stability bifurcation in one mechanical system. Izv. USSR Acad. Sci. OTN, 1958, No. 8, pp. 53–61.
3. Ishlinsky, A. Yu., Storozhenko, V. A., and Temchenko, M. E. On stationary motions of an axisymmetric rigid body rotating on a string. In: Dynamics and stability of controlled systems. Inst. Mat. Ukr. SSR Acad. Sci., Kiev, 1977 (in Russian).
4. Ishlinsky, A. Yu., Storozhenko, Y. A., and Temchenko, M. E. On the motion of a string suspended symmetric rigid body. Izv. USSR Acad. Sci. MTT, 1979, No. 6, pp. 3–16.
5. Storozhenko, V. A. Stationary motion stability of a body of revolution suspended by a string from an arbitrary point. Prepr. Inst. Mat. Ukr. SSR Acad. Sci., 1981, No. 81.13 (in Russian).
6. Storozhenko, V. A. and Temchenko, M. E. Survey of the stationary motions of an axisymmetric rigid body. In: Analytical methods for complex system dynamics and stability studies. Prepr. Inst. Mat. Ukr. SSR Acad. Sci., Kiev, 1982 (in Russian).
7. Chetaev, N. G. Stability of motion: studies in analytical mechanics. USSR Acad. Sci., Moscow, 1962 (in Russian).
8. Rumyantsev, V. V. On the stability of stationary motions. PMM, 1966, Vol. 30, No. 5, pp. 922–933.
9. Skimel, V. N. On the motion of a spring suspended gyrostat. In: Proc. Conf. on Anal. Mech. and Appl. Stability Theory, Kazan, Kazan Aviation Institute, 1964 (in Russian).
10. Storozhenko, V. A. Stability of one form of stationary motion of a spring suspended elongated body of revolution. Izv. USSR Acad. Sci. MTT., 1983, No. 4, pp. 32–40.
11. Ishlinsky, A. Yu. et al. Balancing methods for string drive rotation. Ibid, 1979, No. 5, pp. 3–18.
12. Storozhenko, V. A. Determination of principal axes of inertia in an arbitrarily shaped body. Prepr. Inst. Mat. Ukr. SSR Acad. Sci., Kiev, 1986, No. 86.32 (in Russian).
13. Ishlinsky, A. Yu., Vasilenko, V. P., Storozhenko, V. A. et al. On one form of stationary oscillations of a heavy rigid body. Izv. USSR Acad. Sci. MTT, 1985, No. 2, pp. 3–18.
14. Ishlinsky, A. Yu., Storozhenko, V. A., and Temchenko, M. E. On the stability of the pendular oscillations of a rigid body with a single fixed point. Ibid, 1986, No. 1, pp. 18–26.
15. Ishlinsky, A. Yu., Malashenko, S. V., and Storozhenko, V. A. On stationary motions of a string suspended body with a vertical principal central axis of inertia. Ibid., 1980, No. 2, pp. 34–45.
16. Temchenko, M. E. On stationary motions of two connected bodies. Inst. Mat. Ukr. SSR Acad. Sci., 1981, pp. 95–105.
17. Storozhenko, V. A. On the rotation stability of a body suspended by means of connected rods. Izv. USSR Acad. Sci. MTT, 1985, No. 1, pp. 45–52.
18. Contensou, P. Interrelation between sliding friction and pivoting friction and its role in the theory of a top. In: Problems of gyroscopy. Moscow, Mir., 1967 (in Russian).
19. Magnus, K. Kreisel: Theorie und Anwendungen. Berlin, Springer, 1971.
20. Ishlinsky, A. Yu. and Temchenko, M. E. On the string rotation stability of a rigid body with an ellipsoidal cavity filled with an ideal noncompressible liquid. PMM, 1966, Vol. 30, No. 1, pp. 30–41.
21. Temchenko, M. E. Stability of a ball supported top on a frictionless spheric surface. In: Approximate methods in dynamics and stability of multidimensional systems, Inst. Mat. Ukr. SSR Acad. Sci., Kiev, 1983, pp. 123–128.
22. Neumark, Yu. I. On the distribution of polynomial roots. Dokl. USSR Acad. Sci., 1947, Vol. 58, No. 3, pp. 357–360.

RODRIGUES-HAMILTON PARAMETERS IN RIGID BODY DYNAMICS PROBLEMS

V. N. Koshlyakov

The method of Rodrigues-Hamilton parameters offers considerable computational advantages over competitive techniques and is therefore receiving much attention in the theory of motion control and in orientation and navigation problems. In this paper we apply this method to the analytical treatment of rapid rotation around a vertical axis [1–6].

1. As shown in [4, 5], the equations of motion of a heavy solid about a fixed point may be written in the matrix form

$$2ABC\, d^2\lambda/dt^2 + Q\lambda = 0 \tag{1.1}$$

where $\lambda = [\lambda_0, \lambda_1, \lambda_2, \lambda_3]^T$ is the transposed matrix-column of RH parameters, which are considered as the unknowns of the problem; A, B, C are the moments of inertia about the principal axes Ox, Oy, Oz fixed in the body and meeting at the fixed point O; and Q is the matrix

$$Q = \begin{bmatrix} K & L \\ -L^T & K^T \end{bmatrix} \tag{1.2}$$

in which

$$K = \begin{bmatrix} a & b \\ -b & a \end{bmatrix}, \quad L = \begin{bmatrix} a' & b \\ -b' & a' \end{bmatrix} \tag{1.3}$$

are the Kayley matrices with elements defined as follows

$$\begin{aligned} a &= 2ABC(\dot\lambda_0^2 + \dot\lambda_1^2 + \dot\lambda_2^2 + \dot\lambda_3^2), \quad b = BC[M_x - (C - B)qz] \\ a' &= CA[M_y - (A - C)rp], \quad b' = AB[M_z - (B - A)pq] \end{aligned} \tag{1.4}$$

where p, q, r are, respectively the x, y, and z components of the angular velocity of the body, and M_x, M_y, M_z are the same components of the moment of the force of gravity, the only active force in the problem.

In the subsequent analysis, use will be made of relationships between the *RH* parameters and Euler angles φ, ϑ, ψ. We have

$$\lambda_0 = \cos\frac{\vartheta}{2}\cos\frac{\psi+\varphi}{2}, \quad \lambda_1 = \sin\frac{\vartheta}{2}\cos\frac{\psi-\varphi}{2}$$
$$\lambda_2 = \sin\frac{\vartheta}{2}\sin\frac{\psi-\varphi}{2}, \quad \lambda_3 = \cos\frac{\vartheta}{2}\sin\frac{\psi+\varphi}{2} \tag{1.5}$$

$$\lambda_0^2 + \lambda_1^2 + \lambda_2^2 + \lambda_3^2 = 1 \tag{1.6}$$

2. Suppose that at some initial moment of time the body is set in rapid rotation about the principal axis $0z$ coinciding with the vertical 0ζ.

If the $0z$ axis passes through the center of gravity of the body, the Euler-Poisson equations, and hence equations (1.1), have a partial solution [2, 4, 5]

$$\tilde{p} = \tilde{q} = 0, \quad \tilde{r} = \omega = \text{const}$$
$$\tilde{\gamma}_1 = \tilde{\gamma}_2 = 0, \quad \tilde{\gamma}_3 = 1 \tag{2.1}$$

where γ_1, γ_2, γ_3 are the direction cosines of the vertical axis 0ζ of a certain fixed trihedral $0\xi\eta\zeta$ relative to the axes $0x$, $0y$, $0z$ of the moving trihedral fixed in the body. The solution (2.1) is readily interpreted as rotation about the vertical 0ζ with a constant angular velocity ω. We now turn to the case when the $0z$ axis does not pass through the center of gravity and note at once that we will not be concerned with Staude's permanent stationary rotations about the set of vertical axes not coinciding with the principal axes at the point 0 [8, 9].

Let x_c, y_c, z_c be the coordinates of the center of gravity referred to the x, y, z coordinate system. We will assume x_c and y_c to be small as compared with z_c, which means that the magnitudes of the ratios x_c/z_c and y_c/z_c will be much smaller than unity.

Having this in mind, let us take the solution (2.1) as a generating one. Needless to say, the generating solution so chosen cannot be associated with the unperturbed motion at x_c and y_c different from zero because in this case (2.1) is no longer a partial solution to the initial system of equations. The generating solution corresponds to the case $\vartheta = 0$, so that the rotation of the body degenerates into the rotation through an angle $\psi + \varphi$ about the 0ζ axis. If we denote by $\tilde{\psi} + \tilde{\varphi}$ the angle corresponding to the generating solution, then

$$\dot{\tilde{\psi}} + \dot{\tilde{\varphi}} = \omega = \text{const} \tag{2.2}$$

giving

$$\tilde{\psi} + \tilde{\varphi} = 2\chi = \omega t, \quad \chi = \chi(0) + ct \tag{2.3}$$

where $c = \omega/2$, and $\chi(0)$ is the initial value of the angular variable χ. From (1.5), the *RH* parameters for the generating solution are readily found to be

$$\tilde{\lambda}_0 = \cos\chi, \quad \tilde{\lambda}_1 = \tilde{\lambda}_2 = 0, \quad \tilde{\lambda}_3 = \sin\chi \tag{2.4}$$

In our analysis we shall only need equations for the parameters λ_1 and λ_2. From the matrix equation (1.1) we have

$$\begin{aligned} &2ABC\ddot{\lambda}_1 + BC(C - B)qr\lambda_0 - AC(A - C)rp\lambda_3 \\ &- AB(B - A)pq\lambda_2 + 2ABC(\dot{\lambda}_0^2 + \dot{\lambda}_1^2 + \dot{\lambda}_2^2 + \dot{\lambda}_3^2)\lambda_1 + M_1 = 0 \\ &2ABC\ddot{\lambda}_2 + BC(C - B)qr\lambda_3 + AC(A - C)rp\lambda_0 \\ &- AB(B - A)pq\lambda_1 + 2ABC(\dot{\lambda}_0^2 + \dot{\lambda}_1^2 + \dot{\lambda}_2^2 + \dot{\lambda}_3^2)\lambda_2 + M_2 = 0 \end{aligned} \tag{2.5}$$

where

$$\begin{aligned} M_1 &= -BCM_x\lambda_0 + ACM_y\lambda_3 - ABM_z\lambda_2 \\ M_2 &= -BCM_x\lambda_3 - ACM_y\lambda_0 + ABM_z\lambda_1 \end{aligned} \tag{2.6}$$

The projections p, q, and r are related to the parameters λ_s and $\dot{\lambda}_s$ by the well known expressions

$$\begin{aligned} p &= 2(\lambda_0\dot{\lambda}_1 - \lambda_1\dot{\lambda}_0 + \lambda_3\dot{\lambda}_2 - \lambda_2\dot{\lambda}_3) \\ q &= 2(\lambda_0\dot{\lambda}_2 - \lambda_2\dot{\lambda}_0 + \lambda_1\dot{\lambda}_3 - \lambda_3\dot{\lambda}_1) \\ r &= 2(\lambda_0\dot{\lambda}_3 - \lambda_3\dot{\lambda}_0 + \lambda_2\dot{\lambda}_1 - \lambda_1\dot{\lambda}_2) \end{aligned} \tag{2.7}$$

and for the moment of the force of gravity we find, accordingly,

$$\begin{aligned} M_x &= P[2(\lambda_0\lambda_1 + \lambda_2\lambda_3)Z_c - (\lambda_0^2 + \lambda_3^2 - \lambda_1^2 - \lambda_2^2)y_c] \\ M_y &= P[(\lambda_0^2 + \lambda_3^2 - \lambda_1^2 - \lambda_2^2)\,x_c - 2(\lambda_1\lambda_3 - \lambda_0\lambda_2)Z_c] \\ M_Z &= 2P[(\lambda_1\lambda_3 - \lambda_0\lambda_2)\,y_c - (\lambda_0\lambda_1 + \lambda_2\lambda_3)x_c] \end{aligned} \tag{2.8}$$

where P is the weight of the body.

3. Turning to the perturbed problem we write

$$\lambda_s = \tilde{\lambda}_s + \lambda_s' \tag{3.1}$$

where the $\tilde{\lambda}_s$ are those values of the *RH* parameters which correspond to the generating motion, and λ_s' are some functions to be determined later. By (2.4), we have $\lambda_1' = \lambda_1$, $\lambda_2' = \lambda 2$.

The explicit form of the equation for λ_1 and λ_2 is greatly simplified if we perform a simple rotation to change to the variables u_1 and u_2

$$u_1 = -(\lambda_1\cos\chi + \lambda_2\sin\chi), \quad u_2 = \lambda_2\cos\chi - \lambda_1\sin\chi \tag{3.2}$$

where χ is defined in (2.3), in which we set $\chi(0) = 0$. Inverting (2.3) we obtain

$$\lambda_1 = -(u_1\cos\chi + u_2\sin\chi), \quad \lambda_2 = u_2\cos\chi - u_1\sin\chi \tag{3.3}$$

The variables u_1 and u_2 are of the same order that λ_1, λ_2 and it is easy to see that

$$\lambda_1^2 + \lambda_2^2 = u_1^2 + u_2^2 \tag{3.4}$$

Until no specifications have been made about the unperturbed motion, we may retain terms up to the second order in u_1, u_2 and their derivatives. From (1.5) and (3.4),

$$\sin^2\frac{\vartheta}{2} = \lambda_1^2 + \lambda_2^2 = u_1^2 + u_2^2 \tag{3.5}$$

or, for small ϑ

$$|\vartheta| = 2\sqrt{u_1^2 + u_2^2} \tag{3.6}$$

From (1.5) we then find for small ϑ

$$\begin{aligned} \lambda_0 &= \cos\frac{\tilde{\psi} + \tilde{\varphi}}{2}\left(1 - \frac{\vartheta^2}{8}\right) = \cos\chi\left(1 - \frac{\vartheta}{8}\right)^2 = \left[1 - \frac{1}{2}(u_1^2 + u_2^2)\right]\cos\chi \\ \lambda_3 &= \sin\frac{\tilde{\psi} + \tilde{\varphi}}{2}\left(1 - \frac{\vartheta^2}{8}\right) = \sin\chi\left(1 - \frac{\vartheta^2}{8}\right) = \left[1 - \frac{1}{2}(u_1^2 + u_2^2)\right]\sin\chi \end{aligned} \tag{3.7}$$

where use has been made of the fact that at $\vartheta = 0$ the rotation degenerates into that through an angle 2χ about the 0ζ axis. Using (3.3) we can now express λ_1 and λ_2 in terms of u_1 and u_2.

From (3.5) and (3.7) we find that

$$\begin{aligned} \sum_{s=1}^{4}\lambda_s^2 &= \lambda_1^2 + \lambda_2^2 + \lambda_3^2 + \lambda_0^2 = u_1^2 + u_2^2 + \left[1 - \frac{1}{2}(u_1^2 + u_2^2)\right]^2\cos^2\chi \\ &+ \left[1 - \frac{1}{2}(u_1^2 + u_2^2)\right]^2\sin^2\chi = 1 + \frac{1}{4}(u_1^2 + u_2^2)^4 = 1 + \frac{\vartheta^4}{16} \end{aligned} \tag{3.8}$$

showing that in the approximation adopted the relationship (1.6) is accurate to 0(4) in the variables u_1, u_2 or in the nutation angle ϑ. Confining ourselves to this accuracy, expression (2.8) may be rewritten in the form

$$\begin{aligned} M_x &= -P\{2u_1z_c + [1 - 2(u_1^2 + u_2^2)]y_c\} \\ M_y &= P\{2u_2z_c - [1 - 2(u_1^2 + u_2^2)]x_c\} \\ M_z &= 2P(u_1x_c - u_2y_c) \end{aligned} \tag{3.9}$$

which gives

$$\begin{aligned} M_1 &= BCP\left\{2u_1z_c + \left[1 - \frac{5}{2}(u_1^2 + u_2^2)\right]y_c\right\}\cos\chi \\ &+ ACP\left\{2u_2z_c + \left[1 - \frac{5}{2}(u_1^2 + u_2^2)\right]x_c\right\}\sin\chi \\ &- 2PAB(u_1x_c - u_2y_c)(u_2\cos\chi - u_1\sin\chi) \end{aligned} \tag{3.10}$$

$$M_2 = BCP\left\{2u_1z_c + \left[1 - \frac{5}{2}(u_1^2 + u_2^2)\right]y_c\right\}\sin\chi$$

$$- ACP\left\{2u_2z_c + \left[1 - \frac{5}{2}(u_1^2 + u_2^2)\right]x_c\right\}\cos\chi$$

$$- 2PAB(u_2 \sin\chi + u_1 \cos\chi)(u_1x_c - u_2y_c)$$

To proceed further it is desirable that the products $qr\lambda_0$, $qr\lambda_3$, $pr\lambda_3$, $pr\lambda_0$, pq_2 and $pq\lambda_1$ be expressed in terms of the variables u_1u_2 and their derivatives. We find

$$qr\lambda_0 = 4c[(\dot\lambda_2 + c\lambda_1)\cos\chi + (c\lambda_2 - \dot\lambda_1)\sin\chi]\cos\chi$$

$$+ 0(\lambda_s^3, \dot\lambda_s^3) \quad (s = 1, 2) \tag{3.11}$$

(We have mentioned earlier that the variables λ_1, λ_2 are of the same order as u_1, u_2.) In accordance with (3.11) we write

$$qr\lambda_3 = 4c[(c\lambda_2 - \dot\lambda_1)\sin\chi + (\dot\lambda_2 + c\lambda_1)\cos\chi]\sin\chi$$

$$pr\lambda_0 = 4c[(\dot\lambda_1 - c\lambda_2)\cos\chi + (\dot\lambda_2 + c\lambda_1)\sin\chi]\cos\chi \tag{3.12}$$

$$pr\lambda_3 = 4c[(\dot\lambda_2 + c\lambda_1)\sin\chi + (\dot\lambda_1 - c\lambda_2)\cos\chi]\sin\chi$$

$$pq\lambda_2 = pq\lambda_1 = 0$$

and apply the transformation (3.2) to obtain the expressions

$$qr\lambda_0 = 2\omega(\dot u_2 - \omega u_1)\cos\chi, \quad qr\lambda_3 = 2\omega(\dot u_2 - \omega u_1)\sin\chi \tag{3.13}$$

$$pr\lambda_0 = -2\omega(\dot u_1 + \omega u_2)\cos\chi, \quad pr\lambda_3 = -2\omega(\dot u_1 + \omega u_2)\sin\chi$$

accurate to the third order in u_1, u_2 and their derivatives. Changing to the variables u_1, u_2 and retaining the same accuracy the system (2.5) is then converted to the system of two differential equations

$$AC\ddot u_1 - C\omega(C - B - A)\dot u_2 + C[(C - B)\omega^2 - Pz_c]u_1$$

$$+ AP(u_1x_c - u_2y_c)u_2 + \frac{5}{4}CPy_c(u_1^2 + u_2^2) = \frac{1}{2}CPy_c$$

$$BC\ddot u_2 + C\omega(C - B - A)\dot u_1 + C[(C - A)\omega^2 - Pz_c]u_2 \tag{3.14}$$

$$- BP(u_1x_c - u_2y_c)u_1 + \frac{5}{4}CPx_c(u_1^2 + u_2^2) = \frac{1}{2}CPx_c$$

Note that if the axis of rotation $0z$ passes through the center of gravity, then $x_c = y_c = 0$ and equation (3.14) does not contain terms quadratic in u_1, u_2.

The right-hand sides in equations (3.14) represent continually operating perturbations that arise from the center of gravity having x and y coordinates nonzero. Setting $\dot u_1 = \dot u_2 = \ddot u_1 = \ddot u_2 = 0$ in (3.14) yields a system for the partial solutions

u_1^*, u_2^* that correspond to these perturbations. We have

$$C[(C-B)\omega^2 - Pz_c]u_1^* + AP(u_1^* x_c - u_2^* y_c)u_2^* + \frac{5}{4} CPy_c(u_1^{*2} + u_2^{*2}) = \frac{1}{2} CPy_c$$
$$C[(C-A)\omega^2 - Pz_c]u_2^* - BP(u_1^* x_c - u_2^* y_c)u_1^* + \frac{5}{4} CPx_c(u_1^{*2} + u_2^{*2}) = \frac{1}{2} CPx_c \tag{3.15}$$

Equations for the perturbed motion are obtained from (3.14) by changing to the variables x_1, x_2 according to

$$u_1 = u_1^* + x_1, \quad u_2 = u_2^* + x_2 \tag{3.16}$$

with u_1^*, u_2^* given by the system (3.15). The resulting system contains two differential equations and is conveniently rewritten in the matrix form

$$J\ddot{x} + H\dot{x} + Gx = X \tag{3.17}$$

where $x = [x_1, x_2]^T$; J, H and G are second-rank quadratic matrices with constant elements; X is matrix-column with elements quadratic in x_s and $\dot{x}_s$ $(s = 1, 2)$. Explicitly,

$$J = \begin{bmatrix} A & 0 \\ 0 & B \end{bmatrix}, \quad H = \begin{bmatrix} 0 & -h \\ h & 0 \end{bmatrix}, \quad G = \begin{bmatrix} c_1 & p_1 \\ p_2 & c_2 \end{bmatrix} \tag{3.18}$$

where

$$h = (C - B - A)\omega$$
$$c_1 = (C-B)\omega^2 - Pz_c + \frac{AP}{C} u_2^* x_c + \frac{5}{2} Pu_1^* y_c$$
$$c_2 = (C-A)\omega^2 - Pz_c + \frac{BP}{C} u_1^* y_c + \frac{5}{2} Pu_2^* x_c \tag{3.19}$$
$$p_1 = \frac{P}{C}\left[\left(\frac{5}{2} C - 2A\right)u_2^* y_c + Au_1^* x_c\right]$$
$$p_2 = \frac{P}{C}\left[\left(\frac{5}{2} C - 2B\right)u_1^* x_c + Bu_2^* y_c\right]$$

If the elements p_1 and p_2 differ from each other then, provided $H \neq 0$, the structure of the system (3.17) turns out to be unstable, whatever the second-order terms determined by the matrix x [10]. To show this we note that the matrix G can always be represented as the sum of a symmetric and a skew-symmetric matrix

$$G = G_1 + G_2 \tag{3.20}$$

$$G_1 = \begin{bmatrix} C_1 & \dfrac{P_1 + P_2}{2} \\ \dfrac{P_1 + P_2}{2} & C_2 \end{bmatrix}, \quad G_2 = \begin{bmatrix} 0 & \dfrac{P_1 - P_2}{2} \\ -\dfrac{P_1 - P_2}{2} & 0 \end{bmatrix} \tag{3.21}$$

The instability of the problem is due to the component G_2. It is known that a system of the type (3.17) cannot be stable unless dissipative forces are allowed for [10]. Following the notation (3.19), the characteristic equation of the system (3.17) reads

$$ABD^4 + (h^2 + Ac_2 + Bc_1)D^2 + h(p_2 - p_1)D + c_1c_2 - p_1p_2 = 0 \tag{3.22}$$

Now if the coefficient of D is nonzero, equation (3.22) will have at least one root positive (or with a positive real part), which means that the trivial solution $x = 0$ of the system (3.17) will be unstable regardless of the terms of the second and higher orders.

Neglecting second-order terms in (3.15), the variables u_1 and u_2 for the unperturbed problem are found to be

$$u_1^* = \frac{1}{2}\frac{Py_c}{(C - B)\omega^2 - Pz_c}, \quad u_2^* = \frac{1}{2}\frac{Px_c}{(C - A)\omega^2 - Pz_c} \tag{3.23}$$

in full agreement with R. Grammel's results [7].

By the argument given in Section 2, the quantities u_1^* and u_2^* as defined by (3.23) can always be made as small as we please. If $A \neq B \neq C$, this may be achieved by increasing the angular velocity ω. In the degenerate case ($B = C$ or $A = C$), u_1^* and u_2^* are small because so are the ratios x_c/z_c and y_c/z_c (cf. Section 2).

We can try to improve the accuracy of the quantities u_1^* and u_2^* by setting

$$u_1^* = \frac{1}{2}\frac{Py_c}{(C - B)\omega^2 - Pz_c} + \epsilon_1, \quad u_2^* = \frac{1}{2}\frac{Px_c}{(C - A)\omega^2 - Pz_c} + \epsilon_2 \tag{3.24}$$

in (3.15) and by applying a simple iteration procedure to the resulting system. It turns out, however, that the corrections ϵ_1 and ϵ_2 so determined are of the same order of magnitude as the terms we have omitted in deriving (3.17). So we step back to equations (3.23) and turn once more to the characteristic equation (3.22) to find out the conditions under which the term linear in D vanishes. Using (3.19) and (3.23) we get

$$h(p_2 - p_1) = \frac{P^2\omega}{2C}(C - B - A)(B - A)\mu \tag{3.25}$$

where

$$\mu = \frac{\left[\left(\frac{3}{2}C - B - A\right)\omega^2 + Pz_c\right]x_cy_c}{[(C - A)\omega^2 - Pz_c][(C - B)\omega^2 - Pz_c]} \tag{3.26}$$

and we see that the quantity $h(p_2 - p_1)$ goes to zero if one of the following conditions is satisfied:

$$1)\quad C = A + B; \quad 2)\ A = B; \quad 3)\ \mu = 0 \tag{3.27}$$

Under the first of these conditions the elements of the matrix H in (3.17) are all zero and the characteristic equation is therefore biquadratic. The second condition corresponds to a symmetric body whose inertial ellipsoid at point 0 is that of revolution. The third of conditions (3.27) is fulfilled when either x_c or y_c or both are zero, that is, when the center of gravity lies on one of the principal planes of inertia, Oyz or Oxz.

There is a special case which also satisfies the condition $\mu = 0$; this is when the square bracket in the numerator of (3.26) goes to zero. If, for example, $z_c > 0$, this means that

$$\omega = \sqrt{\frac{2Pz_c}{2(A+B) - 3C}} \tag{3.28}$$

where $2(A + B) > 3C$. If the center of gravity is below the point 0, then $z_c < 0$, and it should be assumed that $3C > 2(A + B)$.

As remarked earlier, the angular velocity ω of the generating solution is assumed to be large. As far as equation (3.28) is concerned, this will be the case if the interrelation between the quantities A, B, and C is such that the denumerator under the radical is sufficiently small and positive. This condition can also be expressed by requiring that the mass distribution of the body be close to that of Kovalevskaya's top, $A = C = 2B$ in our notation, because this is precisely the condition which turns the above denominator to zero.

If none of the above three conditions is satisfied, the motion about the vertical will always be unstable because the fact of there being a root with a positive real part in (3.22) will always increase the angle ϑ irrespective of whether the center of gravity is above or below the point of support.

It can readily be shown that this structural instability is associated with the non-stationary character of the motion about the principal axis Oz. Note that in the special case $A = B$, $x_c = y_c = 0$ (known as the Lagrange case) the starting equations of the problem admit of the integral

$$r = \text{const} \tag{3.29}$$

showing that the rotation about the Oz axis is stationary. But it is easy to see that the component G_2 in (3.17) is identically zero in this case because both p_1 and p_2 turn out to be zero. If x_c and y_c are different from zero, (3.29) is no longer an integral, nor is (2.1) a partial solution, so that the z-component of the angular velocity will in general vary in time even though (in the case of rapid rotation) it will have a considerable constant component ω corresponding to the generating solution. (The reader will recall that we leave out of consideration permanent rotations about vertical axes other than the principal axes of the body.)

4. It seems worthwhile discussing here some aspects of the unperturbed motion we have chosen. It has already been noticed that a trivial solution of (3.17) corresponding to this motion is $x = 0$, so that we must set $u_1 = u_1^*$, $u_2 = u_2^*$, where u_1^* and u_2^* are defined in (3.23). From (3.3) and (3.7) we have

$$\lambda_0^* = \left[1 - \frac{1}{2}(u_1^{*2} + u_2^{*2})\right] \cos\chi, \quad \lambda_1^* = -u_1^* \cos\chi - u_2^* \sin\chi \tag{4.1}$$

$$\lambda_2^* = -u_1^* \sin\chi + u_2^* \cos\chi, \quad \lambda_3^* = \left[1 - \frac{1}{2}(u_1^{*2} + u_2^{*2})\right] \sin\chi$$

where the λ_s^* are the *RH* parameters of the unperturbed motion. From (3.6), the nutation angle for the unperturbed motion is

$$\vartheta^* = 2\sqrt{u_1^{*2} + u_2^{*2}} = P\sqrt{\left[\frac{x_c}{(C-A)\omega^2 - Pz_c}\right]^2 + \left[\frac{y_c}{(C-B)\omega^2 - Pz_c}\right]^2} \tag{4.2}$$

We thus see that in the unperturbed motion the $0z$ axis makes a constant angle ϑ^* with the vertical 0ζ. The unperturbed precession angle, ψ^*, can be found from

$$\tan\psi = (\lambda_0\lambda_2 + \lambda_1\lambda_3)/(\lambda_0\lambda_1 - \lambda_2\lambda_3) \tag{4.3}$$

by setting $\lambda_s = \lambda_s^*$ (cf. (4.1)). Within the accuracy of our calculations the result is

$$\tan\psi^* = (u_1^* \sin\omega t - u_2^* \cos\omega t)/(u_1^* \cos\omega t + u_2^* \sin\omega t) \tag{4.4}$$

Setting now $u_1^* = N\cos\varkappa$, $u_2^* = N\sin\varkappa$, where N and $\varkappa$ are certain constants, we get

$$\psi^* = \omega t - \varkappa \tag{4.5}$$

where

$$\varkappa = \operatorname{arctg}\frac{u_2^*}{u_1^*} = \operatorname{arctg}\left[\frac{x_c}{y_c}\cdot\frac{(C-B)\omega^2 - Pz_c}{(C-A)\omega^2 - Pz_c}\right] \tag{4.6}$$

In the same fashion we use the expression

$$\operatorname{tg}\varphi = \frac{\lambda_1\lambda_3 - \lambda_0\lambda_2}{\lambda_0\lambda_1 + \lambda_2\lambda_3} \tag{4.7}$$

to obtain

$$\varphi^* = \varkappa = \text{const} \tag{4.8}$$

with $\varkappa$ as defined in (4.6). We thus see that

$$\psi^* + \varphi^* = \omega t$$

giving

$$\dot\varphi^* = 0, \quad \dot\psi^* = \omega \tag{4.9}$$

Note that the same value of ψ^* can be found in a more direct manner by putting $\lambda_s = \lambda_s^*$, $\dot\lambda_s = \dot\lambda_s^*$ in the representation

$$\dot\psi = \frac{\dot\lambda_3\lambda_1 - \dot\lambda_0\lambda_2 + \dot\lambda_2\lambda_0 - \dot\lambda_1\lambda_3}{\lambda_0\lambda_1 + \lambda_2\lambda_3} \tag{4.10}$$

In the unperturbed motion the $0z$ axis rotates about the 0 axis with an angular velocity ω and describes a right circular cone whose vertex is at the point 0 and whose generator makes an angle ϑ^* with the vertical 0ζ.

Using (1.5) and (4.1) it is possible to calculate the unperturbed values p^*, q^*, r^* of the projections of the angular velocity on the axes fixed in the body. Within the accuracy of calculation these are

$$
\begin{aligned}
p^* &= -2\omega u_2^* = \frac{P\omega x_c}{Pz_c - (C - A)\omega^2} \\
q^* &= -2\omega u_1^* = \frac{P\omega y_c}{Pz_c - (C - B)\omega^2} \\
r^* &= \omega\left(1 - \frac{\vartheta^{*2}}{2}\right)
\end{aligned}
\tag{4.11}
$$

The important question that remains is, how close to exact solutions are the approximate expressions (4.1) and (4.11) derived for the unperturbed motion? The answer is best obtained by direct use of Euler's dynamic equations. Using (2.8) and noting that the projections p^*, q^*, r^* in equations (4.11) are constants, we obtain

$$
\begin{aligned}
&(C - B)q^*r^* - M_x^* = (C - B)q^*r^* - P[2(\lambda_0^*\lambda_1^* + \lambda_2^*\lambda_3^*)z_c \\
&- (\lambda_0^{*2} + \lambda_3^{*2} - \lambda_1^{*2} - \lambda_2^{*2})y_c] = -3P(u_1^{*2} + u_2^{*2})u_1^* z_c
\end{aligned}
\tag{4.12}
$$

where the λ_s^* and M_x^* correspond to the unperturbed motion, and u_1^* and u_2^* are defined by (3.23).

Similarly,

$$
\begin{aligned}
&(A - C)r^*p^* - M_y^* = -3P(u_1^{*2} + u^{*2}{}_2)u_2^* z_c \\
&(B - A)p^*q^* - M_z^* = P(u_1^{*2} + u_2^{*2})(u_1^* x_c - u_2^* y_c)
\end{aligned}
\tag{4.13}
$$

If $A \neq B$, $A \neq C$, and $B \neq C$, it is seen that the right-hand sides in (4.12) and (4.13) are of the order of ω^{-6} (cf. (3.23)) and may therefore be neglected at large ω. If the problem is a degenerate one, that is, either $A = B$ or $A = C$, or $B = C$, the right-hand sides are again small, as they are of the third order of the small quantities x_c/z_c and y_c/z_c.

We thus conclude that the above formulas, although obtained for the unperturbed motion, satisfy to high accuracy the exact equations of motion, provided of course that the rotation is rapid and the center of gravity is sufficiently close to the axis of rotation.

The study of the perturbed motion of our system requires the construction of the Cauchy integral for (3.17) which may be considered as a separate problem.

REFERENCES

1. Koshlyakov, V. N. On the motions of a heavy rigid body about a fixed point. Ukr. Fiz. Zhurn., 1973, Vol. 25, No. 5, pp. 677–691.
2. Koshlyakov, V. N. Rodrigues-Hamilton and Klein-Cayley parameters in the problem of a heavy solid moving about a fixed point. Ibid, 1974, Vol. 26, No. 2, pp. 179–187.

3. Koshlyakov, V. N. Unitary and Hermitian matrices in the problem of a heavy solid rotating about a fixed point. Ibid, 1981, Vol. 33, No. 1, pp. 9–16.
4. Koshlyakov, V. N. Rigid body rotating about a fixed point: Equations of motion in terms of Rodrigues-Hamilton parameters. Izv. USSR Acad. Sci. MTT, 1983, No. 4, pp. 16–25.
5. Koshlyakov, V. N. Solid body dynamics and applied gyroscope theory. Moscow, Nauka, 1985 (in Russian).
6. Koshlyakov, V. N. and Boguslavskaya, E. S. Heavy solid motion equations in terms of Rodrigues-Hamilton parameters. In: Guidance and inertial navigation systems. Inst. Mat. Ukr. SSR Acad. Sci., Kiev, 1985 (in Russian).
7. Grammel, R. Der Kreisel. Seine Theorie und Seine Anwendunger. Berlin, Springer, 1950.
8. Rumyantsev, V. V. Stability of permanent rotations of a heavy solid. PMM, 1956, Vol. 29, No. 1, pp. 51–55.
9. Rubanovskii, V. N. On bifurcation and stability of the permanent rotations of a solid with the mass center close to the principal plane of the inertial ellipsoid. Ibid, 1982, Vol. 46, No. 5, pp. 762–770.
10. Merkin, D. R. Introduction into motion stability theory. Moscow, Nauka, 1971 (in Russian).

STATIONARY MOTIONS OF A ROD-SUSPENDED BODY: BIFURCATION AND STABILITY

V. N. Rubanovskii

The object we are considering is a dynamically symmetric gyrostat suspended from a fixed point by a weightless rod. For four different formulations of the problem, the bifurcation and stability of the set of all regular precessions of the gyrostat is studied. The results obtained are presented in the form of bifurcation diagrams. (The reader is referred to [1–3] for similar studies of the string suspension case).

1. A dynamically symmetric gyrostat has its center of gravity C on its symmetry axis and is suspended by a weightless undeformable rod from a fixed point 0_1. The point 0 at which the rod is linked to the gyrostat lies on the symmetry axis of the gyrostat. The axis of rotation of the statically and dynamically balanced rotor (gyrowheel) is fixed in the body and coincides with the symmetry axis of the body. At the point 0_1 the rod is linked by means of a cylindric hinge with the vertical shaft of a motor rotating with a constant angular velocity Ω; the other end of the rod is fixed in the body by means of a ball-and-socket joint (point 0). We denote by α the angle that 0_10 makes with the vertical (assumed to be directed downward).

We will employ two rectangular coordinate systems in our discussion. One of these, $Cx_1x_2x_3$ is fixed in the body and has its axes coinciding with the principal central axes of inertia of the gyrostat. The other, $0_1y_1y_2y_3$, rotates with an angular velocity Ω around the axis of y_3 directed positively upwards; the rod is assumed to lie in the plane $y_1 0_1 y_3$. We will assume that the x_3 axis coincides with the axis of symmetry of the body and that for the point 0 we have $x_3 = a > 0$. Let φ, ϑ, Ψ be the Euler direction angles for the system $Cx_1x_2x_3$ relative to the system $0_1y_1y_2y_3$. As usual, φ is the proper rotation, ϑ the nutation measured from the upgoing vertical, and Ψ precession. Let φ_1 be the angle the rotor makes with the body.

The Lagrangian function L of the mechanical system we consider is

$$L = \frac{m}{2}\{l^2(\dot\alpha^2 + \Omega^2 \sin^2 \alpha) + a^2[\dot\vartheta^2 + (\Omega + \dot\Psi)^2 \sin^2 \vartheta]$$

$$+ 2la[\dot\alpha\dot\vartheta(\sin \alpha \sin \vartheta + \cos \alpha \cos \vartheta \sin \Psi) + \dot\alpha(\Omega + \dot\Psi)\cos \alpha \sin \vartheta \cos \Psi$$

$$- \Omega\dot\vartheta \sin \alpha \cos \vartheta \cos \Psi + \Omega(\Omega + \dot\Psi)\sin \alpha \sin \vartheta \sin \Psi]\}$$

$$+ \frac{1}{2} J_1[\dot\vartheta^2 + (\Omega + \dot\Psi)^2 \sin^2 \vartheta] + \frac{1}{2} J_3' [\dot\varphi + (\Omega + \dot\Psi)\cos \vartheta]^2$$

$$+ \frac{1}{2} J_3''[\dot\varphi_1 + \dot\varphi + (\Omega + \dot\Psi)\cos \vartheta]^2 + mg(l \cos \alpha + a \cos \vartheta)$$

where m is the mass of the gyrostat; $J_1 = J_2$, $J_3 = J_3' + J_3''$ are the principal central moments of inertia of the gyrostat; J_3' and J_3''—the axial moments of inertia of respectively the body and rotor; l—the length of the rod; g—the free fall acceleration.

Let Q_φ and Q_{φ_1} be the nonconservative generalized forces corresponding to the coordinates φ, φ_1.

We will be concerned here with four special cases defined as follows:

1) $\dot\varphi = r, \quad \dot\varphi_1 = r_1, \quad Q_\varphi = \left(\dfrac{d}{dt}\dfrac{\partial L}{\partial \dot\varphi}\right)_{\dot\varphi = r,\ \dot\varphi_1 = r_1}, \quad Q_{\varphi_1} = \left(\dfrac{d}{dt}\dfrac{\partial L}{\partial \dot\varphi_1}\right)_{\dot\varphi = r,\ \dot\varphi_1 = r_1}$

2) $\dot\varphi = r, \quad Q_\varphi = \left(\dfrac{d}{dt}\dfrac{\partial L}{\partial \dot\varphi}\right)_{\dot\varphi = r}, \quad Q_{\varphi_1} = 0$ (φ_1 – cyclic coordinate)

3) $Q_\varphi = Q_{\varphi_1} = 0$ (φ, φ_1 – cyclic coordinates)

4) $\dot\varphi_1 = r_1, \quad Q_{\varphi_1} = \left(\dfrac{d}{dt}\dfrac{\partial L}{\partial \dot\varphi_1}\right)_{\dot\varphi_1 = r_1}, \quad Q_\varphi = 0$ (φ – cyclic coordinate)

where r and r_1 are constant.

The modified potential energy W of the corresponding reduced system has the same form for all the four problems, viz.,

$$W_s = \Pi - \frac{1}{2}\Omega^2(mR^2 - A_s \cos^2 \vartheta) - \Omega k_s \cos \vartheta \quad (s = 1, 2, 3, 4)$$

where Π is the potential energy of the force of gravity; R is the distance of the point C from the axis of y_3, and A_s and k_s are constants defined by

$$A_1 = J_1 - J_3, \quad A_2 = J_1 - J_3', \quad A_3 = J_1, \quad A_4 = J_1$$

$$k_1 = J_3 r + J_3' r_1, \quad k_2 = J_3' r + J_3'' \omega_1, \quad k_3 = J_3' \omega + J_3'' \omega_1 \cdot k_4 = J_3 \omega + J_3'' r_1$$

where the constant parameter ω(resp. ω_1) is the x_3 projection of the absolute instantaneous angular velocity of the body (resp. rotor).

Let τ_j, ν_j be the y_j projections of the unit vectors of the x_3 axis and of the direction $0_1 0$ of the rod ($j = 1, 2, 3$; $\nu_2 \equiv 0$). To simplify notation we drop the subscript in W_s, A_s, k_s. We then have

$$W = mg(l\nu_3 - a\tau_3) - \Omega k \tau_3 - \frac{1}{2}\Omega^2\{m[(l\nu_1 - a\tau_1)^2 + a^2\tau_2^2] - A\tau_3^2\}$$

Instead of W we will consider the function

$$2W_*(\lambda, \sigma, \tau_1, \nu_1, \tau_3, \nu_3, \tau_2) = 2W + \Omega^2[(\lambda + ma^2)\, V_1 + (\sigma + ml^2) V_2]$$

$$V_1 = \tau_1^2 + \tau_2^2 + \tau_3^2 - 1 = 0, \quad V_2 = \nu_1^2 + \nu_3^2 - 1 = 0$$

where λ and σ are Lagrange's undeterminate multipliers.

2. Let us consider the conditions of stability of the function W_* towards the variables introduced. We have

$$\begin{aligned}
&\frac{\partial W_*}{\partial \lambda} = \frac{1}{2}\Omega^2 V_1 = 0, \quad \frac{\partial W_*}{\partial \sigma} = \frac{1}{2}\Omega^2 V_2 = 0 \\
&\frac{\partial W_*}{\partial \tau_1} = \Omega^2(\lambda\tau_1 + mal\, \nu_1) = 0, \quad \frac{\partial W_*}{\partial \nu_1} = \Omega^2(mal\, \tau_1 + \sigma\nu_1) = 0 \\
&\frac{\partial W_*}{\partial \tau_3} = \tau_3(\lambda + A_*)\Omega^2 - \Omega k - mga = 0, \quad A_* = A + ma^2 \\
&\frac{\partial W_*}{\partial \nu_3} = \Omega^2(\sigma + ml^2)\nu_3 + mgl = 0, \quad \frac{\partial W_*}{\partial \tau_2} = \lambda\Omega^2\tau_2 = 0
\end{aligned} \tag{2.1}$$

From the last of these we find $\tau_2 = 0$ showing that to stationary values of W_*, there correspond steady-state motions in which the rod and the symmetry axis of the body both lie in the vertical plane $y_1 0_1 y_3$.

Equations (2.1) have four families of solutions which are as follows:

$$\begin{aligned}
&\tau_1 = \tau_2 = \nu_1 = 0, \quad \tau_3 = \pm 1, \quad \nu_3 = \pm 1 \\
&\lambda = A\frac{\mu - \beta\tau_3 p^2}{\tau_3 p^2}, \quad \sigma = -\frac{ml^2}{\nu_3 p^2}(1 + \nu_3 p^2)
\end{aligned} \tag{2.2}$$

where we have introduced the dimensionless parameters

$$p = \Omega\sqrt{\frac{l}{g}}, \quad \beta = \frac{A_*}{A}, \quad \mu = \left(pk\sqrt{\frac{l}{g}} + mal\right)A^{-1}, \quad \varkappa = \frac{a}{l}$$

In the rotations described by (2.2) both the rod and the axis of symmetry are vertical; the point 0 may be either below ($\nu_3 = -1$) or above ($\nu_3 = 1$) the point 0_1; and the center of gravity C may be either below ($\tau_3 = 1$) or above ($\tau_3 = -1$) the point 0.

Equations (2.1) also admit of a two-parameter family of solutions,

$$\nu_1 = \sqrt{1-\nu^2}, \quad \nu_3 = \nu, \quad \tau_1 = -\frac{mal}{\lambda}\nu_1, \quad \tau_2 = 0, \quad \tau_3 = \frac{\mu(1+\nu p^2)}{p^2(\beta + \nu p^2)}$$

$$\frac{mal}{\sigma} = \frac{\lambda}{mal} = -\frac{\varkappa \nu p^2}{1+\nu p^2}, \quad \Phi(p^2, \nu) = \varkappa^2\nu^2 p^4 - (1-\nu^2)(1+\nu p^2)^2 \tag{2.3}$$

$$\mu^2 = \frac{(\beta + \nu p^2)^2}{\varkappa^2\nu^2(1+\nu p^2)^2}\Phi(p^2, \nu)$$

These solutions are interpreted as regular precessions in which the body participates in two types of motion simultaneously. First, it rotates about the axis of symmetry with an angular velocity $\dot\varphi(\dot\varphi = r$ for problems 1, 2 and $\dot\varphi = \omega - \Omega\tau_3$ for problems 3, 4). Second, the rod and the axis of symmetry rotate as one solid body around a vertical with an angular velocity Ω, both lying in the vertical plane $y_1 0_1 y_3$ and forming constant angles α and ϑ with downgoing and upgoing verticals in the respective ranges $0 \leqslant \alpha \leqslant \pi$ and $0 \leqslant \vartheta \leqslant 2\pi$. The rotation of the rotor relative to the body occurs with an angular velocity $\dot\varphi_1$ ($\dot\varphi_1 = r_1$ for problems 1, 4; $\dot\varphi_1 = \omega_1 - r - \Omega\tau_3$ for problem 2; $\dot\varphi_1 = \omega_1 - \omega$ for problem 3).

3. It is of interest to discuss the geometric interpretation of the motions (2.2) and (2.3). It is seen from equations (2.3) that to each set μ, p, ν_3 satisfying the last of equations (2.3) there corresponds one and only one motion of the system. This means that all the motions (2.3) may be represented geometrically as points on the surface S in the space of the parameters μ, p, ν_3, the surface being defined by the last of equations (2.3). On the other hand, one cannot define the motions (2.2) completely by specification of the parameters μ, p, $\nu_3 = +1$ because the values $\tau_3 = \pm 1$ must also be specified. To remedy this situation let us introduce the space of the parameters μ, p, ν_3, τ_3. In this space all the motions (2.2), (2.3) will be represented as points of the variety Σ composed of the variety Σ' defined by equations $\nu_3 = \pm 1$, $\tau_3 = \pm 1$, plus the "cylindrical" variety Σ'' defined by the last of equations (2.3) and having the surface S as its "directrix".

Let us investigate the sections S_p of the surface S by planes $p =$ const. This is best done by considering a certain region G in the plane (p^2, ν) in the band D: $p^2 \geqslant 0$, $-1 \leqslant \nu \leqslant 1$. We assume that, $\Phi(p^2, \nu) > 0$ and write $G = G_1 + G_2$, where the regions G_1 and G_2 have their bounds defined by the equations

$$p^2 = F_{1,2}(\varkappa, \nu) = \pm\frac{\sqrt{1-\nu^2}}{\nu(\varkappa \mp \sqrt{1-\nu^2})}$$

the upper (lower) signs referring to $F_1(F_2)$. Fig. 1 shows regions G_1 and G_2 for the cases a) $\varkappa < 1$ and b) $\varkappa > 1$, with

$$\nu^{*,**} = \pm\sqrt{1-\varkappa^2}, \quad \nu_* = -\sqrt{1-\varkappa^{2/3}}, \quad p_*^2 = F_1(\varkappa, \nu_*) = (1-\varkappa^{2/3})^{-3/2} > 1$$

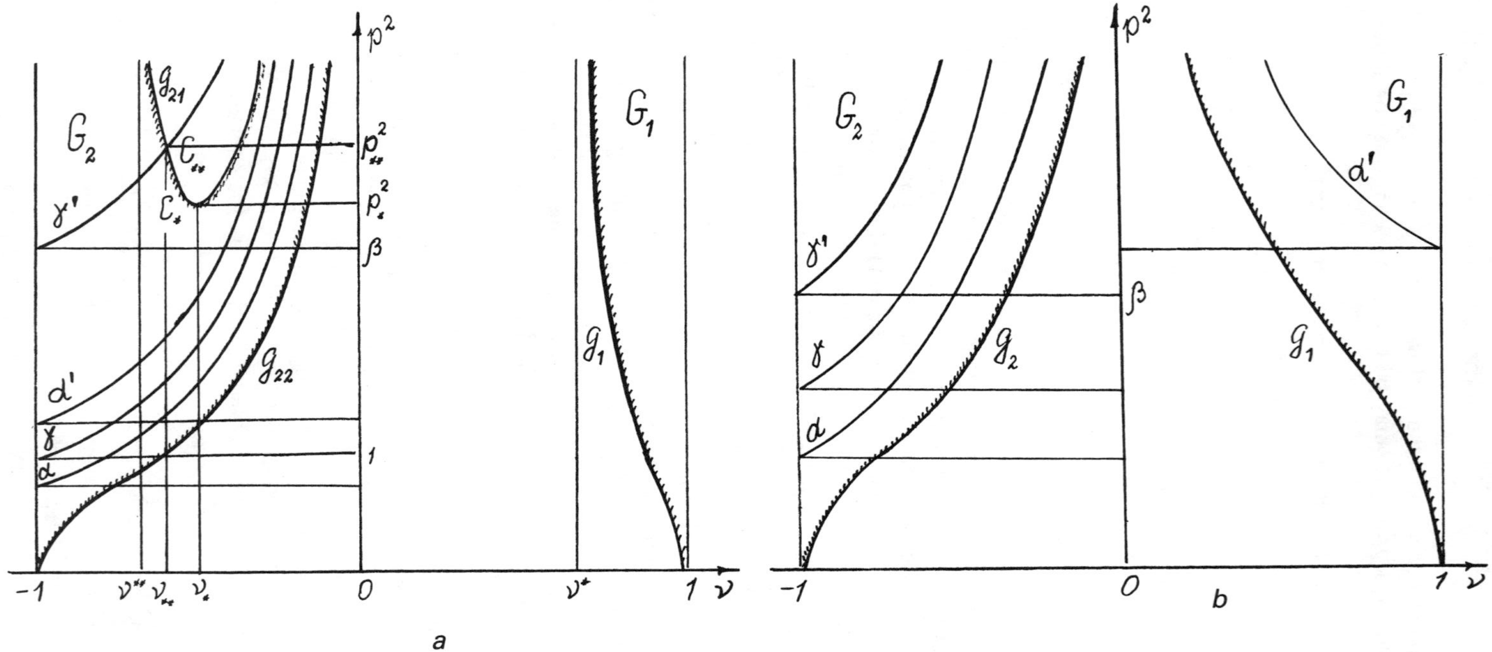
p^2
G_2
g_{21}
g_{22}
γ'
α'
γ
α
β
C_{**}
C_*
p_{**}^2
p_*^2
1
-1
ν^{**}
ν_{**}
ν_*
0
G_1
g_1
ν^*
1
ν
a
g_2
b

Figure 1

Let us consider in the band D two typical curves γ and γ' defined by the respective equations $p^2 = -1/\nu$ and $p^2 = -\beta/\nu$. In those points of the hyperbola γ (resp. γ') that lie in G, the curve S_p have an asymptote $\nu = -1/p^2$ (resp. a self-intersection point on the ν axis). The shape of the curve S_p depends on the interrelation of the curves γ and γ' in G at given values of the parameters $\varkappa$ and β.

On the curve γ we have $\Phi = \varkappa^2 > 0$, which means that this curve is always in the region G_2. For γ' we have $\Phi = \varkappa^2\beta^2 - (\beta - 1)^2\,(1 - \nu^2)$, so that the location of γ' in G depends on which of the inequalities $\varkappa^2\beta^2 \gtreqless (\beta - 1)^2$ is satisfied for given values of $\varkappa$ and β.

Referring to Fig. 2, it is seen that curves *1* and *2* defined by the equations $\varkappa = \varkappa_{1,2}(\beta) = \pm(\beta - 1)/\beta$ in the $(\varkappa, \beta)$-plane divide the half-plane $\varkappa > 0$ into the regions

$$B_1 = b_3 + b_4 + b_{10} + b_{11}, \quad B_2 = b_{19} + b_{20}, \quad B_3 = b_5 + \ldots + b_9,$$

$$B_4 = b_1 + b_2, \quad B_5 = b_{12} + \ldots + b_{18}$$

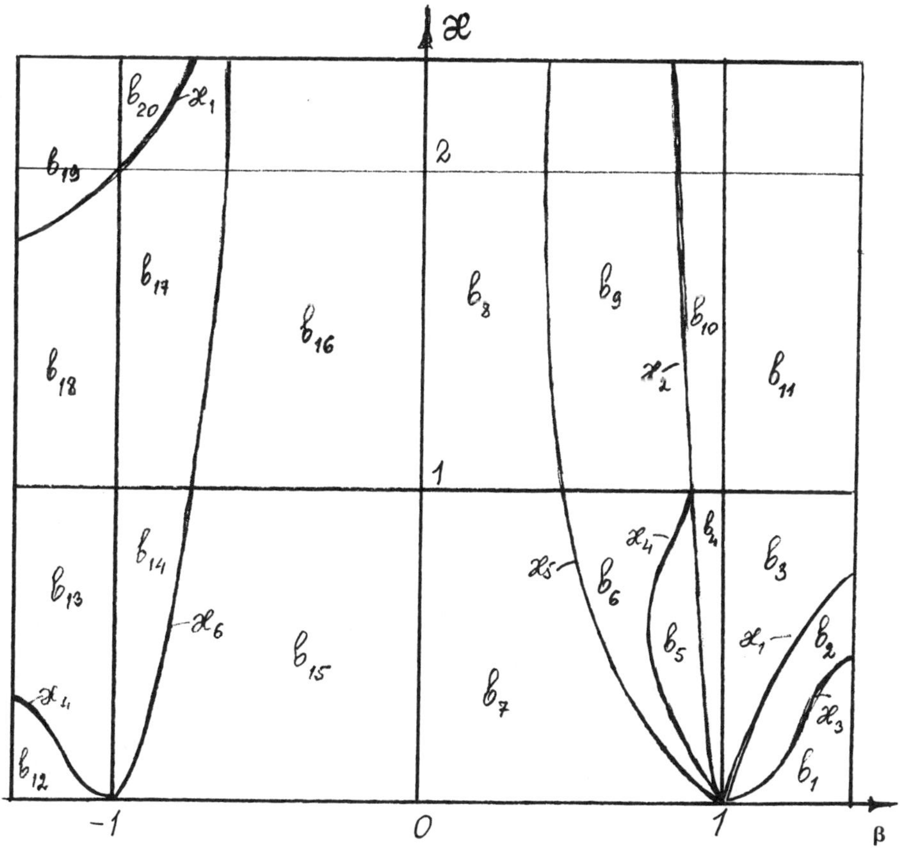

Figure 2

If $(\varkappa, \beta) \in B_1(B_2)$, then the curve γ' lies wholly in region $G_2(G_1)$; if $(\varkappa, \beta) \in B_3$, the curve γ' lies partly in region G_2 and cuts its boundary g_2 (if $\varkappa > 1$) or g_{22} (if $\varkappa < 1$). If $(\varkappa, \beta) \in B_4$, then γ' lies partly in G_2 and crosses the boundary g_{21}. Finally, if $(\varkappa, \beta) \in B_5$, γ' lies partly in G_1 and crosses the boundary g_1. In the cases when γ' cuts the boundary of the region G, the coordinates $\nu = \nu_{**}$ and $p^2 = p^2_{**}$ of the crosspoint C are given by

$$\nu_{**} = -\left[1 - \frac{\varkappa^2\beta^2}{(\beta - 1)^2}\right]^{1/2} \operatorname{sign}\beta, \quad p^2_{**} = \left[1 - \frac{\varkappa^2\beta^2}{(\beta - 1)^2}\right]^{-1/2} \beta \operatorname{sign}\beta \ (\varkappa < 1)$$

Let us find in the $(\varkappa, \beta)$ plane the locus of points for which the points C_* and C_{**} (see Fig. 1*b*) lie on the same line parallel to the axis of β. This locus is described by the equation $p^2_* = p^2_{**}$, which leads to the equation

$$\beta^2(1 - \beta)^2(1 - w)^3 + \beta^2 w^3 - (1 - \beta)^2 = 0 (w = \varkappa^{2/3})$$

with the roots

$$W_1 = W_2 = \frac{\beta - 1}{\beta}, \quad W_3 = \frac{1 - \beta^2}{\beta(2 - \beta)}$$

The required locus is therefore determined by the equations

$$\varkappa = \varkappa_3(\beta) = \frac{\beta - 1}{\beta}\sqrt{\frac{\beta - 1}{\beta}}, \quad \varkappa = \varkappa_4(\beta) = \frac{1 - \beta^2}{\beta(2 - \beta)}\sqrt{\frac{1 - \beta^2}{\beta(2 - \beta)}}$$

(see Fig. 2, curves *3*, *4*) implying that the point C_{**} is above (resp. below) C_* if $(\varkappa, \beta) \in b_1, \ldots b_5, b_{12}$ (resp. $b_6, b_7, b_{13}, b_{14}, b_{15}$).

Let us now investigate the relative position of the curves γ and γ'. If $0 < \beta < 1$ (resp. $\beta > 1$), both curves are in the region G_2, γ being above (resp. below) γ'; if $\beta < -1$ (resp. $-1 < \beta < 0$), γ' and γ lie in G_1 and G_2 respectively, the right extremity of γ' being above (resp. below) the left extremity of γ. It is of interest to find in the $(\varkappa, \beta)$-plane the locus of points for which $p^2_{**} = -1$ and hence the point C_{**} and the left extremity of the curve γ lie on the same straight line $p^2 = 1$. The condition $p^2_{**} = 1$ is fulfilled along the curves

$$\varkappa = \varkappa_{5,6} = \pm\frac{1 - \beta}{\beta}\sqrt{1 - \beta^2}$$

showing that for regions between the curves *5* and *6*, the point C_{**} is below, and elsewhere in the $(\varkappa, \beta)$-plane above, the straight line $p^2 = 1$.

The curve $\varkappa = \varkappa_j(\beta)(j = 1, \ldots, 6)$ and the straight lines $\varkappa = 1$ and $\beta = \pm 1$ divide the $(\varkappa, \beta)$-plane into 20 regions $b_1, \ldots, b_{20}$ differing in the position of the curves γ and γ' in the region G. The following inequalities determine the ordinates of the extremities of these curves as well as those of the points C_* and C_{**}:

b_1: $1 < \beta < p^2_* < p^2_{**}(\nu_{**} < \nu_*)$; $\quad b_3$: $1 < \beta < p^2_*$

b_2: $1 < \beta < p^2_* < p^2_{**}(\nu_* < \nu_{**})$; $\quad b_4$: $0 < \beta < 1 < p^2_*$

b_5: $0 < \beta < 1 < p_*^2 < p_{**}^2$; b_8: $0 < \beta < p_{**}^2 < 1$

b_6: $0 < \beta < 1 < p_{**}^2 < p_*^2$; b_9: $0 < \beta < 1 < p_*^2$

b_7: $0 < \beta < p_{**}^2 < 1 < p_*^2$; b_{10}: $0 < \beta < 1$; b_{11}: $1 < \beta$

b_{12}: $1 < -\beta < p_*^2 < p_{**}^2$; b_{16}: $0 < -\beta < p_{**}^2 < 1$

b_{13}: $1 < -\beta < p_{**}^2 < p_*^2$; b_{17}: $0 < -\beta < 1 < p_{**}^2$

b_{14}: $0 < -\beta < 1 < p_{**}^2 < p_*^2$; b_{18}: $1 < -\beta < p_{**}^2$

b_{15}: $0 < -\beta < p_{**}^2 < 1 < p_*^2$; b_{15}: $1 < -\beta$; b_{20}: $0 < -\beta < 1$

These inequalities make it an easy task to draw curves γ and γ' for given values of $\varkappa$ and β. Fig. 1*a*, *b* shows the positions of these curves for regions b_1 and b_{11} respectively.

The shape of the curve S_p for a given region b_j depends not only on the relative position of the curves γ and γ' in the region G, but also on the given value of p^2. More specifically, this is as follows. If we think of the p^2 axis as being divided by the ordinates of the points C_*, C_{**}, and by those of the curves γ and γ', what is important is precisely which of these intervals contains the given p^2 value. It follows then that in order to construct the curve S_p, 85 possible cases must be considered.

Fig. 3 shows the shape of the curve S_p for region b_1 for the following cases: *a*) $0 < p^2 < 1$, *b*) $1 < p^2 < \beta$, *c*) $\beta < p^2 < p_*^2$, *d*) $p_*^2 < p^2 < p_{**}^2$, *e*) $p_{**}^2 < p^2$.

Our next concern is the geometrical representation of the motions (2.2) on the (μ, ν)-plane. We will represent these motions by points on doubled ("two-rimmed") straight lines $\nu = \pm 1$ with right (resp. left) rims corresponding to those of the motions (2.2) for which $\tau_3 = 1$ (resp. $\tau_3 = -1$).

Fig. 3 shows the geometrical representation on the (μ, ν)-plane of all the motions (2.2), (2.3) for which $(\varkappa, \beta) \in b_1$.

4. Conditions sufficient for the stability of the motions (2.2), (2.3) will be derived from Routh's theorem by requiring that the second variation $\delta^2 W_*$ be definite positive on the linear variety defined by the equations $\delta V_1 = \delta V_2 = 0$. Let ξ_j, $\eta_j (i = 1, 3; j = 1, 2, 3)$ be the variations of the variables ν_i, τ_j and let $(\delta^2 W_*)$ be the value the quantity $\delta^2 W_*$ assumes when $\delta V_1 = \delta V_2 = 0$. For the motions (2.2) we may then write

$$\Omega^{-2}(\delta^2 W_*) = \lambda\eta_1^2 + 2mal\eta_1\xi_1 + \sigma\xi_1^2 + \lambda\eta_2^2$$

The conditions that this quadratic form should be definite positive are

$$\lambda > 0, \quad \Delta = \lambda\sigma - (mal)^2 > 0 \qquad (4.1)$$

Denoting by χ the degree of instability of the unperturbed motion, we see

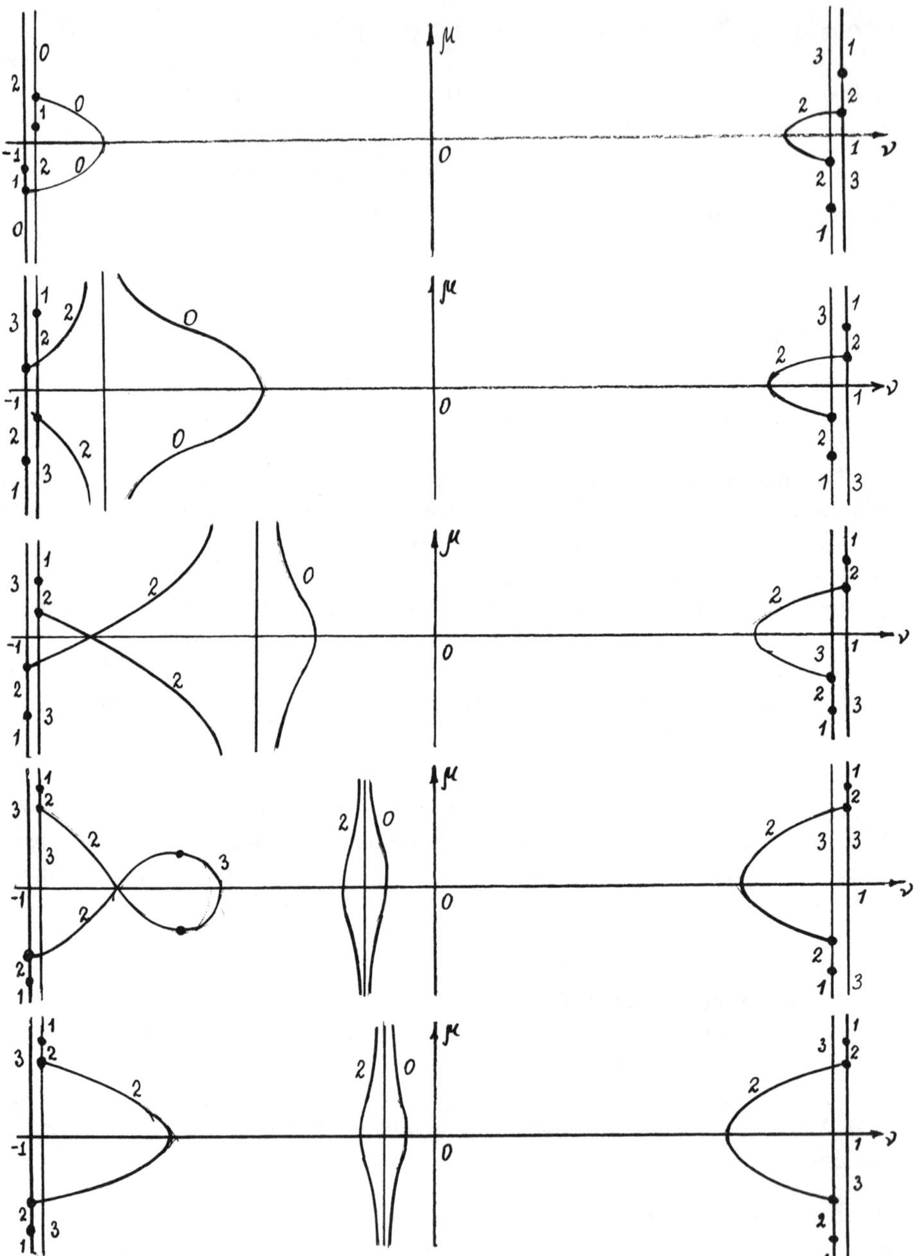

Figure 3

that $\chi = 0$ if $\lambda > 0$, $\Delta > 0$; $\chi = 2$ if $\lambda < 0$, $\Delta < 0$; $\chi = 1$ if $\lambda > 0$, $\Delta < 0$; $\chi = 3$ if $\lambda < 0$, $\Delta > 0$.

Substituting for λ and σ from (2.2), inequalities (2.2) become

$$Q = \tau_3(\beta - 1)(\mu - \mu_*) > 0, \quad Q' = \tau_3(1 - \beta)(p^2 + \nu_3)(\mu - \mu_{**}) > 0$$

$$\mu_* = \tau_3\beta p^2, \quad \mu_{**} = \frac{\tau_3 p^2(p^2 + \nu_3\beta)}{p^2 + \nu_3}$$

Analysis of these conditions leads to the following conclusions.

For the solution $\nu_3 = \tau_3 = 1$ with $\beta > 1$ (resp. $\beta < 1$) we have $\chi = 1$ if $\mu > \mu_*$ (resp. $\mu < \mu_*$); $\chi = 2$ if $\mu_{**} < \mu < \mu_*$ (resp. $\mu_{**} > \mu > \mu_*$); and $\chi = 3$ if $\mu < \mu_{**}$ (resp. $\mu > \mu_{**}$).

For the solution $\nu_3 = -\tau_3 = 1$ with $\beta > 1$ (resp. $\beta < 1$) we have $\chi = 1$ if $\mu < \mu_*$ (resp. $\mu > \mu_*$); $\chi = 2$ if $\mu_* < \mu < \mu_{**}$ (resp. $\mu_* > \mu > \mu_{**}$); and $\chi = 3$ if $\mu > \mu_{**}$ (resp. $\mu < \mu_{**}$).

For the solution $-\nu_3 = \tau_3 = -1$ with $\beta > 1$ (resp. $\beta < 1$) we have $\chi = 0$ if $p^2 < 0$ and $\mu > \mu_{**}$ (resp. $\mu < \mu_{**}$); $\chi = 1$ if $p^2 < 1$ and $\mu_* < \mu < \mu_{**}$ (resp. $\mu_* > \mu > \mu_{**}$), or $p^2 > 1$ and $\mu > \mu_*$ (resp. $\mu < \mu_*$); $\chi = 2$ if $p^2 < 1$ and $\mu < \mu_*$ (resp. $\mu > \mu_*$), or $p^2 > 1$ and $\mu_{**} < \mu < \mu_*$ (resp. $\mu_{**} > \mu > \mu_*$); $\chi = 3$ if $p^2 > 1$ and $\mu < \mu_{**}$ (resp. $\mu > \mu_{**}$).

Finally, for the solution $\nu_3 = \tau_3 = -1$ with $\beta > 1$ (resp. $\beta < 1$) we have $\chi = 0$ if $p^2 < 1$ and $\mu < \mu_{**}$ (resp. $\mu > \mu_{**}$); $\chi = 1$ if $p^2 < 1$ and $\mu_{**} < \mu < \mu_*$ (resp. $\mu_{**} > \mu > \mu_*$), or $p^2 > 1$ and $\mu < \mu_*$ (resp. $\mu > \mu_*$); $\chi = 2$ if $p^2 < 1$ and $\mu > \mu_*$ (resp. $\mu < \mu_*$), or $p^2 > 1$ and $\mu_* < \mu < \mu_{**}$ (resp. $\mu_* > \mu > \mu_{**}$); $\chi = 3$ if $p^2 > 1$ and $\mu > \mu_{**}$ (resp. $\mu < \mu_{**}$).

The curves shown in Figs. 4 and 5 correspond to the solutions $\nu_3 = \tau_3 = 1$ and $-\nu_3 = \tau_3 = 1$ respectively; in both Figs., the cases $\beta > 1$, $0 < \beta < 1$, and $\beta < 0$ are considered. Numbers 0, 1, 2, and 3 indicate the degree of instability for a given pair μ, p. Results for the solutions $\nu_3 = -\tau_3 = 1$ and $\nu_3 = \tau_3 = -1$ are obtained from Figs. 4 and 5 by mirror reflection about the p^2 axis.

Turning to the motions (2.3), it is necessary to consider the Hessian $D = \partial^2 W_*/\partial q^2$ of the function W_* with respect to the variables $q = (\lambda, \sigma, \tau_1, \nu_1, \tau_3, \nu_3, \tau_2)$. More specifically, the quantities we will need are the three last principal diagonal minors of this Hessian, D_5, D_6, and $D_7 = D$; it turns out that for the solutions (2.3) to be stable, it is sufficient that these three be positive [4]. This condition leads to the inequalities

$$D_5 = \nu_1^2(\lambda + A_*\tau_1^2) > 0, \quad \lambda > 0, \quad D_6 = D_* > 0 \tag{4.2}$$

To proceed with the analysis, it proves advantageous to represent D_* in some special form. Fixing the value of p, we may consider the solutions (2.3) as a family of solutions that depends on the parameter μ. Substitution of (2.3) into the first six of equations (2.1) then leads to identities which when differentiated with respect to μ give rise to a system of linear equations for the derivatives λ', σ', τ_1', ν_1', τ_3', ν_3'. By Kramer's rule we have $\nu_3' = d\nu_3/d\mu = D_*'/D_*$, where D_* is the determinant of the system, and D' differs from D in having its last column

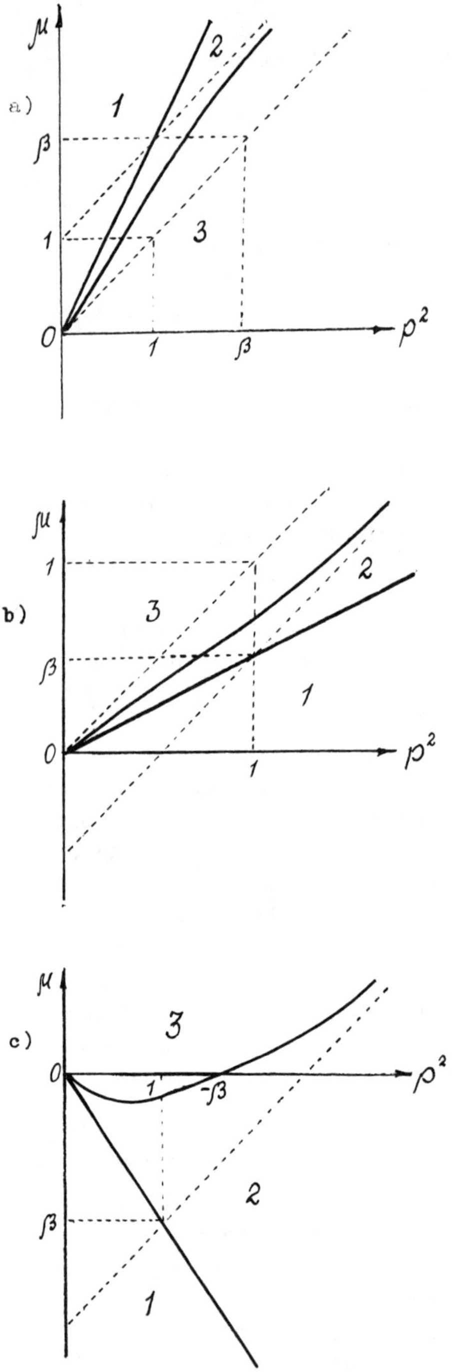

Figure 4

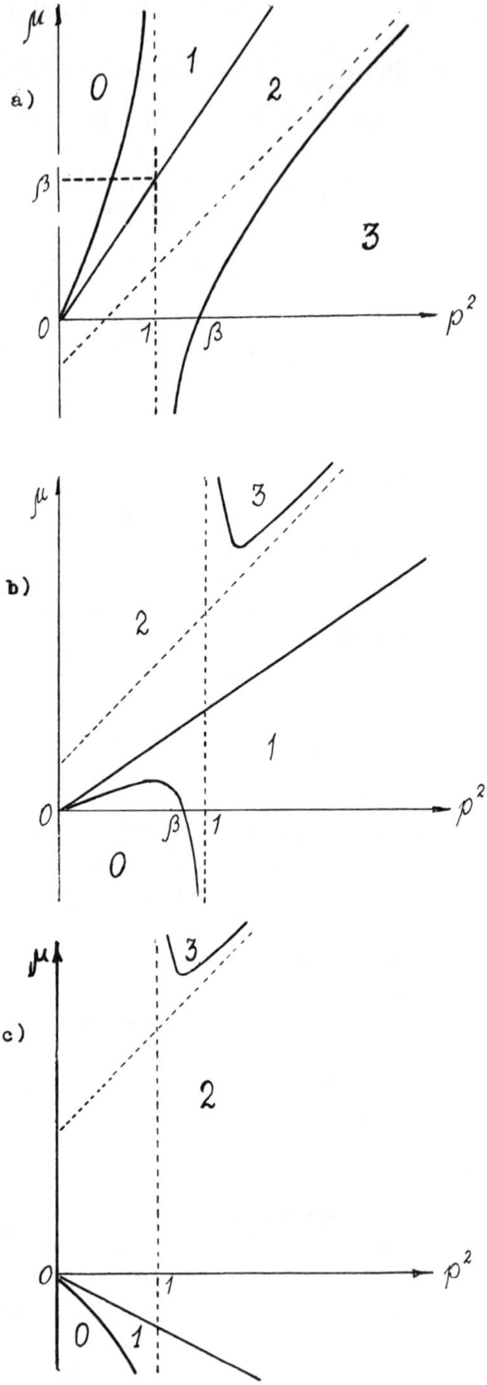

Figure 5

replaced by the column of free terms. The required representation is therefore given by

$$D_* = \frac{m^2a^4\tau_1^2\nu^2\mu}{(1-\beta)p^2(\beta+\nu p^2)}\frac{d\mu}{d\nu} \tag{4.3}$$

We note that the expression for $\delta^2 W_*$ and conditions $\delta V_1 = \delta V_2 = 0$ take the form

$$\Omega^{-2}\delta^2 W_* = \delta^2 W_*^{(1)} + \delta^2 W_*^{(2)}, \ \frac{1}{2}\delta V_1 = \tau_1\eta_1 + \tau_3\eta_3 = 0, \ \frac{1}{2}\delta V_2 = \nu_1\xi_1 + \nu_3\xi_3 = 0$$

$$\delta^2 W_*^{(1)} = \lambda\eta_1^2 + 2mal\eta_1\xi_1 + \sigma\xi_1^2 + (\lambda + A_*)\eta_3^2 + (\sigma + ml^2)\xi_3^2, \ \delta^2 W_*^{(2)} = \lambda\eta_2^2$$

for the motions (2.3), and we are thus led to the conclusion that the inequalities $D_5, D_* > 0$ and $\lambda > 0$ in (4.2) express the conditions that the respective quadratic forms $(\delta^2 W_*^{(1)})$ and $(\delta^2 W_*^{(2)})$ be positive.

Let us now investigate the stability of the motions (2.3) in region b_1, see Fig. 3.

If, say, $0 < p^2 < 1$ (Fig. 3*a*), it follows from (2.3) that $\lambda > 0$ for that branch of S_p where $\nu < 0$. From (4.3) and Fig. 3*a* we conclude that $D_* > 0$ for this branch. Since, further, $\tau_1 \approx 0$ and $D_5 \approx \lambda\nu_1^2 > 0$ for $\nu \approx -1$, and $D_* > 0$ for the whole of this branch, it follows also that $D_5 > 0$ on it. We thus see that if $\nu < 0$ for some of the motions (2.3), these particular motions are stable and have $\chi = 0$. The branch for which $\nu > 0$ has $\lambda < 0$ in view of (2.3), and it follows from (4.3) and Fig. 3*a* that $D_* < 0$ for this branch. We thus see that $\chi = 2$ for these motions.

A similar analysis can be carried out for the motions which correspond to Fig. 3*b–e*. The results obtained from the analysis of the stability conditions (4.1), (4.3) of the solutions (2.2), (2.3) are represented for region b_1 in Fig. 3, where numbers from 0 to 3 on the branches of the curve S_p indicate the degree of in stability of the corresponding motions, and small circles mark bifurcation points, at which the degree of instability changes its value. These bifurcation diagrams exhibit the validity of the law of change of stability at fixed values of the parameter μ.

5. A body with a triaxial central ellipsoid of inertia is attached to a rod in a point *0* lying on the principal central axis of inertia of the body, x_3; the rod hinges on a fixed point 0_1; we are interested in permanent motions of the body in the absence of a rotor.

The system we are considering has five degrees of freedom and admits of the energy and area integrals. The modified potential energy of the system is $W = (1/2)k^2J^{-1} + \Pi$, where k is the constant of the area integral, and J is the moment of inertia of the body about the vertical through the point 0_1.

Let the coordinate system $0_1y_1y_2y_3$ rotate with an angular velocity $\Omega = kJ^{-1}$ about the axis of y_3 directed positively downwards. Let α_j, β_j, γ_j, and $\nu_j (j = 1, 2, 3)$ be the y_j-projections of the unit vectors respectively the axes x_1, x_2, x_3 and of the direction of the rod, 0_10. We have

$$\pi_\alpha = \bar{\alpha}^2 - 1 = 0, \quad \pi_\beta = \bar{\beta}^2 - 1 = 0, \quad \pi_\gamma = \bar{\gamma}^2 - 1 = 0 \tag{5.1}$$

$$\pi_{\alpha\beta} = \bar{\alpha} \cdot \bar{\beta} = 0, \quad \pi_{\beta\gamma} = \bar{\beta} \cdot \bar{\gamma} = 0, \quad \pi_{\gamma\alpha} = \bar{\gamma} \cdot \bar{\alpha} = 0, \quad \pi_\nu = \bar{\nu}^2 - 1 = 0$$

In the system $0_1y_1y_2y_3$

$$\Pi = -mg(l\nu_3 - a\gamma_3), \quad J = J_1\alpha_3^2 + J_2\beta_3^2 + J_3\gamma_3^2 + m[(l\nu_1 - a\gamma_1)^2 + (l\nu_2 - a\gamma_2)^2]$$

Instead of W it is convenient to consider the function

$$2W_* = 2W + \Omega^2(\lambda_2\pi_\nu + \lambda_\alpha\pi_\alpha + \lambda_\beta\pi_\beta + \lambda_\gamma\pi_\gamma + 2\lambda_{\alpha\beta}\pi_{\alpha\beta} + 2\lambda_{\beta\gamma}\pi_{\beta\gamma} + 2\lambda_{\gamma\alpha}\pi_{\gamma\alpha})$$

where $\lambda_\nu, \lambda_\alpha, \ldots, \lambda_{\gamma\alpha}$ are Lagrange's indeterminate multipliers.

6. Combining (5.1) with the condition $\delta W_* = 0$ necessary that the function W_* be stationary with regard to the above variables, we obtain a system for determining the stationary motions of the body. These are uniform rotations about the vertical axis with an angular velocity $\Omega = k_0J_0^{-1}$, the subscript zero indicating that the values refer to the stationary motion.

The following are the families of solutions consistent with the equations of stationary motions:

$$\alpha_1 = \beta_2 = 1, \quad \gamma_3 = \pm 1, \quad \nu_3 = \pm 1$$

$$\alpha_2 = \alpha_3 = \beta_3 = \beta_1 = \gamma_1 = \gamma_2 = \nu_1 = \nu_2 = 0$$

$$\lambda_\nu = ml^2(\sigma + ma^2)\sigma^{-1}, \quad \lambda_\alpha = \lambda_\beta = \lambda_{\alpha\beta} = \lambda_{\beta\gamma} = \lambda_{\gamma\alpha} = 0$$

$$\sigma = \frac{ma^2\Omega^2 l}{\nu_3 g - \Omega^2 l}, \quad \lambda_\gamma = J_3 - \frac{\gamma_3 mga}{\Omega^2}, \quad k = J_3\Omega \tag{6.1}$$

$$\alpha_1 = \alpha_3 = 0, \quad \alpha_2 = 1, \quad \beta_1 = \gamma_3, \quad \beta_2 = 0, \quad \beta_3 = -\gamma_1$$

$$\gamma_1^2 = 1 - \gamma_3^2, \quad \gamma_2 = 0, \quad \gamma_3 = -\frac{mga}{\Omega^2(\sigma + J_2^* - J_3)} \qquad (J_2^* = J_2 + ma^2)$$

$$\nu_1 = -\frac{\sigma}{mal}\gamma_1, \quad \nu_2 = 0, \quad \nu_3 = \frac{g\sigma}{\Omega^2 l(\sigma + ma^2)}$$

$$\lambda_\alpha = \lambda_{\alpha\beta} = \lambda_{\alpha\gamma} = 0, \quad \lambda_\beta = J_2\beta_3^2, \quad \lambda_{\beta\gamma} = J_2\beta_3\gamma_3, \quad \lambda_\nu = ml^2(\sigma + ma^2)\sigma^{-1}$$

$$\lambda_\gamma = (\sigma + ma^2)\left\{1 + \frac{J_2(\sigma + ma^2)[(mal)^2 - \sigma^2]}{(J_2 - J_3)\sigma^2(2\sigma + ma^2 + J_2^* - J_3)}\right\}, \quad k = J\Omega$$

$$\Omega^4 = \frac{(J_2 - J_3)(mga)^2\sigma^2(2\sigma + ma^2 + J_2^* - J_3)}{(\sigma + ma^2)^2(\sigma + J_2^* - J_3)^2[(mal)^2 - \sigma^2]} \tag{6.2}$$

$$J = J_3 + \left[J_2 - J_3 + \frac{(\sigma + ma^2)^2}{ma^2}\right]\beta_3^2$$

$$\beta_3^2 = \frac{\sigma^2(\sigma + J_2^* - J_3)^2 - (mal)^2(\sigma + ma^2)^2}{(J_2 - J_3)\sigma^2(2\sigma + ma^2 + J_2^* - J_3)}$$

$$\beta_1 = \beta_3 = 0, \quad \beta_2 = 1, \quad \alpha_1 = \gamma_3, \quad \alpha_2 = 0, \quad \alpha_3 = -\gamma_1$$

$$\gamma_1^2 = 1 - \gamma_3^2, \quad \gamma_2 = 0, \quad \gamma_3 = -\frac{mga}{\Omega^2(\sigma + J_1^* - J_3)} \qquad (J_1^* = J_1 + ma^2)$$

$$\nu_1 = -\frac{\sigma}{mal}\gamma_1, \quad \nu_2 = 0, \quad \nu_3 = \frac{g\sigma}{\Omega^2 l(\sigma + ma^2)}$$

$$\lambda_\beta = \lambda_{\beta\alpha} = \lambda_{\beta\gamma} = 0, \quad \lambda_\alpha = J_1\alpha_3^2, \quad \lambda_{\alpha\gamma} = J_1\alpha_3\gamma_3, \quad \lambda_\nu = ml^2(\sigma + ma^2)\sigma^{-1}$$

$$\lambda_\gamma = (\sigma + ma^2)\left\{1 + \frac{J_1(\sigma + ma^2)[(mal)^2 - \sigma^2]}{(J_1 - J_3)(2\sigma + ma^2 + J_1^* - J_3)}\right\}, \quad k = J\Omega \qquad (6.3)$$

$$\Omega^4 = \frac{(J_1 - J_3)(mga)^2\sigma^2(2\sigma + ma^2 + J_1^* - J_3)}{(\sigma + ma^2)^2(\sigma + J_1^* - J_3)^2[(mal)^2 - \sigma^2]}$$

$$J = J_3 + \left[J_1 - J_3 + \frac{(\sigma + ma^2)^2}{ma^2}\right]\alpha_3^2$$

$$\alpha_3^2 = \frac{\sigma^2(\sigma + J_1^* - J_3)^2 - (mal)^2(\sigma + ma^2)^2}{(J_1 - J_3)\sigma^2(2\sigma + ma^2 + J_1^* - J_3)}$$

Solutions (6.1) describe four one-parameter families of uniform rotations in which both the rod and x_3 axis are vertical. The point 0 may be below ($\nu_3 = 1$) or above ($\nu_3 = -1$) the point 0_1, and the center of gravity C may be either below or above 0 ($\gamma_3 = -1$ or $\gamma_3 = 1$, respectively). Fig. 6 shows the relative positions of the points 0_1, 0, and C in the uniform rotations (6.1).

Solutions (6.2) and (6.3) describe the one-parameter families of uniform rotations about the vertical with an angular velocity Ω. The rod and the x_3 axis lie

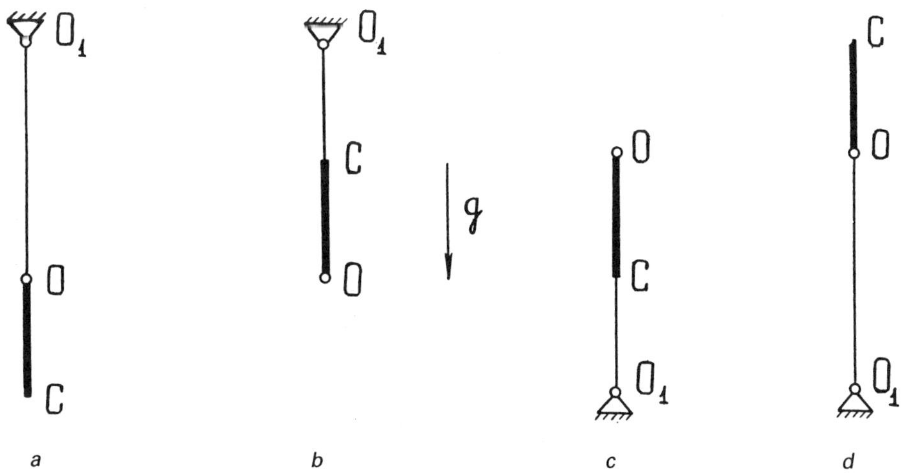

Figure 6

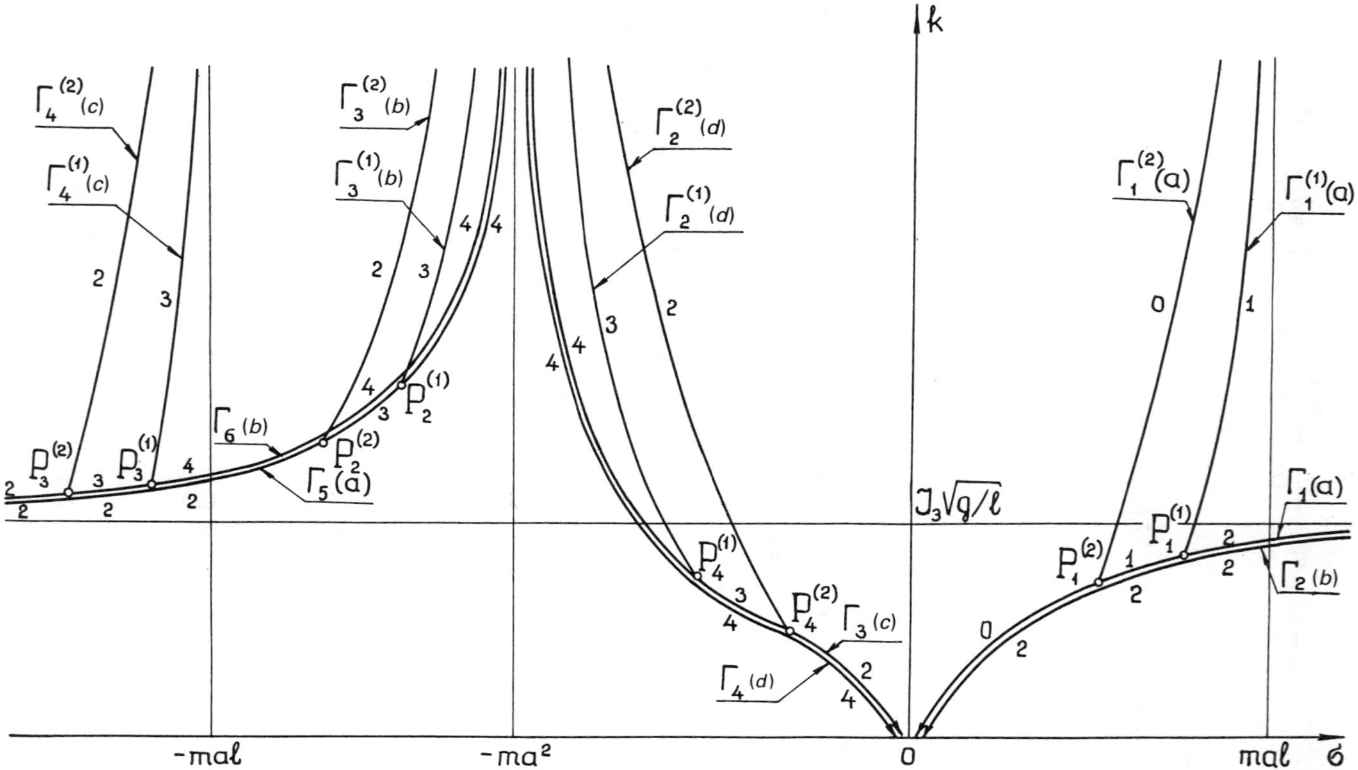
k
$J_3\sqrt{g/l}$
$-mal$
$-ma^2$
0
mal
σ
$\Gamma_4^{(2)}(c)$
$\Gamma_4^{(1)}(c)$
$\Gamma_3^{(2)}(b)$
$\Gamma_3^{(1)}(b)$
$\Gamma_2^{(2)}(d)$
$\Gamma_2^{(1)}(d)$
$\Gamma_1^{(2)}(a)$
$\Gamma_1^{(1)}(a)$
$\Gamma_6(b)$
$\Gamma_5(a)$
$\Gamma_3(c)$
$\Gamma_4(d)$
$\Gamma_1(a)$
$\Gamma_2(b)$
$P_3^{(2)}$
$P_3^{(1)}$
$P_2^{(2)}$
$P_2^{(1)}$
$P_4^{(1)}$
$P_4^{(2)}$
$P_1^{(2)}$
$P_1^{(1)}$

Figure 7

both in the $0_1y_1y_3$ plane, and of the remaining two axes one is perpendicular to this plane: the x_1 axis in (6.2) and the x_2 axis in (6.3).

7. The stationary motions (6.1)–(6.3) may be represented by points on a certain curve Γ in the $(k, \sigma, \lambda_\gamma)$-space. For the motions (6.1), the corresponding branch of Γ is defined by the parametric equations $k = (\Omega)$, $\sigma = \sigma(\Omega)$, $\lambda_\gamma = \lambda_\gamma(\Omega)$. For the motions (6.2) and (6.3), the branches are given by the equations $k = k(\sigma) \equiv J(\sigma)\Omega(\sigma)$, $\lambda_\gamma = \lambda_\gamma(\sigma)$. Fig. 7 shows the projection of Γ onto the plane $\lambda = 0$ for the case defined by the conditions

$$J_3 < J_2 < J_1, \quad ma^2 < J_2^* - J_3 < J_1^* - J_3 < mal \tag{7.1}$$

These are satisfied, in particular, for a homogeneous body elongated in the x_3 direction and suspended by a sufficiently long rod.

The branches $\Gamma_1(a)$, $\Gamma_2(b)$, $\Gamma_3(c)$, $\Gamma_4(d)$, $\Gamma_5(a)$, and $\Gamma_6(b)$ in Fig. 7 correspond to the motions (6.1), the letters a, b, c, d referring to the geometric conditions shown in Fig. 6. The motions (6.2) are represented by the branches $\Gamma_1^{(1)}(a)$, $\Gamma_2^1(d)$, $\Gamma_3^{(1)}(b)$, $\Gamma_4^{(1)}(c)$ with the x_1-axis horizontal, and the motions (6.3) by the branches $\Gamma_1^{(2)}(a)$, $\Gamma_2^{(2)}(d)$, $\Gamma_3^{(2)}(b)$, $\Gamma_4^{(2)}(c)$ with the x_2-axis horizontal, the letters a, b, c, d denoting in these two cases the mutual arrangements of the rod and the portion CO of the x_3-axis as shown in Fig. 8. It will be observed that Fig. 8a–d refer to the case of a finite Ω, and Fig. 8a'–d' to the limiting case $\Omega = \infty$. Coming back to Fig. 7 and the (6.1)-motions, we note that the projections of the corresponding branches of Γ onto the $\lambda_\gamma = 0$ plane should be thought of as doubled curves the two rims of which correspond to the various types of (6.1)-motions shown in Fig. 6.

8. The problem we address next is the stability of the motions (6.1)–(6.3) towards the variables $\boldsymbol{v}$, $\boldsymbol{\omega}$, v_j, α_j, $\beta_j\gamma_j (j = 1, 2, 3)$, where the vectors $\boldsymbol{v}$ and $\boldsymbol{\omega}$ are, respectively, the velocity of the center of gravity and the instantaneous angular velocity of the body. By Routh's theorem, the condition sufficient for these motions to be stable is that the second variation $\delta^2 W_*$ be definite positive on the linear variety defined by the relations

$$\delta\pi_\alpha = \delta\pi_\beta = \delta\pi_\gamma = \delta\pi_{\alpha\beta} = \delta\pi_{\beta\gamma} = \delta\pi_{\gamma\alpha} = \delta\pi_\nu = 0 \tag{8.1}$$

In the study of the perturbed problem, we will retain the previous notation for the variables equal to zero in the unperturbed motion. Denoting by $(\delta^2 W_*)$ the value of $\delta^2 W_*$ on the variety (8.1), we then have

$$\Omega^{-2}(\delta^2 W_*) = \sum_{j=1}^{2} [(\lambda_\nu - ml^2)v_j^2 + 2malv_j\gamma_j + (\lambda_\gamma - J_j^*)\gamma_j^2]$$

for the motions (6.1). This quadratic form will be definite positive if

$$\lambda_\nu - ml^2 > 0, \quad (\lambda_\nu - ml^2)(\lambda_\gamma - J_j^*) - (mal)^2 > 0 \quad (j = 1, 2)$$

or, substituting for λ_ν and λ_γ from (6.1),

$$\sigma > 0, \quad (J_3 - J_j^* - \gamma_3\Omega^{-2}mga)\sigma^{-1} - 1 > 0 \quad (j = 1, 2)$$

which leads in turn to

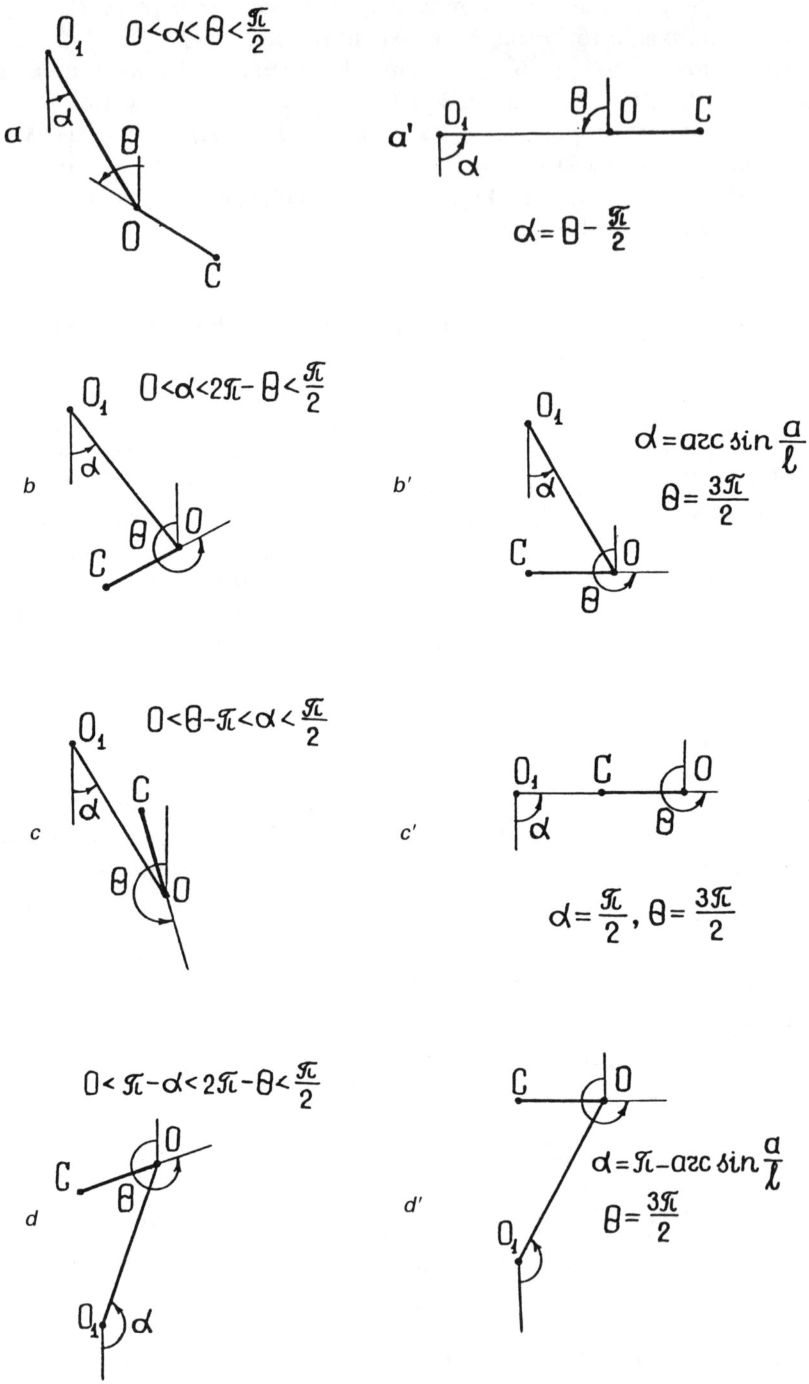

a
O_1
$0<\alpha<\theta<\frac{\pi}{2}$
α
θ
O
C
a'
O_1
θ
O
C
α
$\alpha=\theta-\frac{\pi}{2}$
b
O_1
$0<\alpha<2\pi-\theta<\frac{\pi}{2}$
α
θ
O
C
b'
O_1
$\alpha=\arcsin\frac{a}{l}$
$\theta=\frac{3\pi}{2}$
α
C
O
θ
c
O_1
$0<\theta-\pi<\alpha<\frac{\pi}{2}$
α
C
θ
O
c'
O_1
C
O
α
θ
$\alpha=\frac{\pi}{2}$, $\theta=\frac{3\pi}{2}$
d
$0<\pi-\alpha<2\pi-\theta<\frac{\pi}{2}$
O
C
θ
O_1
α
d'
C
O
$\alpha=\pi-\arcsin\frac{a}{l}$
$\theta=\frac{3\pi}{2}$
O_1

Figure 8

$$\sigma = ma^2\Omega^2 l(\nu_3 g - \Omega^2 l)^{-1} > 0, \quad Q_j(\nu_3, \gamma_3, \Omega^2) > 0 \quad (j = 1, 2)$$
$$Q_j = (J_j - J_3)l\Omega^4 - \nu_3 g(J_j^* - J_3 - \gamma_3\nu_3 mal)\Omega^2 \tag{8.2}$$
$$- \nu_3\gamma_3 g^2 ma \quad (\nu_3 = \pm 1, \gamma_3 = \pm 1)$$

after substituting σ from (6.1). We will assume in discussing these conditions that the parameters of the system obey inequalities (7.1).

Let $P_i^{(1)}$ and $P_i^{(2)}$ $(i = 1, 2, 3, 4)$ denote these points on the curve Γ (Fig. 7) at which the motions (6.1) bifurcate into the motions (6.2) or (6.3), respectively. For these points, the values σ_{ji} and $\Omega_{ji}^2 (j = 1, 2)$ of the parameters σ and Ω^2 are determined by the equations

$$\nu_3 = \frac{\nu_3 ma^2\Omega^2 l}{\Omega^2 l(\sigma + ma^2)}, \quad \gamma_3 = -\frac{mga}{\Omega^2(\sigma + J_j^* - J_3)} \quad (\nu_3 = \pm 1, \gamma_3 = \pm 1)$$

or

$$\sigma = \frac{\nu_3 ma^2\Omega^2 l}{g - \nu_3\Omega^2 l}, \quad Q_j(\nu_3, \gamma_3, \Omega^2) = 0 \quad (\nu_3 = \pm 1, \gamma_3 = \pm 1, j = 1, 2)$$

Let us consider the following 4 types of the motions (6.1). First, let the point 0 be below $0_1(\nu_3 = 1)$ and the point C be below $0(\gamma_3 = -1)$, see Fig. 6*a*. We denote as $\Omega_{kl}^2(k, l = 1, 2)$ the positive roots of the equations $Q_j(1, -1, \Omega^2) = 0 (j = 1, 2)$, it being assumed that $0 < \Omega_{21}^2 < \Omega_{11}^2 < g/l < \Omega_{22}^2 < \Omega_{12}^2$. From (8.2) we then conclude that

$$\sigma > 0, \quad Q_1 > 0, \quad Q_2 > 0, \quad \chi = 0 \quad \text{if} \quad 0 < \Omega^2 < \Omega_{21}^2$$
$$\sigma > 0, \quad Q_1 > 0, \quad Q_2 < 0, \quad \chi = 1 \quad \text{if} \quad \Omega_{21}^2 < \Omega^2 < \Omega_{11}^2$$
$$\sigma > 0, \quad Q_1 < 0, \quad Q_2 < 0, \quad \chi = 2 \quad \text{if} \quad \Omega_{11}^2 < \Omega^2 < g/l$$
$$\sigma < 0, \quad Q_1 < 0, \quad Q_2 < 0, \quad \chi = 2 \quad \text{if} \quad g/l < \Omega^2 < \Omega_{22}^2$$
$$\sigma < 0, \quad Q_1 < 0, \quad Q_2 > 0, \quad \chi = 3 \quad \text{if} \quad \Omega_{22}^2 < \Omega^2 < \Omega_{12}^2$$
$$\sigma < 0, \quad Q_1 > 0, \quad Q_2 > 0, \quad \chi = 4 \quad \text{if} \quad \Omega_{12}^2 < \Omega^2 < \infty$$

Second, let the point 0 be below $0_1(\nu_3 = 1)$ and the point C above $0(\gamma_3 = 1)$, see Fig. 6*b*. Let Ω_{23}^2 and Ω_{13}^2 be the positive roots of the equations $Q_j(1, 1, \Omega^2) = 0 (j = 1, 2)$. Assuming $g/l < \Omega_{23}^2 < \Omega_{13}^2$ we have

$$\sigma > 0, \quad Q_1 < 0, \quad Q_2 < 0, \quad \chi = 2 \quad \text{if} \quad 0 < \Omega^2 < g/l$$
$$\sigma < 0, \quad Q_1 < 0, \quad Q_2 < 0, \quad \chi = 2 \quad \text{if} \quad g/l < \Omega^2 < \Omega_{23}^2$$
$$\sigma < 0, \quad Q_1 < 0, \quad Q_2 > 0, \quad \chi = 3 \quad \text{if} \quad \Omega_{23}^2 < \Omega^2 < \Omega_{13}^2$$
$$\sigma < 0, \quad Q_1 > 0, \quad Q_2 > 0, \quad \chi = 4 \quad \text{if} \quad \Omega_{13}^2 < \Omega^2 < \infty$$

Third, the point 0 is above $0_1(\nu = -1)$ and the point C is below $0(\gamma_3 = -1)$,

Fig. 6*c*. Suppose Ω_{24}^2 and Ω_{14}^2 are the positive roots of $Q_j(-1, -1, \Omega^2) = 0 (j = 1, 2)$, and $0 < \Omega_{24}^2 < \Omega_{14}^2$. Then

$$\sigma < 0, \quad Q_1 < 0, \quad Q_2 < 0, \quad \chi = 2 \quad \text{if} \quad 0 < \Omega^2 < \Omega_{24}^2$$

$$\sigma < 0, \quad Q_1 < 0, \quad Q_2 > 0, \quad \chi = 3 \quad \text{if} \quad \Omega_{24}^2 < \Omega^2 < \Omega_{14}^2$$

$$\sigma < 0, \quad Q_1 > 0, \quad Q_2 > 0, \quad \chi = 4 \quad \text{if} \quad \Omega_{14}^2 < \Omega^2 < \infty$$

Finally, let the point 0 be above $0_1(\nu_3 = -1)$ and the point C be above $0(\gamma_3 = 1)$, Fig. 6*d*. Then $\sigma < 0$, $Q_1 > 0$, $Q_2 > 0$, $\chi = 4$ for all $\Omega^2 > 0$.

9. In discussing the stability of the motions (6.2) and (6.3) it will be assumed, without loss of generality, that the rod is at all times in the plane $y_2 = 0$, and ν_3 is therefore identically zero.

Let us introduce the function

$$2W^*(\mu, \lambda, \varkappa, \gamma_1, \nu_1, \gamma_3, \nu_3, \alpha_3, \beta_3, \gamma_2) = 2W + \Omega^2(\mu V_1 + \lambda V_2 + \varkappa V_3)$$

$$V_1 = \nu_1^2 + \nu_3^2 - 1 = 0,\ V_2 = \gamma_1^2 + \gamma_2^2 + \gamma_3^2 - 1 = 0,\ V_3 = \alpha_3^2 + \beta_3^2 + \gamma_3^2 - 1 = 0$$

where $\Omega = kJ^{-1}$, and μ, λ, $\varkappa$ are Lagrange's indeterminate multipliers. The values obtained for the variables $q = (\mu, \lambda, \varkappa, \lambda_1, \nu_1, \gamma_3, \nu_3, \alpha_3, \beta_3, \gamma_2)$ from equation $\partial W^*/\partial q = 0$ (which is the condition that W^* be stationary) are the same as those in the solutions (6.2) and (6.3); in particular, $\mu = \lambda_1$, $\lambda = \sigma + ma^2$, and $\varkappa$ equals J_2 for (6.2) and J_1 for (6.3).

We may derive sufficient conditions for the stability of the motions (6.2) and (6.3) by requiring that the second variation $\delta^2 W_*$ be definite positive on the linear variety $\delta V_1 = \delta V_2 = \delta V_3 = 0$. Introducing the Hessian $D(W^*) = \delta^2 W^*/\partial q^2$ with respect to all the arguments of the function W^*, these conditions are reduced (see [4]) to the requirement that the four principal diagonal minors of the Hessian should be positive.

For (6.2) sufficient stability conditions are

$$\Omega^2(J_2 - J_1) > 0, \quad \Omega^2\sigma > 0$$

$$\Delta_1' = ml^2\Omega^2\gamma_1^2\gamma_3^2[ma^2\sigma^{-1} + \nu_1^2 + 4J^{-1}m(l\nu_1 - a\gamma_1)^2\nu_3^2] > 0 \tag{9.1}$$

$$\Delta_1 = \frac{\nu_1^2\beta_3^2 ml^2\Omega^7(\sigma + ma^2)^3(\sigma + J_2^* - J_3)^3[(mal)^2 - \sigma^2]}{J(J_2 - J_3)(mga)^2\sigma^4(2\sigma + ma^2 + J_2^* - J_3)} \frac{dk}{d\sigma} > 0$$

with $D(W^*) = \Omega^4\sigma(J_1 - J_2)\Delta_1$, and for (6.3) we have

$$\Omega^2(J_1 - J_2) > 0, \quad \Omega^2\sigma > 0$$

$$\Delta_2' = ml^2\Omega^2\gamma_1^2\gamma_3^2[ma^2\sigma^{-1} + \nu_1^2 + 4J^{-1}m(l\nu_1 - a\gamma_1)^2\nu_3^2] > 0 \tag{9.2}$$

$$\Delta_2 = \frac{\nu_1^2\alpha_3^2 ml^2\Omega^7(\sigma + ma^2)^3(\sigma + J_1^* - J_3)^3[(mal)^2 - \sigma^2]}{J(J_1 - J_3)(mga)^2\sigma^4(2\sigma + ma^2 + J_1^* - J_3)} \frac{dk}{d\sigma} > 0$$

with $D(W^*) = \Omega^4\sigma(J_2 - J_1)\Delta_2$.

Let us investigate the motions (6.2). For the branch $\Gamma_1^{(1)}$, see Fig. 7, we have $\sigma > 0$, and $dk/d\sigma > 0$, so that the last three conditions in (9.1) are fulfilled, but the first is not, which means that the motions in this branch are unstable ($\chi = 1$). The motions represented by the branches $\Gamma_2^{(1)}$, $\Gamma_3^{(1)}$, and $\Gamma_4^{(1)}$ are also unstable (with $\chi = 1$) because in $\Gamma_2^{(1)} \sigma < 0$, $\sigma + ma^2 > 0$, $dk/d\sigma < 0$, so $\Delta_1 < 0$; in $\Gamma_3^{(1)} \sigma + ma^2 < 0$, $2\sigma + ma^2 + J_2 - J_3 > 0$, $dk/d\sigma > 0$ and $\Delta_1 < 0$; in $\Gamma_4^{(1)} 2\sigma + ma^2 + J_i - J_3 < 0$, $dk/d\sigma > 0$ and $\Delta_1 < 0$.

The stability of the (6.3)-motions is investigated in a similar fashion by analyzing conditions (9.2). The results obtained for both types of motion are illustrated in Fig. 7, where the numbers 0 through 4 on the branches of the curve Γ refer to the degree of instability of the corresponding motions. Note that the distribution of the stable and unstable motions on the branches of Γ obeys the principle of change of instability for fixed values of the parameter k, and that the change of the degree of instability occurs only at the bifurcation points.

10. Taking $0_1y_1y_2y_3$ as a reference system (cf. Section 1), the problem of the relative equilibrium of the gyrostat we are considering is reduced to that of analyzing the potential energy $U = \Pi - (^1/_2)J\Omega^2$ due to the force of gravity and centrifugal forces.

Let U_* denote the function that results from replacing W by U in the expression for W_*. These two functions must satisfy the same conditions to be stationary, and the families of all relative equilibrium states of the system will be described by formulas (6.1)–(6.3) in which the relationships of the form $k = k(\sigma)$ should be dropped.

The set of relative equilibrium states admits of an easy geometrical interpretation as points on a curve B defined by the equations $\Omega = \Omega(\sigma)$, $\lambda_\gamma = \lambda_\gamma(\sigma)$ in the $(\Omega, \sigma, \lambda_\gamma)$-space. Fig. 9 illustrates the shape of the projection of this curve on the $\lambda_\gamma = 0$ plane for the case (7.1). To relative equilibria (6.1), there correspond the branches $B_1(a)$, $B_2(b)$, $B_3(c)$, $B_4(d)$, $B_5(a)$, $B_6(b)$; the letters a, b, c, d referring to the geometrical conditions shown in Fig. 6. For the equilibria (6.2), we have the branches $B_1^{(1)}(a)$, $B_2^{(1)}(d)$, $B_3^{(1)}(b)$, $B_4^{(1)}(c)$ with the x_1 axis horizontal, and for the equilibria (6.3) we have the branches $B_1^{(2)}(a)$, $B_2^{(2)}(d)$, $B_3^{(2)}(b)$, $B_4^{(2)}(c)$, with the x_2 axis horizontal; for these two types of equilibrium, the letters a, b, c, d refer to the geometric conditions shown in Fig. 8. In Fig. 9 we represent by double curves the projections on the $\lambda_\gamma = 0$ plane of the branches that correspond to the equilibrium states (6.1). Each "rim" of such curves corresponds to different types of equilibrium states (6.1)—see Fig. 6.

The stability of the relative equilibrium states is best discussed in terms of the function U^* obtained by replacing W by U in the definition of W^*. The conditions $\partial U^*/\partial q = 0$ securing that the function U^* is stationary yield the same values for the variables $q = (\mu, \lambda, \varkappa, \gamma_1, \nu_1, \gamma_3, \nu_3, \alpha_3, \beta_3, \gamma_2)$ as those obtained for the families of relative equilibrium states (6.2) or (6.3), with $\mu = \lambda_\gamma$, $\lambda = \sigma + ma^2$, and $\varkappa$ equal to J_2 for the former and to J_1 for the latter case.

The sufficient stability conditions for the relative equilibria (6.2) and (6.3) are reduced to the requirement that the Hessian $D(U^*) = \partial^2 U^*/\partial q^2$ have its four principal diagonal minors positive. For (6.2) we obtain the inequalities

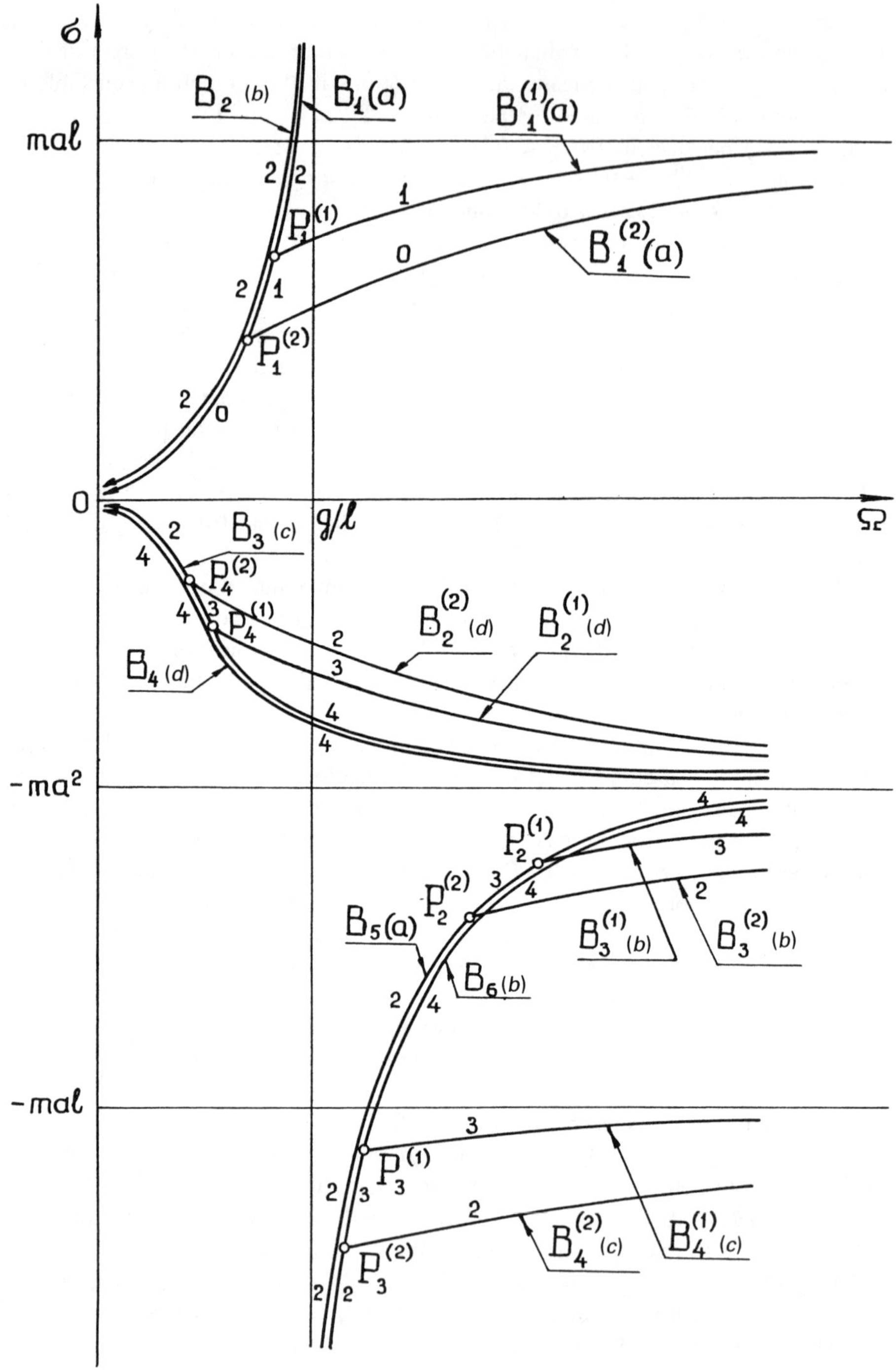

σ
mal
0
$-ma^2$
-mal
g/l
Ω
B_2 (b)
B_1(a)
$B_1^{(1)}$(a)
$B_1^{(2)}$(a)
$P_1^{(1)}$
$P_1^{(2)}$
B_3 (c)
$P_4^{(2)}$
$P_4^{(1)}$
B_4 (d)
$B_2^{(2)}$ (d)
$B_2^{(1)}$ (d)
$P_2^{(1)}$
$P_2^{(2)}$
B_5(a)
B_6(b)
$B_3^{(1)}$ (b)
$B_3^{(2)}$ (b)
$P_3^{(1)}$
$P_3^{(2)}$
$B_4^{(2)}$ (c)
$B_4^{(1)}$ (c)

Figure 9

$$\Omega^2(J_2 - J_1) > 0, \quad \Omega^2\sigma > 0, \quad ml^2\Omega^{14}\gamma_1^2\gamma_3^2\sigma^{-1}(ma^2 + \sigma\nu_1^2) > 0$$

$$\Delta_1 = \frac{(J_2 - J_3)\gamma_1^4\Omega^8 mg^2\sigma^2(2\sigma + ma^2 + J_2^* - J_3)}{2(\sigma + ma^2)(\sigma + J_2^* - J_3)} \frac{d\Omega^4}{d\sigma} > 0 \tag{10.1}$$

with $D(U^*) = \Omega^4\sigma(J_1 - J_2)\Delta_1$, and for (6.3)

$$\Omega^2(J_1 - J_2) > 0, \quad \Omega^2\sigma > 0, \quad ml^2\Omega^{14}\gamma_1^2\gamma_3^2\sigma^{-1}(ma^2 + \sigma\nu_1^2) > 0$$

$$\Delta_2 = \frac{(J_1 - J_3)\gamma_1^4\Omega^8 mg^2\sigma^2(2\sigma + ma^2 + J_1^* - J_3)}{2(\sigma + ma^2)(\sigma + J_1^* - J_3)} \frac{d\Omega^4}{d\sigma} > 0 \tag{10.2}$$

with $D(U^*) = \Omega^4\sigma(J_2 - J_1)\Delta_2$.

The results of an analysis of the condition (10.1) and (10.2) are given in Fig. 9, where the numbers from 0 to 4 on the branches of the curve B indicate the degree of instability of the corresponding relative equilibrium states (6.1)–(6.3).

REFERENCES

1. Skimel, V. N. On the motion of a string suspended gyrostat. Proc. Conf. on Anal. Mech. & Motion Stability Theory, Kazan, 1964, pp. 118–122.
2. Vozlinskii, V. I. On bifurcations of the stationary motions of conservative systems with two cyclic coordinates. PMM, 1967 Vol. 31, No. 5, pp. 841–847.
3. Rubanovskii, V. N. and Rumyantsev, V. V. On stationary motions of a heavy string suspended solid. Izv. USSR Acad. Sci., MTT, 1985, No. 5, pp. 3–7.
4. Rubanovskii, V. N. and Stepanov, S. Ya. Routh's theorem and the method of Chetaev for deriving the Lyapunov function from motion integrals. PMM, 1969, Vol. 33, No. 5, pp. 904–912.

A SOLID BODY MOVING IN A RESISTANT MEDIUM: QUALITATIVE ANALYSIS

B. Ya. Lokshin, V. A. Privalov, V. A. Samsonov

The analysis of motions in a resistant medium is a relatively unexplored subject, even though its practical importance is evident and some of its aspects—those for example pertaining to aircraft design—have been given considerable attention in the last decades. The major difficulty that slows down the progress in this field is the lack of finite dimensional models capable of describing the effects of the medium on the moving body. On the one hand, the development of such models calls for intensive, and often non-traditional, experimentation with moving objects. On the other, the very arrangement of such experiments presupposes the qualitative analysis of the system under study as well as some prior knowledge about the possible behavior of the moving body.

Let us assume that a "plane" body performs a translational motion as shown in Fig. 1. The main feature which is characteristic of motions of real bodies in real media is that the instantaneous velocity $\boldsymbol{V}$ of a body is generally not colinear with the force the medium exerts on the body. In addition to the opposing force S acting in the $-\boldsymbol{V}$ direction as in point mass problems, a real body also experiences a so-called lateral, or lift, force $\boldsymbol{P}$. It is usually assumed that

$$S = s(\alpha)V^2, \quad P = p(\alpha)V^2$$

where α is the angle of attack that specifies the orientation of the vector $\boldsymbol{V}$ relative to the body; occasionally it proves more convenient to employ the vector $\boldsymbol{V}_c = -\boldsymbol{V}$ which is, of course, the velocity of the medium with respect to the body.

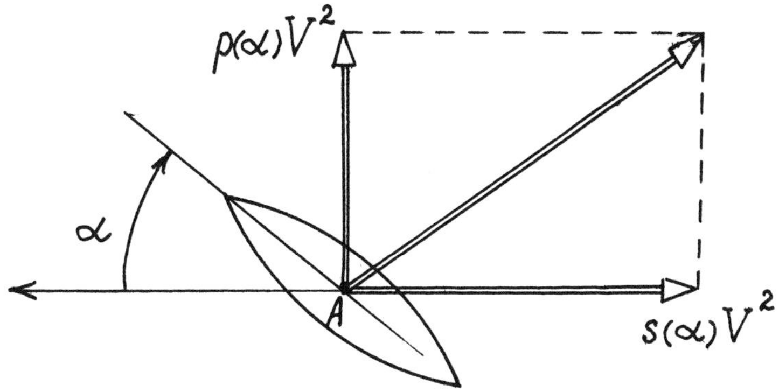

Figure 1

Although the role of the lift force is known very well as far as practical aircraft design is concerned, we are still far from a full theoretical understanding of this phenomenon. In the model we will consider in this paper it is assumed that the medium offers no resistance to the motion of the body ($s(\alpha) \equiv 0$) and that its only effect is to produce a lift force. For all its simplicity, the model turns out to provide useful physical insight into the properties of the lift force.

1. Referring to Fig. 2, we thus consider a rigid-sailed iceboat sailing perpendicular to the direction of the wind. We neglect the friction of the runners on the ice and assume the force of the wind on the sail to be the only external influence in the problem. The position of the sail relative to the iceboat will be

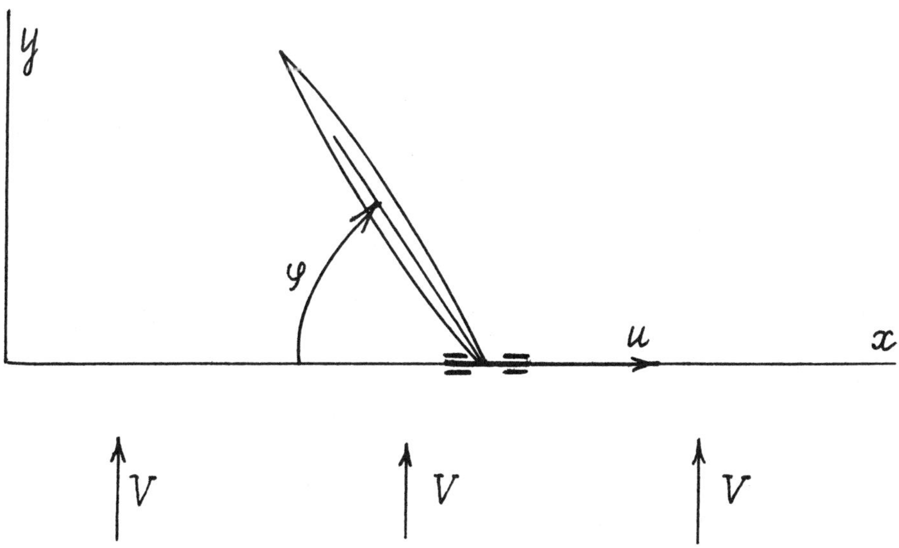

Figure 2

determined by specification of the (constant) incidence φ, which represents one of the structural parameters of the system. The medium moves at a velocity $\boldsymbol{V}_c$ relative to the sail, and we take as an angle of attack the angle of the sail with the vector $\boldsymbol{V}_c$.

The motion of the iceboat is governed by the equation

$$mu^{\cdot} = p(\alpha)V_c^2 \sin(\alpha + \varphi) \tag{1.1}$$

in which the angle α and the quantity V_c are found from the kinetic relations

$$V_c \cos \alpha = V \sin \varphi + u \cos \varphi \tag{1.2}$$

$$V_c \sin \alpha = V \cos \varphi - u \sin \varphi \tag{1.3}$$

and the coefficient $p(\alpha)$ is an odd function of α which is positive on the segment $[0, \pi/2]$ and vanishes at its ends. We will require that this function be continuous and differentiable only in some neighborhood about the points $\alpha = 0$ and $\alpha = \pi/2$.

The steady state velocity of the iceboat is determined by the equation

$$p(\alpha) \sin (\alpha + \varphi) = 0 \tag{1.4}$$

If $\sin 2\varphi \neq 0$, (1.4) has four solutions, two of which correspond to the zeroes of the function $p(\alpha)$

$$\alpha_1 = 0, \quad u_1 = V \cot \varphi, \quad \alpha_2 = \pi/2, \quad u_2 = -\cot \varphi$$

While the first of these two describes a quite natural mode of motion, the second is rather an exotic one. So too are the two remaining solutions of (1.4),

$$\alpha_3 = -\varphi, \quad u_3 = +\infty, \quad \alpha_4 = \pi/2 - \varphi, \quad u_4 = -\infty$$

The stability of the steady-state motion will be best discussed by considering the first-order approximation equation in variations. We obtain

$$m(\delta u)^{\cdot} = p'(\alpha_i)\, V_{ci}^2 \sin(\varphi + \alpha_i)\delta\alpha \tag{1.5}$$

(*a*) Let $i = 1$. Then

$$V_{ci}(\delta\alpha) = -\delta u \sin \varphi, \quad m(\delta u)^{\cdot} = -p'(0)\, V_{ci} \sin^2\varphi\, \delta u$$

Since $p'(0) > 0$, it follows that the dependence of the thrust force on velocity variation is dissipative in nature, so that the natural mode of motion for $\alpha = \alpha_1 = 0$ is asymptotically stable as regards velocity variations

(*b*) Let $i = 2$. Then

$$V_{c2}(\delta\alpha) = -(\delta u) \cos \varphi$$

$$m(\delta u)^{\cdot} = -p^1\left(\frac{\pi}{2}\right) V_{c2} \cos^2\varphi(\delta u)$$

Since $p(\pi/2) < 0$, the dependence of the thrust force on velocity variation is now of accelerating rather than of dissipative character, which means that the exotic regime $\alpha = \alpha_2 = \pi/2$ is unstable. The velocity of the iceboat changes in such a way that if $u < u_2 (\delta u < 0)$, the device will slide in the direction opposite to the natural one, and the velocity u will tend to minus infinity showing that one of the exotic modes, that with $\alpha = \alpha_4$, will be attractive in nature.

The phase space of equation (1.5) is the straight line u shown in Fig. 3. The line contains four steady-state points $(-\infty, u_2, u_1, +\infty)$ that divide it into three intervals, $(-\infty, u_2)$, (u_2, u_1), and $(u_1, +\infty)$. The arrows on the line indicate the direction in which the representative points moves in each particular trajectory.

We may thus ascribe the following properties to the lift force viewed as an independent force factor. First, the generalized forces may be either dissipative or accelerating with respect to the generalized velocities that produce them. Second, the effect of the lift is to decrease the angle between the direction of the velocity V_c and the plane of the sail.

The first property appears to be a rather important one, especially in situations when the structure of stabilizing influences is to be determined. It turns out that a natural force exists which possesses a property of acceleration rather than that of dissipation. Note also that lift may cause energy dissipation even when the medium presents no resistance to the motion.

2. Clearly a real medium does present a resistance. Let us first assume that the resistance is small in the sense that $s(\alpha) \ll \max p(\alpha)$. The equation of motion of the iceboat will then take the form

$$mu^{\cdot} = [p(\alpha) \sin (\alpha + \varphi) - s(\alpha) \cos (\alpha + \varphi)] V_c^2 \tag{2.1}$$

and its phase space (phase line rather) must be very much the same as that of equation (1.1).

Since the function $s(\alpha)$ is small in magnitude, the "perturbation" it introduces in (2.1) should be insignificant in those points of the phase line where the right-hand side of (1.1) assumes finite values. If we determine the phase flow direction at $-\infty$, u_1, u_2, $+\infty$, as shown in Fig. 4, we see that one of the exotic regimes, $u = +\infty$, disappears, whereas the second, $u = -\infty$, shifts toward very large—but finite—negative velocities. The two remaining special points will move to the left. If the resistance of the media is increased—say, if its viscosity becomes higher—it is clear that both the exotic negative-velocity regimes get closer together, merge, and finally disappear thus leaving us with only one—natural—steady-state regime.

It must be kept in mind that in some cases (air, for example) the functions $s(\alpha)$ and $p(\alpha)$ may be of the same order of magnitude, which of course complicates qualitative analysis. If we rewrite (2.1) in the form

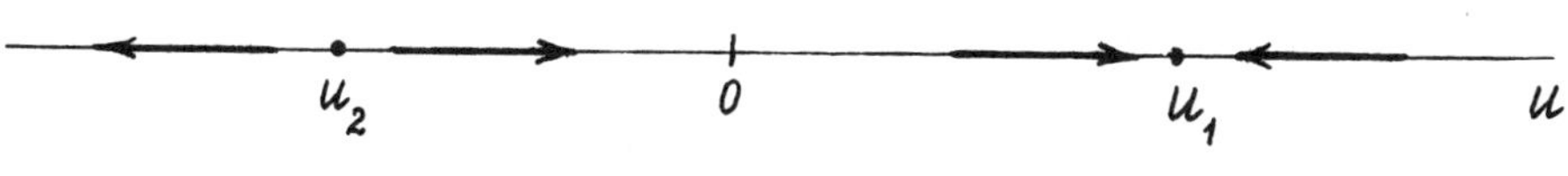

Figure 3

Figure 4

$$mu^{\cdot} = s(\alpha)\sin(\alpha + \varphi)[k(\alpha) - \cot(\alpha + \varphi)]V_c^2 \tag{2.2}$$

$$k(\alpha) = p(\alpha)/s(\alpha)$$

it is easy to see that the stationary regimes are now determined by the equation

$$k(\alpha) = \cot(\alpha + \varphi) \tag{2.3}$$

If α_* is one of the solutions to (2.2), the sliding velocity of the iceboat is found from the kinematic relation

$$u_* = V\cot(\alpha_* + \varphi)$$

The total number of solutions of equation (2.2) is found by simultaneously plotting the function cot α and possible shapes of the function $k(\alpha)$. Let us consider, for example, the case when the action of the air on an airplane profile is characterized by the inequality $\alpha_1 < |\alpha| < \pi/2$ ($\alpha_1 \approx \pi/6$) [1]. We have then

$$k(\alpha) = \cot\alpha \tag{2.4}$$

It should be remarked, however, that if for short plates $k(\alpha) < \cot\alpha$ at $|\alpha| < \alpha_1$, for long plates (see Fig. 5) there exists a certain value $\tilde{\alpha}$ such that on the interval $(0, \tilde{\alpha})$ we have $k(\alpha) < \cot\alpha$, whereas for $\tilde{\alpha} < \alpha < \alpha_1$ the reverse inequality is true.

We thus conclude that if the sail is long and perpendicular to the iceboat ($\varphi = 0$), a nonisolated set of stationary regimes must exist for which

$$u < u_1 = V\cot\alpha_1$$

There also exist two isolated attracting regimes among the stationary motions of the system, $\alpha_1 = \tilde{\alpha}$ and $\alpha_2 = \pi - \tilde{\alpha}$, in which the velocities are equal in magnitude but opposite in direction.

It is a well-known fact from the qualitative theory of differential equations that the sets of the nonisolated singular points do not possess the property of "robustness" and may be destroyed by even the slightest modifications in the equations of motion. It is therefore impossible, from this point of view, to improve our model in terms of the forces acting on the iceboat. The isolated attracting points do possess the above property and the corresponding regimes consequently admit of experimental study.

It should be noted that for some sail geometries the sliding velocity may happen to be a maximum precisely in one of the exotic regimes (i.e., when the sail setting angle φ assumes a certain small value), but in this case no transition into the attraction region will occur unless a certain amount of acceleration is achieved in the direction opposite to the natural one.

The above identification of possible stationary regimes and the stability analysis of these regimes have thus revealed some new properties of the lift force, which seems to be a rather interesting result, especially when the practical importance of this phenomenon is taken into account. We also believe that this result may suggest new approaches to certain well-known problems concerned with motions in a resistant medium.

3. The object we consider next is the straight blade of a wind wheel. Let the axis rotation of the wheel be colinear with the wind direction, as shown in Fig. 6, and let ω be the angular velocity of rotation. When the blade turns about the axis, so does the vector $\boldsymbol{V}_c$ of the relative velocity of the medium, and the plane of the angle of attack formed by $\boldsymbol{V}_c$, and the normal to the blade therefore retains its orientation relative to the blade throughout the motion even though it changes its "absolute" position. If we assume also that the blade is small in size compared with the length L of the connecting element, and that this latter does not interact with the medium, the wheel will be describable in terms of the model we have discussed in the preceding sections provided only that $L\omega$ replaces u in all the results obtained. In a special case of a long plate [1] at least three stationary regimes of rotation prove to be possible. The phase space of the problem is the

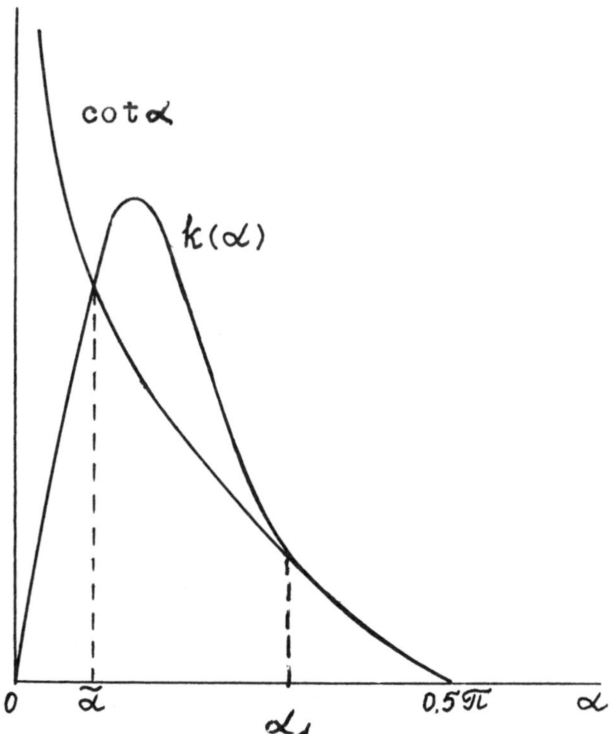

Figure 5

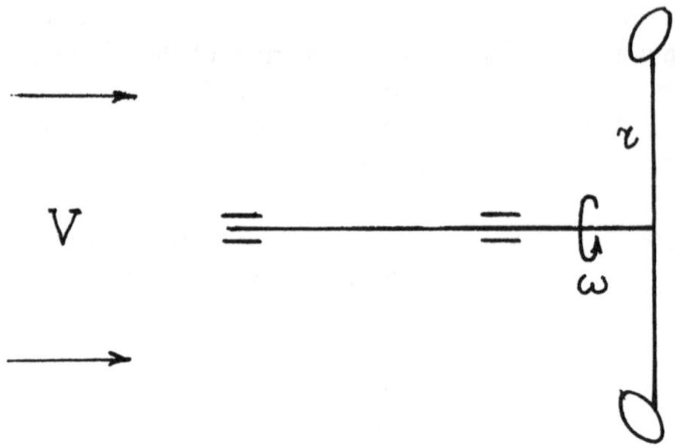

Figure 6

straight line ω with three singular points corresponding to stationary rotations, and three intervals representing the trajectories of transient processes. The maximum and minimum values of the autorotation angular velocity are asymptotically stable, whereas those in-between are not. The maximum value is attained in an exotic regime, with rotation occurring in the direction opposite to the natural one. Note that the attraction region of this steady-state regime can only be attained after a certain angular speed has been imported to the top. This remark is only a theoretical interest, though.

For a series of plates, $s(\alpha)$ is a continuous monotonically increasing function on the interval $[0, \pi/2]$. For blades of this type, the resistance along the axis of rotation of the wind wheel is at its greatest at attack's angles larger than those maximizing the angular velocity. Such an autorotation regime may prove unstable.

4. A feature which is common to the transient regimes we have discussed is that the angle of attack α is a monotonically varying quantity, and there is therefore every reason to believe that the above models must be reasonably close to reality both in qualitative and quantitative terms. There are many problems, however, in which transient regimes are accompanied by oscillatory variations of the angle of attack. In such cases the quantitative adequacy of a model is often very difficult to justify, which makes the "rough" conclusions based on qualitative analysis all the more valuable.

Let us consider a physical pendulum with axis of oscillation perpendicular to the flow of the medium (Fig. 7). The straight blade A of the pendulum is attached to the holder OA in such a way that the line OA coincides with the normal to the plane of the blade. It will be assumed that the blade A is the only element of the system to be affected by the medium and that the transverse dimension of the blade is much smaller than the length of the holder OA.

If ϑ is the angle the holder makes with the direction of the flow, the equation of motion of the pendulum may be written in the form

$$J\ddot{\vartheta} = LV_c^2\,[p(\alpha)\sin(\alpha) - s(\alpha)\cos(\alpha)] \tag{4.1}$$

The kinematic relations relevant to the problem are

$$V_c \sin \alpha = V \cos \alpha, \; V_c \cos \alpha = V \sin \alpha + r\dot{\vartheta}$$

Since the right-hand side of (4.1) depends only on the instantaneous value of α for its sign, it proves possible to apply the phase plane method for qualitatively analyzing the motion. To do this, let us draw the curves $\alpha(\vartheta, \dot{\vartheta}) = \text{const}$ in the $(\vartheta, \dot{\vartheta})$-plane. It is easy to see that these lines form a family of sinusoids differing in their amplitudes and phases.

It is not within the scope of this discussion to explore all the possible types of the functions $s(\alpha)$ and $p(\alpha)$. We will confine our attention here to the special case of a long blade with the $p(\alpha)/s(\alpha)$ ratio as in Fig. 5. The lines $\alpha(\vartheta, \dot{\vartheta}) = \pm\tilde{\alpha}$, $\alpha(\vartheta, \dot{\vartheta}) = \pm\alpha_1$ divide the phase plane $(\vartheta, \dot{\vartheta})$ into a number of regions, each of which is given an arrow to indicate the direction of the phase flow. It does not present any difficulties to analyze the vicinity of the special points in the problem, and it is clear from the results obtained, see Fig. 8, that if a proper set $\tilde{\alpha}$, α_1, $\max k(\alpha)$ is found, a periodic phase path may exist which corresponds to the rotary motion of the pendulum. This path possesses the attraction property and is likely to correspond to the operation regime of the Darrieus Rotor [2].

5. Although there is a considerable amount of literature available (see [3, 4], for example) that treats the classical plate flutter problem, it seems not out of place here to consider the effects of fixing conditions on the low-amplitude oscillations of a plate. (Panovko and Gubanova [5] point out, for example, that an analysis of this kind requires that account should be taken of the effects that the proper motion of the plate has on the forces the plate experiences in the flow).

Let a plate be fixed by means of a massive holder AG in such a way (see

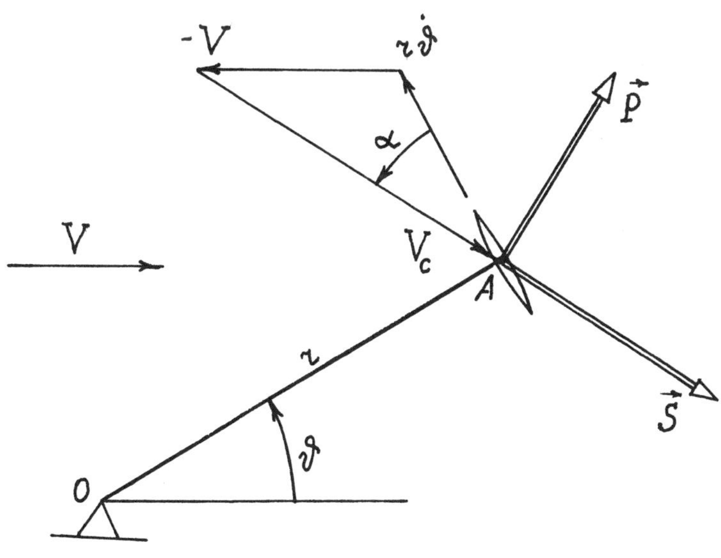

Figure 7

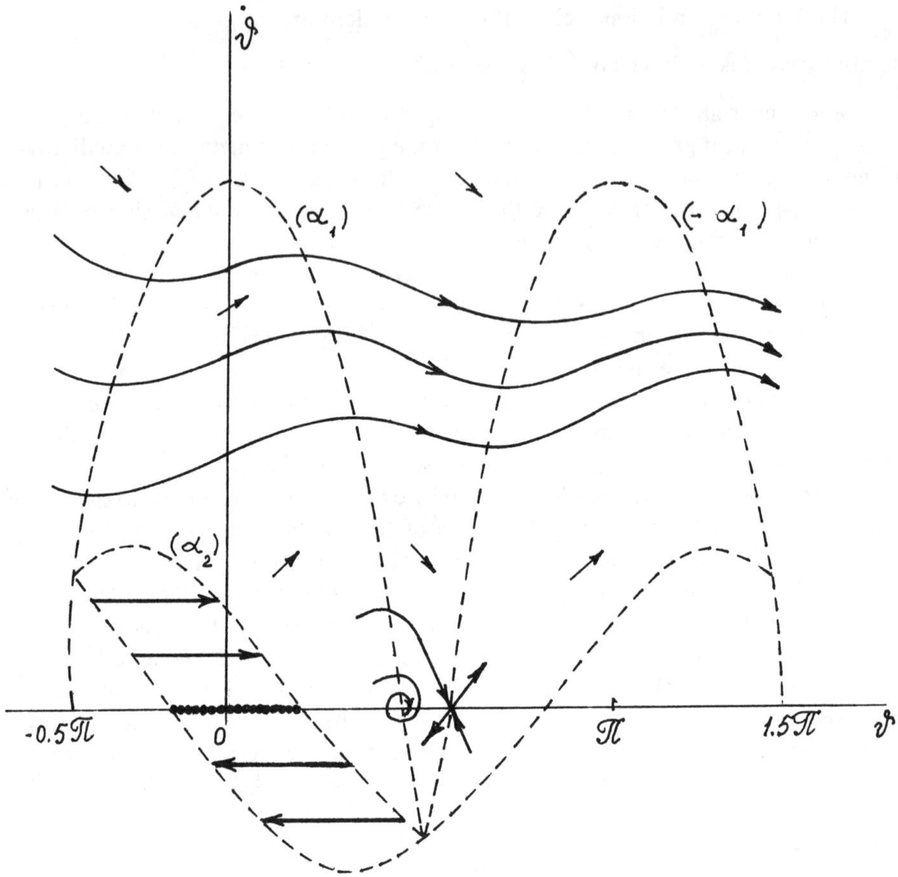

Figure 8

Fig. 9) that the center of gravity of the plate-holder system may only move along the vertical (which will be taken as the x axis) and the holder itself is allowed to turn relative to the direction of the flow (through an angle ϑ). The plate is assumed to be weightless, and its center of pressure to lie at point A. It is also assumed that the quantities x and ϑ are sufficiently small that the deformation forces arising in the compliant elements be linear in these variables. This gives us the restoring force $F = -kx$ and the restoring moment $M = -c\vartheta$.

Taking into account the kinematic relation

$$V\alpha = -\dot{x} - L\dot{\vartheta} + V\vartheta$$

the equations of small oscillations may be written

$$m\ddot{x} + [p'(0) + s(0)](V\dot{x} + lV\dot{\vartheta}) + p'(0)V^2\,\vartheta + kx = 0 \tag{5.1}$$

$$J\ddot{\vartheta} + [p'(0) + s(0)]Vl^2\,\dot{\vartheta} + [p'(0) + s(0)]Vl\dot{x} + \{C - [p'(0) + s(0)]Vl\}\vartheta = 0$$

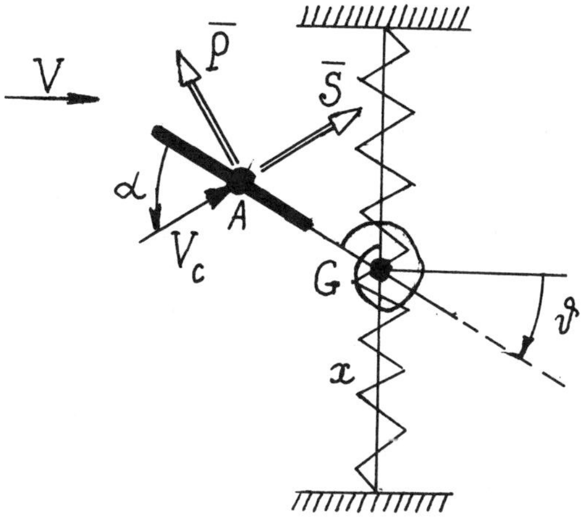

Figure 9

In the situation that has been discussed in the literature, the plate is fixed along the holder. For long plates we have $p'(0) \gg s(0)$, and the conditions of stability of the trivial solution to (5.1) will take the form of the inequalities

$$V^2 < c/Lp'(0) \tag{5.2}$$

$$V^2 < (-kJ - cm)/(J + mL^2)p'(0) \tag{5.3}$$

for $s(0) = 0$. Now let the parameters of the plate be such that $kJ < cm$, $1 > 0$ (the plate is shifted toward the flow). The plots drawn in the (V^2, l)-plane for these conditions (Fig. 10) show the domain of validity for the inequality (5.2) (below the curve *I*) and that for (5.3) (below the curve *II*).

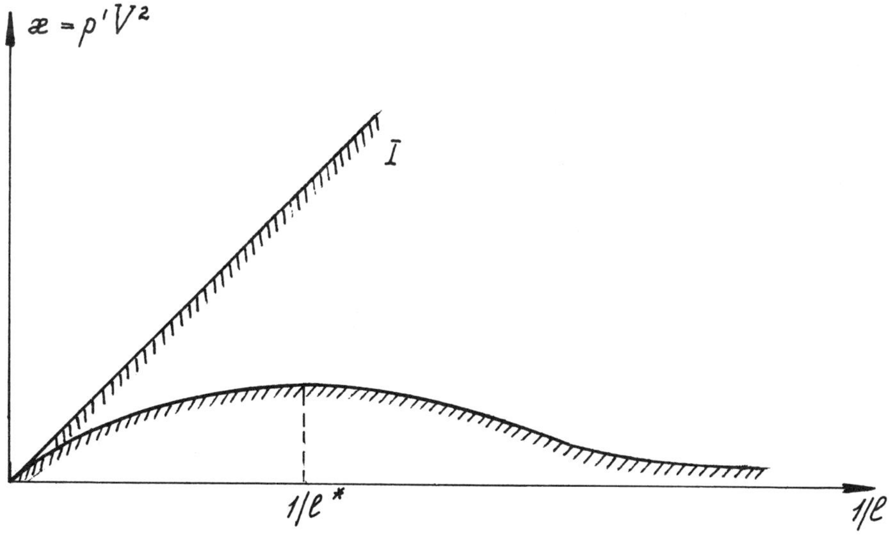

Figure 10

It is seen that as the flow velocity is increased the inequality (5.3) is the first of the two to be invalidated, and an oscillatory form of instability, or flutter, results. We note also that a certain optimum position of the plate exists (l_*) at which the critical value of velocity is a maximum.

6. We have thus discussed a number of remarkable types of motion which have been discovered theoretically and admit of experimental observation. When the global analysis is too difficult to carry out, it often proves helpful to investigate the stability boundaries for the stationary regimes of the problem [6]. Useful insight into the effects of the medium on a moving body may be gained by experimentation, especially for perturbations building up with time. Experiment is of particular importance for the situations when rotation (autorotation) is accompanied by so-called Magnus effects [7, 8].

To conclude, the problem of motion in a resisting medium is by no means a newcomer in the field of theoretical mechanics: suffice it to remember the works of Zhukovskii and Chetaev [9, 10]. The problem is a vast and promising application area for the methods of theoretical mechanics, and it is a pity it has not yet received the attention it deserves.

REFERENCES

1. Tabachnikov, V. G. Stationary low velocity wing properties for all angles of attack. Tr. TsAGI, 1974, No. 1621.
2. Vittecog, P. and Laneville, A. The aerodynamic forces for a Darrieus rotor with straight blades. Wind tunnel measurements. J. Wind Eng. and Ind. Aerodyn, 1983, No. 12, pp. 381–388.
3. Keldysh, M. G. Selected works. Mechanics, Moscow, Nauka, 1985 (in Russian).
4. Nekrasov, A. I. Collected Works. Moscow, Nauka, Vol. 1, 1961 (in Russian).
5. Panovko, Ya. G. and Gubanova, I. I. Stability and the oscillations of elastic systems. Moscow, Nauka, 1979 (in Russian).
6. Privalov, B. A. Stability of a vertically parachuting load. Proc. Inst. Mech., Moscow State Univ., 1975, No. 40.
7. Lokshin, B. Ya. On the stability of the stationary motions of a symmetric solid rapidly rotating in air. Izv. USSR Acad. Sci. MTT, 1974, pp. 18–24.
8. Samsonov, V. A. Qualitative analysis of motions in an air flow. MGU, Moscow, 1980, pp. 56–66.
9. Zhukovskii, N. E. Theoretical fundamentals of aeronautics. In: Collected Works, Moscow-Leningrad, Vol. 6, 1950 (in Russian).
10. Chetaev, N. G. Stability of motion. Moscow-Leningrad, Gostekhizdat, 1946 (in Russian).

MOTION SEPARATION IN ELECTROMECHANICAL SYSTEMS

Yu. G. Martynenko, I. V. Novozhilov

Because of the great practical importance of the title problem a great deal of effort is being devoted to its analysis and there is a large literature on the subject.

1. There is a wide class of mechanisms among those studied in mechanics that are usually referred to as rigid systems. They include gyroscopes, robotic control systems, magnetic suspension control systems, etc. The rigid systems usually depend for their characterization on so-called "large" parameters, such as stiffness coefficient, friction coefficient, angular momenta or discrepancy control coefficients. The rigidity of such systems is usually conditioned by high precision requirements on their performance.

A feature which seems to be common to all the rigid systems is that they have rapid and slow components in their motions and that neither may generally be neglected if a reasonable degree of accuracy is needed. If one also remembers the large frequency diversity characteristic of such problems, it comes as no surprise that computational difficulties that result are hardly, if at all, surmountable, even with a high-speed computer available.

It turns out, however, that problems of this kind may be rendered tractable if the diversity of the characteristic times of a system is properly disposed of in the solution. As a result, certain motion separation methods can be employed, and approximate equations for rapid and slow motions can be derived separately.

The first two steps in carrying out such a program is the normalization of the equations of the problem and the introduction of a small parameter into them [1, 2]. It is also necessary that for each of the quantities chosen to describe the system, such as the phase variables X_i, time T, moments M, rigidities K, etc., a characteristic value for the given problem should be indicated ($X_{i*}, T_*, M_*, K_*, \ldots$). Specification of these values determines the class of motions to be used in the subsequent analysis and is the least formalizable stage in the search for an approximate solution. It may be based on experimental data, analogies, rough estimates of the solutions of the initial system, etc., and depends mainly on the skill and intuition of the researcher for its success.

The step to be taken next is to change to the dimensionless variables $x_i = X_i/X_{i*}$, $t = T/T_*$, $m = M/M_*$, $k = K/K_*$, etc. The characteristic values X_{i*}, M_*, K_*, ... have been chosen properly if the variables x_i, m, k, ... do not exceed unity in the class of motions being investigated. For each of the equations of the problem, a certain combination of the characteristic values is now devised of the same dimension as the equation, and each equation is made dimensionless by being divided by its respective combination. The small parameters that should be introduced into the system are thought of as ratios of partial time constants for rapid and slow components of the motion.

Experience shows that the success or otherwise of a particular investigation depends to a great extent on the manner in which the small parameter of the problem has been chosen. To make this point clearer, the reader will be reminded of the separation of rapid (nutational) and slow (precessional) motions of gyroscopic systems, which is a major problem in the precessional theory of the gyroscope [3–5]. As remarked earlier, we may take as a small parameter of this problem the ratio $\mu = T_1/T_2$, where the time $T_1 = I_*/H$ is of the order of one nutational period of the gyroscope, I_* and H being the characteristic values of the moment of inertia and of the angular momentum of the gyroscope; $T_2 = H/M_*$ is the time of the precession of the gyroscope axis under the action of a disturbing moment of the order of M_*. Alternatively, the dimensional quantity $\epsilon = 1/H$ may be chosen as a small parameter. In this case, the asymptotics derived from a finite interval of the dimensionless time $t = T/T_2$ proves to be valid on a dimensional-time interval $T \sim T_2 = H/M_* \sim 1/\epsilon$ which is asymptotically large with respect to H. It turns out that it is by far a simpler task to justify precessional equations on a finite interval of the dimensional time than on an asymptotically large time interval with $\epsilon = 1/H$ as a small parameter. Note also that the small parameter $\mu = T_1/T_2 = I_*M_*/H^2$ is of the second order in $\epsilon = 1/H$ so that for the same accuracy, fewer approximations are needed with the former than with the latter parameter.

It will be understood that depending on the particular choice of characteristic values, there are various possibilities of introducing a small parameter into the normalized equations [2]. After changing to the dimensionless variables the equations of motion are reduced to the form

$$\frac{T_1}{T_*}\frac{dx_1}{dt} = \mathbf{f}_1(\mathbf{x}_1, \mathbf{x}_2, \ldots)$$
$$\frac{T_2}{T_*}\frac{d\mathbf{x}_2}{dt} = \mathbf{f}_2(\mathbf{x}_1, \mathbf{x}_2, \ldots), \quad T_1 \ll T_2 \tag{1.1}$$

where T_1, T_2 are the partial time constants for the rapid and slow components of motion with respect to the vector variables $\mathbf{x}_1$, $\mathbf{x}_2$.

If we choose $T_* \ll T_1, T_2$, small parameters emerge in the right-hand sides of the equations, and the approximate solution is represented as an expansion in powers of the small parameters. This expansion makes it possible to bring out those details of the solution that develop on a characteristic time interval considerably smaller than both the partial time constants.

When $T_* = T_1$, the system (1.1) becomes

$$d\mathbf{x}_1/d\tau = \mathbf{f}_1(\mathbf{x}_1, \mathbf{x}_2, \ldots) \tag{1.2}$$
$$d\mathbf{x}_2/d\tau = \mu\mathbf{f}_2(\mathbf{x}_1, \mathbf{x}_2, \ldots), \quad \tau = T/T_1, \quad \mu = T_1/T_2 \ll 1$$

and if the solution is expanded in a series of powers of μ, it is known that the expansion coefficients are generally unbounded functions of the rapid time τ.

If, further, $T_* = T_2$, the system (1.1) converts to

$$\mu\, dx_1/dt = f_1(x_1, x_2, \ldots)$$
$$dx_2/dt = f_2(x_1, x_2, \ldots), \quad t = T/T_2 = \mu\tau \tag{1.3}$$
$$\mu = T_1/T_2$$

Finally, if $T_* > T_2$, a system of equations arises in which the derivatives involved have as their coefficients small parameters of various orders, and the analysis of the degenerate equations turns out to be equivalent to the problem of finding the stationary solutions of the system.

If the right-hand side of a (1.2)-type system depends in a regular manner on a small parameter, the system is usually termed regularly perturbed.

Systems of the form (1.3), with small parameters in front of derivatives, are known as singularly perturbed. There exist several small parameter techniques suitable for analyzing both types of systems, such as the Tikhonov-Vasilyeva method of boundary functions [6], the averaging method in a number of versions [7, 8], and multi-scale (or regularization) methods [9].

The Tikhonov-Vasilyeva method turns out to be effective for the system (1.1) written in the form (1.3) if there exists an asymptotically stable solution to the zero-th-approximation equations for a rapid variable in the rapid dimensionless time (boundary layer equations relative to the time). From the equations of the zero-th order in μ for the system (1.3) the solution with respect to $\mathbf{x}_2$ is obtained with an error μ on a finite interval of the dimensionless time t, that is, on an interval T_2 in terms of the dimensional time T.

The averaging method is usually applied to equations written in the form (1.2). To assess the effectiveness of this method it is necessary to consider the so-called "generating system" of the averaging method, that is, the equations of the zero-th order in the rapid variable that are obtained by setting $\mu = 0$ in (1.2)

$$d\mathbf{x}_1/d\tau = \mathbf{f}_1(\mathbf{x}_1, \mathbf{x}_2, \ldots), \quad \mathbf{x}_2 = \text{const} \tag{1.4}$$

The method is effective if (1.4) describes undamped oscillations. The equations in slow variables, when averaged over the trajectories of the generating system, yield approximate solutions with an error of order μ on an asymptotically large interval $1/\mu$ of the dimensionless time τ, or, in terms of dimensional time, on an interval of the order of the larger time constant, T_2.

The singularly perturbed problem (1.3) is best discussed in terms of the multiscale (or regularization) method, in which damped and undamped oscillations are considered simultaneously.

The above methods have different application areas. Although the averaging method appears to be the most general and sophisticated investigation tool, requirements it poses on the system under study are rather severe ones. First, there are relatively restrictive conditions concerning the integrability of the generating system (1.4). Second, the right-hand sides of the slow-variable equations must have a limit independently of the initial conditions for the rapid variables.

In the boundary function method, the integrability of (1.4) is not necessary; instead, the asymptotic stability of the system of the boundary layer is required. In this case the Tikhonov-Vasilyeva procedure is easier to carry out than the general averaging scheme [8] because clearly a quantitative analysis of the rapid-motion system is simpler than its actual solution. It should be noted, however, that boundary function methods are of no use when a system with undamped rapid motions is being considered, and in this case the averaging method is needed. A major problem currently on the agenda is the development of a general theory for the asymptotic integration of singularly perturbed systems. An approach that appears to offer some possibilities in this context is the multiscale method, which enables one to reduce a singularly perturbed system to a regular one by introducing additional independent variables into the problem. One important attraction of the multiscale method is that it may be applied directly to the initial system, so that neither the standardization nor other preliminary transformations of the system are needed. A practical application of the method may present considerable computational difficulty, though. Note also that in the resonance regime for the success of this method it is necessary that certain very restrictive stability conditions should be imposed on the spectrum of the system [9].

The method of boundary functions has proved extremely useful in justifying certain limiting models of mechanics, such as holonomic and nonholonomic constraints and the perfectly rigid body, because in problems of this kind a qualitative analysis of rapid motions is all that is needed [10]. The rigid-body and holonomic-constraint models are generally defined as limiting cases in which an infinitely large rigidity is assumed in the corresponding elements of the system. Small parameters to work with are usually introduced as the ratios of small-time elastic-oscillation constants to the large-time constants characteristic of the system as a whole. The equation of the limiting model is then obtained by letting this parameter go to zero. Situations exist, however, particularly in conservative systems, when it is not perfectly clear whether the limiting model being studied is valid on an infinite time interval [10].

Whether a particular holonomic constraint model is valid or not in a given problem depends on the properties of the forces determining constraints in question, Coulomb friction forces for example. A feature characteristic of these forces is that certain intervals exist in which they change their magnitude in an abrupt, very nearly jump-like, manner. A small parameter suitable here is the ratio of a small time constant of the system inside the jump region to a large time constant outside. An equation of the model is again obtained by allowing this parameter to approach zero. This eliminates the kinematic contradiction arising, for example, in problems concerned with the rolling of rigid railway vehicles, when non-holonomic constraints enter as *a priori* restrictions imposed on the system. From the requirement that the zero limit may be taken, the conditions of the correctness of the model are obtained, expressed in terms of the forces involved and of the length of the time interval of the problem. The method just outlined was successfully applied to the problem of the transverse oscillations of a railway train having an arbitrary number of cars. As a result, the model of slow "kinematic wiggles" was developed and some insight into the forced vibrations and stability of a train was gained [11].

2. One of the difficulties encountered in motion separation is the chaotic behavior which the integral curves of the problem may sometimes exhibit even in the absence of random external influences. However deterministic the formulation, it may happen at certain values of the parameters that some features characteristic of random processes appear in the solution of the problem. As an example, a balanced gyro on a gimbal suspension with a variable ungular momentum may be mentioned. Let the gyro be placed on immovable base. Assuming there are no perturbing moments in the axes of the internal and external rings, the equations of motion are

$$[A\dot{\alpha} + H \sin \beta]^{\cdot} = 0, \quad B\ddot{\beta} - H \cos \beta\alpha^{\cdot} = 0 \tag{2.1}$$

where α and β are the turning angles of respectively the external and internal rings; $A = A_2 + A_0 + A_1$, $B = B_1 + A_0$, A_i, B_i, C_i being the traditional notations for the moments of inertia of the rings and the rotor [12] (it has been assumed for the sake of simplicity that $A_1 + A_0 - C_1 = 0$); and we employ the dot notation for differention with respect to the dimensional time. We will not assume the gyromotor to be an ideal one, so that the angular momentum H will be a sum of a constant term plus a small periodic term with a frequency of the order of the angular frequency of the rotor; that is

$$H = H_0 h, \quad h = 1 + \epsilon \cos \Omega T \tag{2.2}$$

where ϵ is a small parameter characterizing the "modulation depth" of the angular momentum.

Using the angular momentum integral associated with the axis of rotation of the external ring and eliminating the angular velocity $\dot{\alpha}$ between the equations

(2.1), we arrive at a single, nonlinear, nonautonomous, differential equation of the second order

$$d^2\beta/d\tau^2 = hk\cos\beta - h^2\sin\beta\cos\beta, \quad h = 1 + \epsilon\cos\nu\tau \tag{2.3}$$

where $\tau = \Omega_n T$ is the dimensionless rapid time; $\Omega_n = H_0/(AB)^{1/2}$ the nutational frequency; $H_0 k$ is the constant of the integral of the angular momentum relative the axis of rotation of the external ring.

Setting $\epsilon = 0$, the potential energy of the system (2.3), $\Pi(\beta) = -hk\sin\beta + (1/2)h^2\sin^2\beta$, does not depend on the time explicitly and has stationary points: (1) when $\sin\beta = k/h$ (Π is a minimum), and (2) when $\beta = \pi/2$, that is the rings of the suspension are folded up, (Π is a maximum). The phase plane $(\beta, d\beta/d\tau)$ has a saddle point at $\beta = \pi/2$ and center points at $\beta = \beta_0$, $\beta = \pi - \beta_0(\beta_0 = \sin^{-1} k)$. If $\epsilon \neq 0$, it can be seen from the numerical integration of (2.3) that at some values of ϵ and ν certain randomness features appear in the solutions of the system (2.1), namely, at certain instants of time, $\tau_1, \tau_2, \ldots$ the system "jumps" from the neighborhood of the point $\beta = \beta_0$ into the neighborhood of $\beta = \pi - \beta_0$. The system thus operates in the regime of chaotic oscillations, the fold-up instants forming in effect a random sequence. It is a matter of considerable importance in the theory of the gyroscope to develop simple algorithms for determining those of system parameters resulting in chaotic regimes of this kind.

3. Methods of motion separation are often usefully applied to the asymptotic integration of the equations of linear-current electromechanical systems. There is a wide class of systems of this kind whose equations may be written in the Lagrange-Maxwell form

$$\frac{d}{dt}\left(\frac{\partial L}{\partial \dot{e}_k}\right) - \frac{\partial L}{\partial e_k} + \frac{\partial \Psi}{\partial \dot{e}_k} = u_k \quad (k = 1, \ldots, r)$$

$$\frac{d}{dt}\left(\frac{\partial L}{\partial \dot{q}_j}\right) - \frac{\partial L}{\partial q_j} + \frac{\partial \Psi}{\partial \dot{q}_j} = Q_j \quad (j = 1, \ldots, n) \tag{3.1}$$

$$L = T - \Pi + W_m - W_e$$

Here e_k is the charge on the capacitor C_k in the k-th circuit of the system; $\dot{e}_k$ is the current; q_j—the generalized coordinate and $\dot{q}_j$—the generalized velocity; Q_j—the generalized force; u_k—the external emf applied to the k-th circuit; r—the number of circuits in the system; n—the number of mechanical degrees of freedom in the system; L—the Lagrangian function; $T = \frac{1}{2}\sum_{i=1}^{n}\sum_{j=1}^{n} a_{ij}q_i q_j$ the kinetic energy of the system (we limit ourselves to the case of stationary constraints) $\Pi = \Pi(q_j)$ is the potential energy; $W_m = \frac{1}{2}\sum_{k=1}^{r}\sum_{s=1}^{r} L_{ks}\dot{e}_k\dot{e}_s$—the energy of the magnetic field; $W_e = \frac{1}{2}\sum_{k=1}^{r} e_k^2/c_k$—the energy of the electric field between the capacitors' plates; a_{ij}—the coefficients of inertia; L_{ks}—the mutual inductness for the k-th and s-th circuits; L_{kk}—the self-inductance of the k-th circuit; Ψ—the dissipative function.

In many cases of interest electrical processes in a system proceed far more rapidly than mechanical ones, so when the normalization procedure has been carried out, the vector of the electrical variables turns out to be rapid whereas that of the mechanical variables slow. In order to separate the motions two approaches may be employed. First, we may apply the averaging technique. It turns out, however, that since the rapid motions are not oscillatory in this problem, the general averaging scheme is too difficult to carry through, and some modification of the averaging method is therefore needed. Khodzhaev and colleagues [13, 14] performed this task for capacitance-free systems and were able to provide some insight into the structure of the slow motion equations arising in this case. Alternatively, we may use the multiscale method to analyze the problem (3.1). To give an illustration, let us consider a simple model of a passive resonant electrostatic suspension (see fig. 1). This is a device which is employed in a diversity of instruments whenever a conducting body must be suspended to be isolated electrically from other elements. The electric circuit of the system contains an inductunce coil L_1, a parallel-plate capacitor $C(y)$, and an ohmic resistor R, and is connected to an external emf $u = A \sin \Omega t$.

The upper of the capacitors is at rest whereas the lower movable one must be maintained at such a distance from the upper that its weight is balanced by the attracting ponderomotive force known to act on any conductor placed in the electric field between conductor plates. The parameters of the circuit are chosen in such a way that the voltage across the plates increases with the inter-plate gap, so that the ponderomotive force exceeds the weight of the plate and returns the plate to its equilibrium position.

The Lagrangian of the system is

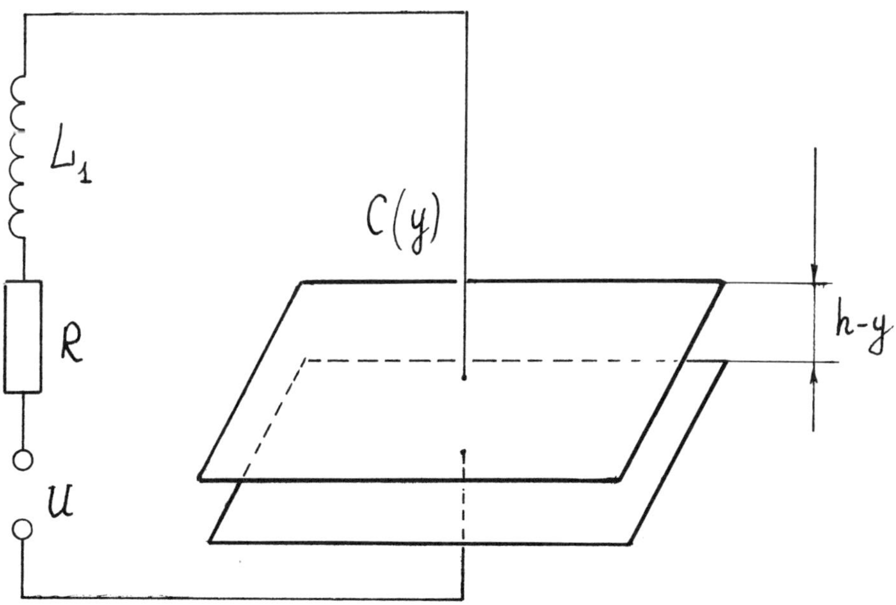

Figure 1 The distribution of movements in the electromechanics of the systems of hard bodies.

$$L = \frac{1}{2} M\dot{y}^2 - Mgy + \frac{1}{2} L_1\dot{e}^2 - e^2(h - y)/(2S) \tag{3.2}$$

where M is the mass of the plate; y is the displacement of the plate from the equilibrium position; $\dot{y}$—the velocity of the plate; L_1—the inductance of the circuit; e—the charge on the capacitor plate; S—the area of the plate; h—the equilibrium distance between the plates. The dissipative function we will need further on is $\Psi_e = \frac{1}{2} R\dot{e}^2$.

After normalization, the equations of motion (3.1) for a passive electrostatic suspension are

$$\mu \frac{dz}{dt} = A(y)z, \quad \frac{dy}{dt} = f(y, z)$$

$$z|_{t=0} = z^0, \quad y|_{t=0} = y^0$$

$$y = \begin{Vmatrix} y_1 \\ y_2 \end{Vmatrix}; \quad f = \begin{Vmatrix} y_2 \\ z_1^2 = 1 \end{Vmatrix} \tag{3.3}$$

$$z = \begin{Vmatrix} z_1 \\ z_2 \\ z_3 \\ z_4 \end{Vmatrix}, \quad A(y) = \begin{Vmatrix} 0 & 1 & 0 & 0 \\ y_1 - 1 & -2\delta & 1 & 0 \\ 0 & 0 & 0 & \nu \\ 0 & 0 & -\nu & 0 \end{Vmatrix}$$

where y_1 is the dimensionless displacement of the lower plate; y_2—its velocity; z_1, z_2—the dimensionless charge and current; z_3—the voltage applied to the circuit; 2δ—the dimensionless resistance of the circuit; ν—the frequency of the source.

By the multiscale method [9, 15], the singularly perturbed system (3.3) is replaced by the system of partial differential equations

$$\mu \frac{\partial z}{\partial t} + Dz = A(y)z$$

$$\mu \frac{\partial y}{\partial t} + Dy = \mu f(y, z), \quad D = \sum_{j=1}^{4} \lambda_j \frac{\partial}{\partial \vartheta_j} \tag{3.4}$$

To solve the Cauchy problem for this system is to find an integral surface passing through a given initial curve. The solution of the extended problem (3.4) is the solution of the initial problem for

$$\vartheta_j = \frac{1}{\mu} \int_0^t \lambda_j(\tau)\, d\tau \tag{3.5}$$

It is shown in [15] that the functions $\lambda_j(\tau)$ involved in the definition of the regularizing functions (3.5) may be expanded regularly in powers of the small parameter μ. The coefficients of the expansion are found in the process of constructing the asymptotic series

$$\lambda_i = \lambda_i^{(0)} + \mu\lambda_i^{(1)} + \mu^2\lambda_i^{(2)} + \dots$$

and the operator D is represented in the form

$$D = \sum_{k=0}^{\infty} \mu^k D^{(k)}, \quad D^{(k)} = \sum_{j=1}^{4} \lambda_j^{(k)} \frac{\partial}{\partial \vartheta_j}$$

To a first approximation we may replace the functions λ_i by the eigenvalues of the matrix $A(y)$ and consider the slow variables y as parameters. We have then

$$\lambda_1^{(0)} = -\delta + i\sqrt{1 - \delta^2 - y_1}; \quad \lambda_2^{(0)} = -\delta - i\sqrt{1 - \delta^2 - y_1}; \quad \lambda_3^{(0)} = i\nu; \quad \lambda_4^{(0)} = i\nu$$

To find the solution of (3.4) we may suppose it expanded in a series of powers of the small parameter μ, the coefficients of the series being determined by systems of differential equations. To extend the definition of these systems and to solve them it proves convenient to introduce a certain system of function spaces, namely, of spaces comprised of the elementary solutions to the continuous approximations equations. The necessary and sufficient solvability conditions for these equations in the elementary-solution spaces are expressed as the requirement that the right-hand side of the equation be orthogonal to the kernel of the corresponding conjugate operator [9, 15]. By separating rapid and slow motions we arrive at the system of equations

$$dy_1/dt = y_2, \quad dy_2/dt = -\partial\Pi/\partial y_1 - \mu\partial\psi/\partial y_2$$

for the slow motions of the lower plate. Here

$$\Pi = y_1 - \frac{a^2}{4\nu\delta} \tan^{-1} \frac{y_1 + \nu^2 - 1}{2\nu\delta}$$

is the potential energy and

$$\psi = -\frac{a^2 y_2^2 \delta[4\nu^2\delta^2 + (\nu^2 - y_1 - 1)(3\nu^2 + 1 - y_1)]}{[(1 - y_1 - \nu^2)^2 + 4\nu^2\delta^2]^3}$$

is the dissipative function.

We thus see that if we calculate to a zero-th order in the small parameter μ, the system turns out to be conservative and its potential energy is $\Pi(y_1)$. In the next-order approximation, small non-conservative forces with a negative dissipative function will appear. In the vicinity of the zero-th-order—and hence automatically stable—equilibrium position, these non-conservative forces act as "negative friction" forces, that is, they "pump" energy into the system and lead to a slow build-up of the oscillations. To circumvent this difficulty, the passive electrostatic suspension may be equipped with some mechanical dampers or with special electrical circuits so as to make the dissipative function positive and thus to eliminate the instability of the kind just discussed.

4. The problem we address next is the motion of a three-dimensional solid body in a magnetic field. The system of equations we must deal with in this case includes the equations of electrodynamics in the quasi-stationary approximation and the equations of motion of the body about its mass center. Let us refer the motion to the set x of rectangular axes fixed in the body, and let the magnetic

permeability of the body be unity. The magnetic flux density **B** within the volume G occupied by the body is then determined by the equations

$$\Delta \mathbf{B} = \frac{4\pi\lambda}{c^2} \mathbf{B}^{\cdot}, \quad \operatorname{div} \mathbf{B} = 0, \quad \mathbf{B}(\mathbf{r}, t)|_{t=0} = \mathbf{B}^0(\mathbf{r}) \tag{4.1}$$

where c the velocity of light in a vacuum, λ the conductivity, and a dot denotes differentiation with respect to the dimensional time in the trihedral x. Outside G the magnetic flux density vector is equal to ∇U, U being a harmonic function satisfying the boundary conditions

$$\mathbf{B}|_S = \nabla U|_S, \lim_{r\to\infty} \nabla U = \mathbf{B}^\infty = \sum_{i=1}^{3} \mathbf{B}_i^\infty e_i$$

where the e_i are the unit vectors of the axes x_i; $\mathbf{B}^\infty = \Gamma \mathbf{B}_\xi^\infty$; $\mathbf{B}_\xi^\infty$ is the magnetic flux density vector referred to a fixed coordinate system ξ and produced by external sources at points remote from the surface S of the body; Γ is the matrix of the direction cosines of the system x relative the axes ξ.

The kinematic equations for the matrix Γ are

$$\Gamma^{\cdot} = \Omega\Gamma, \quad \Omega = \left\| \begin{matrix} 0 & \omega_{x_3} & -\omega_{x_2} \\ -\omega_{x_3} & 0 & \omega_{x_1} \\ \omega_{x_2} & -\omega_{x_1} & 0 \end{matrix} \right\| \tag{4.2}$$

where $\boldsymbol{\omega} = \Sigma_{i=1}^3 \omega_{x_i}\mathbf{e}_i$ is the vector of the angular velocity of the body.

The equations of motion of the solid body relative the mass center are

$$I\boldsymbol{\omega}^{\cdot} + \boldsymbol{\omega} \times I\boldsymbol{\omega} = \mathbf{P} \times \mathbf{B}^\infty \tag{4.3}$$

where

$$\mathbf{P} = \frac{1}{8\pi} \int_G r \times \operatorname{curl} \mathbf{B}\, dv$$

is the vector of the magnetic moment induced in the conductor, and I the inertial tensor of the conductor.

An important special case arises when the magnetic field does not penetrate deep into the conductor, that is, when the magnetic field attenuation time $T_f = 4\pi\lambda L^2/c^2$, L being the dimension of the body, is much greater than the characteristic time of magnetic field variations in the trihedral x. In this case the normalized equations (4.3) have a small parameter at the Laplacian operator, thus

$$\epsilon^2\, \Delta \mathbf{B} = \partial \mathbf{B}/\partial t \tag{4.4}$$

According to [16], the variety of all the slow motions of the solid body is described by the integrodifferential equation

$$I\boldsymbol{\omega}^{\cdot} + \boldsymbol{\omega} \times I\boldsymbol{\omega} = \left[E^0 \mathbf{B}_x^\infty + \frac{\epsilon}{\sqrt{\pi}} E^1 \int_0^t \frac{\mathbf{B}_x^\infty(\tau)\, d\tau}{\sqrt{t-\tau}} \right] \times \mathbf{B}_x^\infty \tag{4.5}$$

where E^0, E_1 are constant tensors dependent only on the shape of the body. The elements of E^0, E_1 are defined as functionals with respect to the solutions of stationary problems.

It should be noted that a model with an integral term of the type (4.5) proved very successful in fitting experimental data on the motion of a magnetic suspension train [17].

In the large penetration depth limit, the field attenuation time is considerably less than the characteristic time constant of magnetic field variations outside the body, and a small parameter appears in the normalized equations (4.1) in the time derivative term

$$\Delta \mathbf{B} = \epsilon\, \partial \mathbf{B}/\partial t \tag{4.6}$$

In solving this singularly perturbed electrodynamic problem there arises a boundary layer in time outside which, as Kobrin and Martynenko [18] have shown, the force moment applied to the body is given by the asymptotic expression

$$M = P \times \mathbf{B}^{\infty}, \quad P = \sum_{k=1}^{\infty} (-1)^k \epsilon^k P^{(k)} \frac{d^k \mathbf{B}^{\infty}}{dt^k} \tag{4.7}$$

Here the $P^{(k)}$ are symmetric and positive defined magnetic polarizability tensors whose components depend only on the shape of the body and require for their determination the solution of a number of stationary problems in the region occupied by the body. In the same paper the reader will find expressions relating to the tensors $P^{(k)}$, $E^{(k)}$ and the tensors of associated masses and of associated moments of inertia appearing in ideal fluid hydrodynamics.

Taking projections on the axes of the fixed trihedral ξ, the principal term of the asymptotic expansion (4.7) may be represented in the form

$$\mathbf{M} = \mathbf{M}' + \mathbf{M}'' \tag{4.8}$$

$$\mathbf{M}' = \epsilon \mathbf{B}_{\xi}^{\infty} \times \boldsymbol{\Gamma} P^{(1)} \boldsymbol{\Gamma}^T [\mathbf{B}_{\xi}^{\infty} \times \boldsymbol{\omega}], \quad \mathbf{M}'' = \epsilon \mathbf{B}_{\xi}^{\infty} \times \boldsymbol{\Gamma} P^{(1)} \boldsymbol{\Gamma}^T \frac{d\mathbf{B}_{\xi}^{\infty}}{dt}$$

This formula proved helpful in studying the motion of a conducting satellite in the Earth's magnetic field inducing eddy currents in the satellite's surface [19]. The satellite was assumed to move under the action of the moment of the eddy currents and to have a triaxial inertial ellipsoid and an arbitrary polarizability tensor. The effect of the moment $\mathbf{M}'$ is to damp the component of the satellite's angular velocity perpendicular to the magnetic flux density vector. The moment $\mathbf{M}''$ depends on the derivative of the magnetic field strength and may be called a driving moment because of its constant component capable of increasing the kinetic energy of the satellite in a non-uniform field. At high angular velocities the moment $\mathbf{M}''$ is considerably less than $\mathbf{M}'$ and its effect in the motion is negligible. But if the rotation of the satellite is on the average slower than its orbital motion, the driving moment $\mathbf{M}''$ may become larger than the damping moment, with a resultant increase in the angular velocity. The analysis is of the interplay of the damping and driving components of the moment, equation (4.8), have revealed a number of previously unknown regimes of satellite motion [19]. The averaged equations of motion derived in [19] can be useful in treating tidal effects in the study of the rotation of heavenly bodies in the case of a nonspheric tidal moment tensor.

One further application of the expression (4.8) is found in the paper by Kudin and Martynenko [20] who studied the problem of bringing up to speed a contactless gyroscope in a resisting medium. It was assumed that the moment of resistance of the medium was linear in the angular velocity of the instrument, and that the active moment was due to the action on the conductor of a uniform rotating field produced by the driving system of the gyroscope rotor, in full analogy with the stator of an asynchronous motor. The (linearized) problem the authors were concerned with was the small oscillations of the rotor dynamic symmetry axis about the axis of rotation of the field. It proved possible to derive the stability conditions both for the non-resonance and the resonance case, which in the latter the nutational oscillations of the rotor have their frequency equal to the angular velocity of the magnetic field. In the rapid rotation limit, an analysis was carried out of the simplified system of equations obtained for slow motion variables by averaging over the Euler-Poinsot motion.

5. A number of interesting problems arise when the motion separation method is applied to the dynamics of gyroscopes equipped with contactless rotor suspension systems of electrostatic, electromagnetic, or cryogenic type. These gyroscopes are extremely complex electromechanical devices whose basic element is a rapidly spinning solid body, the rotor, suspended in a vacuum by means of electric and magnetic fields.

In contrast to conventional gimbal suspension gyroscopes, their contactless counterparts have no fixed point in them so that, generally speaking, the equations of motion of the mass center and of the suspension servo system of such gyroscopes are interconnected with those for the angular motion. The particular feature of dynamics of such devices is that the right-hand sides of the slow motion equations must be averaged both over the unperturbed Euler-Poinsot motion and over the motion of the mass center of the rotor. According to [21], the equations for the slow motion of the rotor symmetry axis are conveniently written in the form

$$\boldsymbol{\gamma}^{\cdot} = \boldsymbol{\gamma} \times \nabla U' + \boldsymbol{\gamma} \times [\boldsymbol{\gamma} \times \nabla U'] \tag{5.1}$$

where $\boldsymbol{\gamma}$ is the unit vector along the symmetry axis; U'—the force function of the conservative moments being applied to the gyroscope rotor; U''—the force function of the nonconservative moments. Both U' and U'' turn out to be dependent on the parameter of the suspension servo system, and analysis of equations (5.1) makes it possible to estimate the accuracy of the instrument, and to evaluate the systematic drift of the system for the subsequent correction of the gyroscope's readings.

6. The motion separation method is well suited for analyzing the motions of walking machines. In the supported motion regime, the motion of a chosen point

of the walking machine frame is stabilized by the control system of the device relative to the preset program. At feet change instants, the discrete control system takes over to control the cyclic walk of the device.

If high accuracy requirements are imposed on the instrument concerning the realization of the preset program, very strict mismatch control must be secured. Let $K_1(V - V_s)$, $K_2(y - y_s)$, ... be the actuator control siquals. Here K_1, K_2 are "large" control coefficients; V, y—the velocity and height of the point being stabilized; V_s, y_s—their respective set values.

Prior to the normalization of the equations involved we must change from the initial phase variables, which are the quantities V, y, ... subject to stabilization, to the mismatch variables $U_1 = V - V_s$, $U_2 = y - y_s$, ... In carrying out the normalization procedure account must be taken of the relationships of the form $M_{1*} = K_1U_{1*}$, $M_{2*} = K_2U_{2*}$, M_{i*} being the final value of the variable characterizing the control action in the system. The smaller partial time constants of the system are given by equations of the form $T_1^2 = M_{1*}/K_1$, ... The larger time constant is of the order of the time of one step, $T_2 = L/V_s$, L being the length of the step.

The normalized equations contain a parameter $\mu = T_1/T_2$ in the coefficients of the derivatives, which is small when the coefficients K_1, K_2 ... are large. The equations have the form (1.3), and their zero-th approximation with regard to the slow variables yields the motion of the system in the preprogrammed servo constraints. These equations having been obtained, discrete walk control becomes accessible to analysis. As for the rapid motions, the zero-th-approximation equations are linear and have constant coefficients. They describe the stability of the system's rapid motions with respect to errors in the vicinity of the slowly varying preset "posture" of the system [22, 23].

Similar normalization techniques and subsequent singular-perturbed formulation also prove helpful in the approximate derivation of both slow-motion and rapid-motion equations in control problems concerned with railway train magnetic suspension [24] and spacecraft angular motion [25].

REFERENCES

1. Kline, S. J. Similitude and approximation theory. New York, McGraw-Hill, 1965.
2. Novozhilov, I. V. Lectures on the approximate methods in dynamic studies. Moscow Energ. Inst., Moscow, 1980 (in Russian).
3. Merkin, D. R. Gyroscopic systems. Moscow, Nauka, 1974 (in Russian).
4. Kobrin, A. I., Martynenko, Yu. G., and Novozhilov, I. Y. On the precession equations of gyroscopic systems. PMM, 1976, Vol. 40, No. 2, pp. 230–237.
5. Gorelova, E. Ya. and Strygin, V. V. Singularly perturbed systems near an integral variety: Cauchy problem solution representation on asymptotically large time periods. In: Approximate methods in the study of differential equations, Kuibyshev St. Univ., Kuibyshev, 1983.
6. Vasilyeva, A. B. and Butusov, V. F. Asymptotic expansions of the solutions of singularly perturbed equations. Moscow, Nauka, 1973 (in Russian).
7. Mitropolskii, Yu. A. Averaging method in nonlinear mechanics. Kiev, Nauk. Dumka, 1971 (in Russian).

8. Volosov, V. M. and Morgunov, B. I. Averaging method in the theory of nonlinear oscillatory systems. Moscow St. Univ., Moscow, 1971 (in Russian).
9. Lomov, S. A. Introduction into the general theory of singular perturbations. Moscow, Nauka, 1981 (in Russian).
10. Novozhilov, I. V. On the correctness of certain limiting models in mechanics. Prepr. Inst. Prik. Mat. USSR Acad. Sci., Moscow, No. 253, 1985 (in Russian).
11. Kopylov, I. A. and Novozhilov, I. V. Transverse vibrations of a railway train. Izv. USSR Acad. Sci. MTT, 1985, No. 4, pp. 66–70.
12. Klimov, D. M. and Kharlamov, S. A. Dynamics of the gimbal suspension gyroscope. Moscow, Nauka, 1978 (in Russian).
13. Reimers, N. A. and Khodzhayev, K. Sh. Averaging of quasilinear systems with many fast variables. Differ. Eq., 1978, Vol. 14, No. 8, pp. 1388–1396.
14. Khodzhayev, K. Sh. and Shatalov, S. D. On slow motions of a conducting solid in a magnetic field. Izv. USSR Acad. Sci. MTT, 1981, No. 2, pp. 175–182.
15. Kobrin, A. I. and Martynenko, Yu. G. Asymptotic solution of one nonlinear differential system. In: Approximate methods in the study of differential equations. Kuibyshev St. Univ., Kuibyshev, 1979, No. 5 .
16. Kobrin, A. I. and Martynenko, Yu. G. Dynamics of a conducting body in an HF magnetic field. Dokl. USSR Acad. Sci., 1980, Vol. 255, No. 5, pp. 1063–1065.
17. Rabinovich, B. I. and Rogovoi, V. M. Mathematics of nonstationary eddy currents and vortex fluid motions in spacecraft stabilization/orientation problems. Kosm. Issled., 1984, Vol. 22, No. 5, pp. 683–692.
18. Kobrin, A. I. and Martynenko, Yu. G. Conducting solid moving about its mass center in a slowly varying magnetic field. Dokl. USSR Acad. Sci., 1981, Vol. 261, No. 5, pp. 1070–1073.
19. Martynenko, Yu. G. Eddy current effects on satellite attitude and rotation. Kosm. Issled., 1985, Vol. 23, No. 3, pp. 347–357.
20. Kudin, S. F. and Martynenko, Yu. G. Driving of a contactless gyroscope in a resistant medium. Ibid., No. 6, pp. 14–22.
21. Martynenko, Yu. G. and Medvedev, A. B. To the theory of the contactless gyroscope with a spheric unbalanced rotor. Izv. USSR Acad. Sci. MTT, 1985, No. 4, pp. 3–11.
22. Novozhilov, I. V. Supported phase walking machine foot control. Proc. Riga Traum. Orthoped. Inst., 1975, No. 13, pp. 634–639.
23. Zatsepin, M. F. and Novozhilov, I. V. Four-foot pace control. Izv. USSR Acad. Sci. MTT, 1986, No. 5.
24. Kapustina, O. M. and Novozhilov, I. V. Motion separation in the dynamics of a contactless suspension vehicle. Proc. Moscow Energ. Inst., 1983, No. 16, pp. 36–41.
25. Novozhilov, I. V. On the separation of fast and slow motions in flywheel attitude control. Kosm. Issled., 1970, Vol. 8, No. 5, pp. 693–700.

COMPUTER ALGEBRA IN MECHANICS

D. M. Klimov

The role of high-speed computers in today's research can hardly be overestimated and is usually associated with the ever-increasing volume of information, increased requirements on accuracy and reliability of computation, and the growing complexity of mathematical models. The subject of this paper is analytical machine computing, a comparatively recent innovation in the computer field which has already proved its worth in the study of mechanical systems and in other areas and is receiving increasing attention the world over.

As far as natural sciences are concerned, analytical manipulations play a very important role in any serious investigation, indeed they often account for the major part of all the effort required to solve the problem, and it is becoming increasingly evident today that it would be profitable to enlist the aid of computers to do the job. This is where computer algebra, or symbolic computation, comes into play. The basic element of symbolic computation is a symbolic (analytical) expression which is organized and transformed according to certain present logical rules. Modern symbolic computation makes it possible to integrate and differentiate a symbolic expression, to regroup or rearrange its terms with respect to some of their characteristics, to collect like terms, to perform substitutions, etc.

The high level of automatization in technology, laboratory research, and information-processing systems carries with it the connotation of high accuracy requirements on mathematical models being used and makes symbolic computation all the more important since any increase in the accuracy of a model, be it a machine, a mechanism, or other mechanical system, depends to a large extent on the analytical properties of the model. The reason is that any sophistication of a model implies a higher dimensionality of a linear system involved, an increased number of nonlinear components, and incorporation of additional physical influences into the model. Symbolic computation has already made significant contributions to many areas of theoretical physics and heavenly mechanics and is constantly expanding its application area. One very important use of symbolic computation is in the treatment of kinematically complex mechanical systems ranging from the traditional models considered in the theory of machines and mechanisms to the present-day robots and manipulators. Symbolic computation systems are being used for deriving equations of motion and for developing new computational methods advantageous over previous techniques in taking account of the specific features of the problem being considered.

One of the earliest mechanical applications of symbolic computation was the development of equations of motion for a system of connected bodies. These equations may contain as many as tens of thousands of monomials and their numerical analysis is generally a formidable task. It should also be noted that some of the computational schemes now in use, like the finite-element method or various difference techniques, presuppose certain transformations of the equations, such as the reduction to normal variables or to the normal Cauchy form. Occasionally, a linearization of the equations in the neighborhood of a certain solution or equilibrium state is needed. All these problems may, and some are already being, solved by the help of symbolic computation.

Let us consider an equation of the form

$$\ddot{u} + \omega_0^2 u = f(u, \dot{u}, \epsilon) \tag{1}$$

where dots represent differentiation with respect to the time t, ϵ is a small parameter and f is an analytical function of its arguments.

One way of treating problems like this is to apply Poincaré's perturbation method, which is a closed algorithm capable of approximating as closely as desired the periodic solution of (1). The algorithm was carried out on the EC-1055M machine using the symbolic computation system REDUCE.

Duffing's equation

$$\ddot{u} + u = -\epsilon u^3 \tag{2}$$

for an oscillator with a nonlinear elastic force gives rise to oscillations with amplitude-dependent period. The same computer constructed the periodic solution $u^*(\tau, \epsilon)$ and the frequency $\nu(\epsilon)$ for this equation, both up to the 20-th order in the small parameter, viz.,

$$\begin{aligned} u^*(\tau, \epsilon) &= a \cos \tau + \frac{\epsilon a^3}{32} (\cos \tau - \cos 3\tau) \\ &+ \epsilon^2 a^5 \left(\frac{23}{1024} \cos \tau - \frac{3}{128} \cos 3\tau + \frac{1}{1024} \cos 5\tau \right) + 0(\epsilon^3) \end{aligned} \tag{3}$$

$$\nu(\epsilon) = 1 - \sum_{i=1}^{20} a^{2i} \frac{N_i}{Z_i} \epsilon^i + 0(\epsilon^{21}) \tag{4}$$

where a is the initial amplitude of the oscillator, and the numbers N_i and Z_i are given in Table 1. (We have reproduced only two approximations for $u^*(\tau, \epsilon)$). Note that equations (3) and (4) are identical to analogous formulas of Hayashi [1].

Equations of the type (1) emerge in many physical and mechanical problems and apart from nonconservative nonlinear elasticity systems are also capable of describing self-excited oscillations, a wide and important class of processes in which a system excites itself into an oscillation with a certain period and amplitude. As a example we have selected Van der Pol's equation

$$\ddot{U} + U = \epsilon U(1 - U^2) \tag{5}$$

which makes it possible to illustrate most of the qualitative features of self-excited oscillations. We have applied the perturbation method to derive to the 18-th order in ϵ the periodic solution $u^*(\tau, \epsilon)$ of this equation and its frequency ν. The results are

$$\begin{aligned} u^*(\tau, \epsilon) &= 2 \cos \tau + \epsilon\left(\frac{3}{4} \sin \tau - \frac{1}{4} \sin 3\tau\right) \\ &+ \epsilon^2\left(-\frac{1}{8} \cos \tau + \frac{3}{16} \cos 3\tau - \frac{5}{96} \cos 5\tau\right) \\ &+ \epsilon^3\left(-\frac{7}{256} \sin \tau + \frac{21}{256} \sin 3\tau - \frac{35}{576} \sin 5\tau + \frac{7}{576} \sin 7\tau\right) + 0(\epsilon^4) \end{aligned} \tag{6}$$

Table 1

i	N_i	Z_i	i	N_i	Z_i
1	3	2^3	11	3511276321347	2^{49}
2	21	2^8	12	401225915283063	2^{56}
3	81	2^{11}	13	2892201453147555	2^{59}
4	6549	2^{18}	14	84053106665670789	2^{64}
5	37737	2^{21}	15	614945335384090729	2^{67}
6	936183	2^{26}	16	115819270549999634l141	2^{78}
7	6077907	2^{29}	17	856653848289440l288225	2^{81}
8	2604833685	2^{38}	18	254612814518190043882263	2^{86}
9	17839453041	2^{41}	19	1899627691040292362960331	2^{89}
10	49715865027	2^{46}	20	227596989316436230247319519	

$$\nu(\epsilon) = 1 + \sum_{i=1}^{9} \frac{N_{2i}}{D_{2i}} \epsilon^{2i} + 0(\epsilon^{20}) \tag{7}$$

where the number N_{zi} and Z_{zi} are listed in Table 2. Again our results are in full agreement with those by Hayashi [1].

One further possible application for the perturbation method is the determination of the region of stability for Hill type equations

$$\ddot{U} + \left(a + 2\epsilon \sum_{i=1}^{\infty} (\theta_{is} \sin 2it + \theta_{ic} \cos 2it) \right) U = 0 \tag{8}$$

with periodic coefficients.

For equations of this type it has been shown by Floquet that the stability boundaries of such equations are curves in the plane of the parameters (a, ϵ). These curves define the a–ϵ relationships securing the existence of a π-periodic or 2π-periodic solution to (8) [1, 2]. Using again the REDUCE system we were able to work out a computer algorithm for applying the perturbation method to equations with periodic coefficients. The algorithm enables the determination of stability boundaries in the form of expansions in powers of the small parameter. Applying the program to the Mathieu equation

Table 2

N_2	−1
Z_2	16
N_4	17
Z_4	3072
N_6	35
Z_6	884736
N_8	−678899
Z_8	5096079360
N_{10}	28160413
Z_{10}	2293235712000
N_{12}	16729607288111
Z_{12}	3698530556313600000
N_{14}	−5722795344767278507
Z_{14}	5219366321069752320000000
N_{16}	−1846779765852173498887007
Z_{16}	1473113950458726894796800000000
N_{18}	3202506386889338212533493973243
Z_{18}	4157716813774710787878787448832000000000000

$$\ddot{U} + (a - 2\epsilon \cos 2t)\, U = 0 \tag{9}$$

we obtained power expansions in u for the Mathieu functions $cl_n(t)$ and $sl_n(t)(n = 1, 2, \ldots 5)$ and for the corresponding frequencies a_n. With view of determining higher-order instability regions, the calculation was carried out up to the 20-th power of ϵ. We obtained

$$\begin{aligned}
a_{c1} &= 1 + \epsilon - \frac{1}{8}\epsilon^2 - \frac{1}{64}\epsilon^3 - \frac{1}{1536}\epsilon^4 + \frac{11}{36864}\epsilon^5 \\
&+ \frac{49}{589824}\epsilon^6 + \frac{55}{9437184}\epsilon^7 - \frac{83}{35389440}\epsilon^8 + 0(\epsilon^9) \\
a_{s1} &= 1 - \epsilon - \frac{1}{8}\epsilon^2 + \frac{1}{64}\epsilon^3 - \frac{1}{1536}\epsilon^4 - \frac{11}{36864}\epsilon^5 \\
&+ \frac{49}{589824}\epsilon^6 - \frac{55}{9437184}\epsilon^7 - \frac{83}{35389440}\epsilon^8 + 0(\epsilon^9) \\
a_{c2} &= 4 + \frac{5}{12}\epsilon^2 - \frac{763}{13824}\epsilon^4 + \frac{1002401}{79626240}\epsilon^6 - \frac{1669068401}{458647142400}\epsilon^8 + 0(\epsilon^{10}) \\
a_{s2} &= 4 - \frac{1}{12}\epsilon^2 + \frac{5}{13824}\epsilon^4 - \frac{289}{79626240}\epsilon^6 + \frac{21391}{458647142400}\epsilon^8 + 0(\epsilon^{10}) \\
a_{c3} &= 9 + \frac{1}{16}\epsilon^2 + \frac{1}{64}\epsilon^3 + \frac{13}{20480}\epsilon^4 - \frac{5}{16384}\epsilon^5 \\
&- \frac{1961}{23592960}\epsilon^6 - \frac{609}{104857600}\epsilon^7 + \frac{4957199}{2113929216000}\epsilon^8 + 0(\epsilon^9) \\
a_{s3} &= 9 + \frac{1}{16}\epsilon^2 - \frac{1}{64}\epsilon^3 + \frac{13}{20480}\epsilon^4 + \frac{5}{16384}\epsilon^5 \\
&- \frac{1961}{23592960}\epsilon^6 + \frac{609}{104857600}\epsilon^7 + \frac{4957199}{2113929216000}\epsilon^8 + 0(\epsilon^9) \\
a_{c4} &= 16 + \frac{1}{30}\epsilon^2 + \frac{433}{864000}\epsilon^4 - \frac{5701}{2721600000}\epsilon^6 \\
&- \frac{112236997}{2006581248000000}\epsilon^8 + 0(\epsilon^{10}) \\
a_{s4} &= 16 + \frac{1}{30}\epsilon^2 - \frac{317}{864000}\epsilon^4 + \frac{10049}{2721600000}\epsilon^6 \\
&- \frac{93824197}{2006581248000000}\epsilon^8 + 0(\epsilon^{10})
\end{aligned} \tag{10}$$

$$a_{c5} = 25 + \frac{1}{48}\epsilon^2 + \frac{11}{774144}\epsilon^4 + \frac{1}{147456}\epsilon^5$$

$$+ \frac{37}{891813888}\epsilon^6 - \frac{7}{339738624}\epsilon^7 + \frac{63439}{201364441399296}\epsilon^8 + 0(\epsilon^9)$$

$$a_{s5} = 25 + \frac{1}{48}\epsilon^2 + \frac{11}{774144}\epsilon^4 - \frac{1}{147456}\epsilon^5 + \frac{37}{891813888}\epsilon^6$$

$$- \frac{7}{339738624}\epsilon^7 + \frac{63439}{201364441399296}\epsilon^8 + 9(\epsilon^9)$$

These results agree with those given in [2] and are reasonably accurate even for not too small ϵ: this was established by a numerical comparison with the exact a_n obtained for equation (9) by the Hill method.

Although Poincaré's perturbation method is simple to work with and readily amenable to high-order approximations, it proves of no use and must be replaced by some other scheme if nonstationary periodic oscillations are sought for. We may employ, for example, the asymptotic method of Krylov-Bogolybov-Mitropolskii [3, 4], well-suited for treating transient processes and stability problems.

By this method, the general solution to (1) is expanded in a series of the form

$$U = a\cos\psi + \sum_{i=1}^{\infty} u_i(a, \psi)\cdot\epsilon^i \tag{11}$$

where the functions $u_i(a, \psi)$ are 2π-periodic in ψ, and the quantities a and ψ vary in time according to

$$\begin{cases} \dot{a} = \sum_{i=1}^{\infty} A_i(a)\epsilon^i \\ \dot{\psi} = \omega_0 + \sum_{i=1}^{\infty} \psi_i(a)\epsilon^i \end{cases} \tag{12}$$

To find the unknown functions $u_1(a, \psi)$, $A_i(a)$, and $\psi_i(a)$, the same procedure as in the perturbation method is employed, that is, (11) is substituted into (1), orders are separated, and an infinite system of equations is obtained. To solve this system, that is, to obtain the functions $u_i(a, \psi)$, one starts with calculating the coefficients $A_i(a)$ and $\psi_i(a)$ using the 2π-periodicity of the functions $u_i(a, \psi)$ and the fact of there being no weight terms in them. This simple algorithm is easily realized by means of symbolic computation and yields

$$u_n(a, \psi),\ A_n(a),\ \psi_n(a) \quad (n = 1, 2, 3 \ldots)$$

thus reducing equation (1) to the simpler system (12) suitable for the analysis of nonstationary mechanical processes.

The condition $\dot{a} = 0$ makes it possible to divide out oscillations with constant amplitude and frequency. These oscillations are clearly periodic in nature and their stability analysis may easily be carried out by standard techniques.

Applying the above procedures to Van der Pol's equation (5) and calculating to the 7-th order in ϵ we obtain

$$\dot{a} = a\left(\frac{1}{2} - \frac{1}{8}a^2\right)\epsilon + a^3\left(-\frac{1}{16} + \frac{35}{1024}a^2 - \frac{37}{8192}a^4\right)\epsilon^3$$

$$+ a^5\left(-\frac{55}{16384} - \frac{453}{131072}a^2 - \frac{10757}{9437184}a^4 + \frac{4589}{37748736}a^6\right)\epsilon^5$$

$$+ a^5\left(-\frac{2277}{262144} + \frac{24959}{2097152}a^2 - \frac{3505961}{679477248}a^4 + \frac{3729109}{5435817984}a^6\right.$$

$$\left. + \frac{1465519}{28991029248}a^8 - \frac{4493323}{347892350976}a^{10}\right)\epsilon^7 + 0(\epsilon^9)$$

$$\dot{\psi} = 1 + \left(-\frac{1}{8} + \frac{1}{8}a^2 - \frac{7}{256}a^4\right)\epsilon^2 + \left(-\frac{1}{128}\right.$$

$$\left. - \frac{1}{64}a^2 + \frac{85}{4096}a^4 - \frac{211}{32786}a^6 + \frac{497}{786432}a^8\right)\epsilon^4 \tag{13}$$

$$+ \left(-\frac{1}{1024} - \frac{1}{1024}a^2 + \frac{421}{65536}a^4 - \frac{2063}{524288}a^6\right.$$

$$\left. + \frac{15179}{56623104}a^8 + \frac{34795}{150994944}a^{10} + \frac{9887}{268435456}a^{12}\right)\epsilon^6 + 0(\epsilon^8)$$

Note that the first and the second terms in these expressions are identical to those calculated by the Bogolyubov-Mirtopolskii asymptotic method [4].

We also applied symbolic computation to expand the stationary solution (13) in a series of powers of ϵ. When the frequencies of this solution were subsequently compared with (7), the coefficients of the same powers of ϵ turned out to be identical as indeed might have been expected.

By means of symbolic computation it proves possible to express the final result explicitly—or analytically—in terms of the parameters of a mechanical system, which of course offers a great deal of new possibilities for those involved in mechanical engineering design.

Computer algebra may be incorporated in the most natural way into the finite-element method, difference schemes of various kinds and into higher-order numerical integration methods, including those based on the regularization of the systems of ordinary differential equations. The important point to be made is that symbolic computation not only improves the efficiency of the numerical methods already available but also provides the basis for developing new methods capable of being optimized with respect to one or more parameters, with attention paid to the specifics of the problem. Typical optimization criteria are the accuracy of the solution, reduction of program running time or of the computer memory required for the program, etc. Symbolic computation systems are also very effective in automated programming, where they transform the symbolic expressions of the equation(s) into operators of a certain programming language and in doing so totally exclude program errors so typical of the human programmer. Some of the symbolic computation systems not only function as translators from one algorithmic language to another but can also perform a detailed analysis or, more important, an optimization of the computation program. The optimized version of the program may be superior to the original both in compactness and speed.

Theoretical and applied mechanics provide a vast application field for such a powerful mathematical tool as symbolic computation. Symbolic computation systems have already played a great role in the solution of a number of major problems in theoretical physics and in celestial mechanics [5]. They can process rapidly and without errors massive volumes of analytical information and may therefore be assigned the tedious and time-consuming routine work of analytical transformation. As an example, let us consider a well-known problem concerned with the nutation-induced (or Magnus) drift of a gimbal suspension astatic gyroscope. Of the several asymptotic techniques that have thus far been applied to the analysis of the problem [6–9], we choose the averaging method, one of the most popular and effective in nonlinear mechanics. A complex of programs based on the symbolic computation system REDUCE makes it possible to achieve any desired accuracy when applying this method to a system of nonlinear differential equations.

As a starting point of our investigation we took the nonlinear equations of motion of a mechanical system

$$F(\ddot{x}, \dot{x}, x, t, \epsilon) = 0, \quad F \in R^n, \quad x \in R^n \tag{14}$$

where ϵ is a small parameter.

Before directly applying the averaging method to these equations, it is necessary that they be first reduced to a standard form. The algorithms used in doing so were based on the approach discussed in [10].

The separation of the rapid from the slow motions in equations (14) is a problem in itself and is reduced, in fact, to the frequency analysis of the linear part of the equations. In the general case equations (14) are, however, nonlinear at $\epsilon \neq 0$ and some additional assumptions are therefore necessary. These assumptions are fairly natural for the averaging method, in which solutions in the vicinity of a certain partial solution are investigated, the partial solution being usually determined by the equilibrium configuration of the mechanical system.

We start by forming an equation to describe small deviations from a partial solution of (14). We assume that the deviations are of the order of ϵ and we retain in the right-hand sides of the equations the terms of the corresponding order of small quantities. The order of the terms retained should be equal to the corresponding number of iterations in the averaging method. From the left-hand sides of the resulting equations the operation matrix, the adjugate matrix $P(D)$ and the characteristic determinant $\Delta(D)(D = d/dt)$ are determined. The transformation converting (14) to the standard form, i.e., that changing from the initial variables to the normal ones, will be of the form

$$\frac{d^\nu x_j}{dt^\nu} = \sum_{\sigma=1}^{s'} V_{j\sigma}^{(\nu)} y_\sigma + \sum_{h=1}^{s''} N_{jh}^{(\nu)} y_{s'+h} \cos(\psi_h + \gamma_{jh} + \nu\zeta_h) \tag{15}$$

$$(j = 1, 2, \ldots, n; \quad \nu = 0, 1, 2, \ldots, m_j)$$

Here $V_\sigma^{(\nu)} = V_\sigma x_\sigma^\nu$ and V_σ is the model column corresponding to the σ-th real root of the characteristic determinant; $N_h^{(\nu)} = N_h\rho_h^\nu$, N_h and γ_h, ρ_h and ζ_h are respectively the modulus and the argument of the modal column of the i-th complex root $\epsilon_h + i\omega_h$ of the characteristic equation $\Delta(D) = 0$; s' and s'' are respectively the number of real and complex roots; y and ψ—new variables, m—the highest order of the time derivative of the j-th component of the vector x.

In most applications there are no multiple roots in the characteristic equation $\Delta(D) = 0$. The quantities N_j, N_{jh}, γ_{jh} and the modal rows W_σ and W_h may then be found in the following way.

If a root is a simple one, then at least one of the elements of the adjugate matrix P will be different from zero if the value $D = \varkappa_\sigma$ of the real root is substituted into the matrix. We may take as the modal column the one containing this element in itself and choose the modal rows from the condition

$$P(\varkappa_0) = V_\sigma W_\sigma \quad (\sigma = 1, 2, \ldots, s')$$

A similar procedure applies to the case of complex roots of the equations $\Delta(D) = 0$.

When written in the standard form, the differential equations in the new variables take the form

$$\frac{dy_\sigma}{dt} = \varkappa_\sigma y_\sigma + \left(\frac{d}{d\varkappa}\Delta(\varkappa)\bigg|_{\varkappa_\sigma}\right)^{-1} \sum_{k=1}^{n} W_{k\sigma} FP_k(x, t, \epsilon)$$

$$\frac{dy_{s'+h}}{dt} = \epsilon_h y_{s'+h} + 2\mathrm{Re}\left[\frac{e^{-i\psi_h}}{\Delta'(\epsilon_h + i\omega_h)} \sum_{k=1}^{n} W_{k,s'+h} FP_k(x, t, \epsilon)\right]$$

$$\frac{d\psi_h}{dt} = \omega_h + \frac{2}{y_{s'+h}} Im\left[\frac{e^{i\psi_h}}{\Delta'(\epsilon_h + i\omega_h)} \sum_{k=1}^{n} W_{k,s'+h} FP(x, t, \epsilon)\right]$$

$$(\sigma = 1, \ldots, s'; \quad h = 1, \ldots, s''; \quad s' + s'' = n), \quad \Delta' = \frac{d\Delta}{dD}$$

where $FP_k(x, t, \epsilon)$ is the vector function obtained from (14) by expanding in powers of ϵ in the vicinity of a partial solution.

The derivation of the transformation (15) and of the new system of equations was realized as a set of REDUCE procedures. What is left to the user in this approach is to specify the left-hand sides of equations (14), the partial solution in the neighborhood of which the expansion procedure must be carried out, and to present some information concerning the characteristic roots, namely, which of them are real and which complex; if known, the actual values of the roots should be entered into the program.

In studying the linear parts of the differential equations derived from (14), it proves convenient to print out the characteristic polynomial of the system. If a zero root occurs, a diagnostic message is printed out concerning the multiplicity of the root. Although the eigenvalues of the problem remain unknown, the initial information provided by the user concerning the eigenvalue distribution is quite sufficient to find the transformation necessary to "standardize" the system. Although it is not actually necessary to divide the eigenvalues into real and complex ones, this division greatly simplifies the study of the system, the more so that it is readily obtainable either on physical grounds or from the analysis of the characteristic polynomial.

This having been done, the computer determines a change from $(x_1, x_2, \ldots x_n)$ to $(y_1, y_2, \ldots y_{n-k}; \psi_1, \ldots \psi_k)$, applies it to equations (14), and reduces these to the standard form

$$\begin{cases} \dot{y} = \epsilon Y(y, \psi, \epsilon) \\ \dot{\psi} = \omega + \epsilon G(y, \psi, \epsilon) \end{cases} \tag{16}$$

where $y = (y_1, y_2, \ldots y_{n-k})$ are the slow variables; $\psi = (\psi_1, \ldots, \psi_k)$, the fast; $G = (G_1, \ldots, G_k)$, $\omega = (\omega_1, \ldots, \omega_k)$.

The next stage of the procedure is to directly apply the averaging method to the system of equations of the form (16). It is assumed that the standardized system of equations obtained at the first stage contains only one fast variable, so that ψ, ω, and G in (16) are all scalars, and ω = const. If we take the ordinary time t as a variable ψ, then $\omega \equiv 1$, and $G(y, \psi, \epsilon)$ is identically zero.

The derivation of averaged equations with any prescribed accuracy is best programmed by using the method of Zhuravlev [11] based on the Lie group formalism. In this method, the system (16) is associated with the differential operator

$$A(y, \psi) = \epsilon Y(y, \psi, \epsilon) \frac{\partial}{\partial y} + (\omega + \epsilon G(y, \psi, \epsilon)) \frac{\partial}{\partial \psi} \tag{17}$$

The essence of the averaging method is to find a change of variables

$$(y, \psi) \rightarrow (u, v) \tag{18}$$

which reduces the operator (17) to a simpler form. Let us assume that the required transformations form a Lie group with the infinitesimal operator

$$U(y, \psi) = \xi(y, \psi) \frac{\partial}{\partial y} + \eta(y, \psi) \frac{\partial}{\partial \psi} \tag{19}$$

It then follows from the theory of Lie groups that the new differential operator $\tilde{A}(u, v) = A(y(u, v)), (u, v)$ may be found by Hausdorff's formula, thus

$$\tilde{A}(u, v) = A(u, v) + \tau[A, U] + \frac{\tau^2}{2!}[[A, U], U] + \ldots \tag{20}$$

where τ is the parameter of the group.

Let us now suppose the unknown operators $\tilde{A}$ and U to be expanded in powers of the small parameter, viz.,

$$\tilde{A} = \tilde{A}_0 + \epsilon\tilde{A}_1 + \epsilon^2\tilde{A}_2 + \ldots \tag{21}$$
$$U = U_0 + \epsilon U_1 + \epsilon^2 U_2 + \ldots$$

Substituting in (20) and setting $\tau = \epsilon$, we obtain, after separating orders, an infinite system of the form

$$\begin{cases} \tilde{A}_0 = A_0 \\ \tilde{A}_1 = A_1 + [A_0, U_0] \\ \tilde{A}_2 = A_2 + [A_0, U_1] + [A_1, U_0] + \dfrac{1}{2!}[[A_0, U_0], U_0] \\ \cdots\cdots\cdots\cdots\cdots\cdots\cdots\cdots\cdots\cdots \end{cases} \tag{22}$$

By these equations the differential operator of the initial problem (16) is related to the operator of the system obtained by means of the transformation (18); the unknown operator U is found from the condition that the averaging procedure should eliminate the fast variable from the transformed operator A. The process is carried out successively for the equations (16), each successive approximation yielding equations allowing to find in quadratures the components $\xi_i(u, v)$ and $\eta_i(u, v)$ of the operator U_i.

The above procedure thus represents a closed algorithm allowing to determine the operators $\tilde{A}_i$, $U_{i-1}(u, v)$ in a unique manner at each stage. As a result, a simpler, averaged system is obtained which has $\tilde{A}(u)$ as a corresponding differential operator. Once this system has been investigated, a transition to the old coordinates (y, ψ) is carried out. This algorithm is programmed by the help of procedures developed in the framework of the symbolic computation system REDUCE. The input data for this part of the program are the equations of the system (16), either known prior to the study or obtained at the previous stage. The operator $A(y, \psi)$ is formed by assigning the right-hand sides of these equations to corresponding array elements. Simultaneously, an array is defined which forms the operator U for the procedure. If there is no friction in the problem, the motion of the astatic gyroscope is governed by the system of differential equations (see [16])

$$\begin{cases} [A_2 + (A + A_1)\cos^2\beta + C_1\sin^2\beta]\ddot{\alpha} + H\dot{\beta}\cos\beta \\ -\ 2(A + A_1 - C_1)\dot{\alpha}\dot{\beta}\sin\beta\cos\beta = 0 \\ (A + B_1)\ddot{\beta} - H\dot{\alpha}\cos\beta + (A + A_1 - C_1)\dot{\alpha}^2\sin\beta\cos\beta = 0 \end{cases} \tag{23}$$

where α is the angle the external gimbal makes with the base; β—the angle the external gimbal makes with the internal one; A_2—the moment of inertia of the internal gimbal; A_1, B_1, C_1—the moments of inertia of the internal gimbal; A and H = const are, respectively, the moment of inertia and the angular momentum of the gyroscope rotor. The problem is to investigate the system in the vicinity of the partial solution $\alpha = \alpha_0$, $\beta = \beta_0$, which clearly satisfies (23). The initial equations are taken to be

$$\alpha(0) = \alpha_0, \quad \beta(0) = \beta_0, \quad \dot{\alpha}(0) = \dot{\alpha}_0, \quad \dot{\beta}(0) = 0 \tag{24}$$

and we are assuming that the general solution differs only slightly from the partial solution, so $\alpha = \alpha_0 + x$ and $\beta = \beta_0 + y$.

There are two zero and two pure imaginary roots in the characteristic equation of the linear part of (23), and the linear approximation to the gyroscope motion is represented by small nutational oscillations of the gyroscope gimbals (see [6]) with a frequency

$$\nu = (H \cos \beta_0)/\sqrt{A_0 B_0}$$

where $A_0 = A_2 + (A + A_1) \cos^2 \beta_0 + C_1 \sin^2 \beta_0$; $B_0 = A + B_1$.

This makes it possible to take the quantity $\mu = \dot{\alpha}_0/\nu$ as a small parameter and to express in terms of μ the variables x and y, which we assume to be small, so that $x = \mu\xi$, $y = \mu\eta$, ξ and η being the new variables.

This information is inputed into the computer as the initial information for executing the first stage of the program. The computer expands the initial equations in a series of powers of the small parameter about the equilibrium position $\alpha = \alpha_0$, $\beta = \beta_0$. The expansion is carried out to the prescribed power of μ, the fourth in our case. The corresponding computational procedure then introduces new variables a_0, b_0, and ψ defined by

$$\begin{cases} \xi = a_0 + a \cos \psi \\ \eta = b_0 + a\sqrt{\dfrac{A_0}{b_0}} \sin \psi \end{cases} \tag{25}$$

in terms of which a system of four standard equations is derived. The initial equations (24) imposed on the variables α, β, $\dot{\alpha}$, $\dot{\beta}$ now become

$$\begin{aligned} &a_0(0) = 0, \quad b_0(0) = \frac{\dot{\alpha}_0}{\nu\mu}\sqrt{\frac{A_0}{B_0}} \\ &a(0) = -\frac{\dot{\alpha}_0}{\nu\mu}, \quad \psi(0) = \frac{\pi}{2} \end{aligned} \tag{26}$$

The system of equations for a_0, b_0, a and ψ is now processed by the second part of the program, and three approximations of the averaging method are found given the following system of averaged differential equations:

$$\dot{\tilde{a}} = 0$$

$$\dot{\tilde{a}}_0 = -\mu \frac{H\tilde{a} \sin \beta_0}{2A_0B_0}(A_2 + C_1) + \mu^2 \frac{H\tilde{a}^2\tilde{b}_0}{2A_0^2B_0 \cos \beta_0}[(A_2 + C_1)^2 \sin^2 \beta_0$$

$$- A_0(C_1 + A_2)(1 + 2 \sin^2 \beta_0)] + \mu^3\Phi_3(\tilde{a}, \tilde{b}, H, \beta_0) + 0(\mu^4)$$

$$\dot{\tilde{b}}_0 = 0 \tag{27}$$

$$\dot{\tilde{\psi}} = \nu - \mu\tilde{b}_0 tg\beta_0 \frac{A_2 + C_1}{A_0} + \mu^2 \frac{\nu}{4 \cos^2 \beta_0 A_0^2 B_0}\{\tilde{a}^2[A_0^3(2 - 5 \sin^2 \beta_0)$$

$$+ A_0^2(A_2 + C_1)(7 \sin^2 \beta_0 - 3) - 3A_0(A_2 + C_1)^2 \sin^2 \beta_0]$$

$$+ \tilde{b}_0^2 [A_0B_0(A_2 + C_1)(8 \sin^2 \beta_0 - 2)$$

$$- 4 A_0^2B_0 \sin^2 \beta_0 - 2B_0(A_2 + C_1)^2 \sin^2 \beta_0]\} + \mu^3\psi_3(\tilde{a}, \tilde{b}_0, H, \beta_0) + 0(\mu^4)$$

Expressions for Φ_3 and ψ_3 are too lengthy to be reproduced here.

It follows from the first and the second of equations (27) that $\tilde{a}$ and $\tilde{b}_0$ are constant quantities. Their values must be found from the initial conditions (26) and then substituted as constants into the equations for determining $\dot{\tilde{\psi}}$ and $\dot{\tilde{a}}_0$. The last equation will then yield a correction to the frequency ν of the gyroscope's nutational oscillations. The constant velocity $\dot{\tilde{a}}_0$ obtained from the second equation corresponds to the constant angular velocity of the gyroscopes drift relative the axis of the external gimbal. Substituting the appropriate values of $\tilde{b}_0$, $\tilde{a}$ and μ and reverting to the initial variables α, β we can obtain from the second of equations (27) the average angular velocity of the external gimbal relative the base. Calculating to the fourth order in ϵ,

$$\langle\dot{\alpha}\rangle = -\frac{\dot{\alpha}_0^2 \sin \beta_0}{2H \cos^2 \beta_0}(C_1 + A_2) + \frac{\dot{\alpha}_0^3}{2H^2 \cos^4 \beta_0}[2(A_2 + C_1)^2 \sin^2 \beta_0$$

$$- A_0(C_1 + A_2)(3 \sin^2 \beta_0 + 1)] + \frac{\dot{\alpha}_0^4}{48H^3 \cos^6 \beta_0} \tag{28}$$

$$[87 \sin^2 \beta_0(C_1 + A_2)^3 + (63 - 230 \sin^2 \beta_0)A_0(A_2 + C_1)^2$$

$$+ (197 \sin^2 \beta_0 - 156)A_0^2(A_2 + C_1) - 84 \sin^2 \beta_0 A_0^3]$$

The first term in this expression coincides with the well-known result for the Magnus drift of the gimbal suspension gyroscope [6], the remaining terms representing a higher order correction.

It was interesting to check (28) against the exact solution of the system (23) obtained earlier by Stepanenko and the present author [7]. To do this, the mass distribution of the gyroscope was assumed to satisfy the relation $A + A_1 - C_1 = 0$, implying that $A_0 = A_2 + C_1$. In this special case the exact solution is expressible in terms of elliptical integrals and the average drift velocity of the external gimbal turns out to be

$$\langle \dot{\alpha} \rangle = \frac{v_e m}{2(u_1 + u_2)} \left[4\sqrt{(1 - u_1^2)(1 - u_2^2)}\, \frac{\Pi(h^2, k)}{K(k)} - (\sqrt{1 - u_1^2} + \sqrt{1 - u_2^2})^2 \right]$$

$$v_e = \frac{H}{\sqrt{A_0 B_0}}, \quad m = \sqrt{\frac{B_0}{A_0}}, \quad u_1 = \sin \beta_0, \quad u_2 = u_1 + \frac{2\dot{\alpha}_0}{m v_e} \tag{29}$$

where $K(k)$ is the complete elliptic integral of the first kind; $\Pi(h^2, k)$—the complete elliptic of the third kind; h and k—depending on u_1 and u_2 [7].

To compare with the results obtained by the averaging method the exact solution is expanded in a series of powers of the small parameter $\dot{\alpha}_0/v_e$. Retaining terms to the fourth order in $\dot{\alpha}_0^4$ and performing some manipulation, the expression (29) becomes

$$\langle \dot{\alpha} \rangle = -\frac{\dot{\alpha}_0^2 \sin \beta_0}{2H \cos^2 \beta_0} A_0 - \frac{\dot{\alpha}_0^3 A_0^2}{2H^2 \cos^4 \beta_0} (1 + \sin^2 \beta_0)$$

$$- \frac{\dot{\alpha}_0^4 A_0^3}{16H^3 \cos^6 \beta_0} \sin \beta_0 (31 + 10 \sin^2 \beta_0)$$

which is fully identical to the expression (28) derived by the averaging method for the drift of a gimbal suspension gyroscope.

The Institute of Mechanical Problems of the Academy of Sciences has realized a number of symbolic computation systems, including universal, general-purpose, and special-purpose ones. The universal system REDUCE, the most popular among scientific researchers, is capable of performing a vast diversity of mathematical operations of considerable generality, and its range of possibilities may be easily extended by enlarging the computer software system. Of the general-purpose systems now at work at the Institute, the system FORMAK is realized on the PL/1 and the AMR on the ASSEMBLER. These two systems are superior to the REDUCE both in speed and in reduced memory requirements, and the AMR shares with the REDUCE the ability of changing from a batch to an interactive mode. The specialized system MMANG written in REFAL is intended to realize analytical procedures involved in theoretical mechanics problems and is characterized by high speed and high record compactness. The modules comprising the system enable it to perform the following procedures:

1. derivation of the Lagrangian function for a mechanical system;
2. derivation of Lagrange's equations of the second kind from the Lagrangian function;
3. linearization of equations of motion;
4. reduction of a system of equations to a standard form;
5. analysis of the asymptotic stability of stationary motions;
6. numerical integration of differential equations for given initial conditions;
7. construction of power-series approximate solutions of differential equations;
8. asymptotic motion separation in single-fast variable systems.

The structure of the MMANG system enables it to perform each procedure as a separate stage in the course of analytical manipulation, communication between the stages being the function of external direct access files. Because of this property the MMANG is readily compatible with other algorithmic languages and systems, in particular with the REDUCE.

The above symbolic computation systems operate in conjunction with large computers such as EC-1055 and EC-1055M. The library computation programs are written in FORTRAN.

A factor of considerable importance in both numerical and analytical programming is the manner in which the obtained results are presented. It is a common experience for everyone that computed numerical information is much easier to perceive when presented in the form of tables, plots, or films.[1] Symbolic computing has methods of its own by which analytical information may be represented in a convenient form. A rearrangement of the expression, well-chosen substitution or notation, collecting like terms or a lucky regrouping may sometimes throw a new light on the result obtained, reveal deep and unexpected links between the phenomena being studied and provide a new understanding of the problem. An analytical computation system can expand an expression in a series of powers of a certain parameter, determine a common multiplier or the greatest common divisor for two or more expressions, carry out identical transformations of trigonometric equations, etc.

Interactive regime of computer operation is even more important in symbolic than in numerical computation, allowing as it does to intervene "on-line" in the computation process and to monitor and correct it if need be by means of the interactive graphics system. On the one hand, this mode of computer operation makes it possible to see at an intermediate stage of computation whether the solution proceeds as it should; on the other, the explicit dependence of an intermediate result on one or the other parameter of the system reveals sometimes hardly predictable aspects of the problem which may suggest more effective methods for the subsequent stages of the computation or a more exact specification of the parameters and inputs of the problem. As a whole, an indisputable advantage that analytical representation of results has over their numerical representation is the possibility to analyze the results qualitatively and to optimize and specify at will the parameters involved if the mechanical system being investigated must be given a certain property or a set of properties, such as amplitudes, eigenfrequencies, overloads, etc.

The computation scenario we have thus outlined integrates in a most natural manner analytical and numerical machine computing and has as its ultimate goal the complete automation of the research and development process except perhaps for the initial stage, at which the physical and mathematical aspects of the problem are formulated and appropriate algorithms decided on. We are not unaware, how-

1) The computer graphics system developed at the Institute of Mechanical Problems enables the visual examination of a number of thermal and hydrodynamic processes. A special cine-camera may be employed whenever a color film is required to demonstrate the evolution of a process.

ever, of the difficulties which are encountered in translating this happy picture in practice and which arise from the limited resources and low reliability characteristics of domestic computer technology. The interactive regime we described above is perhaps one possible way to bypass these difficulties.

REFERENCES

1. Hayashi, Ch. Forced oscillations in nonlinear systems. Osaka, Nippon Print. Company, 1953.
2. McLachlan, N. W. Theory and applications of mathieu functions. Oxford, 1947.
3. Nayfeh, A. H. Perturbation methods. Wiley, N.Y., 1973.
4. Bogolyubov, N. N. and Mitrololskii, Yu. A. Asymptotic methods in the theory of nonlinear oscillations. Moscow, Fizmatgiz, 1958.
5. Analytical computing and its application in theoretical physics. In: Proc. of intern. conf. on anal. comp. and its appl. in theor. phys., Dubna, 1985.
6. Klimov, D. M. and Kharlamov, S. A. Dynamics of a gimbal suspended gyroscope. Moscow, Nauka, 1978 (in Russian).
7. Klimov, D. M. and Stepanenko, N. P. On the integration of equations of motion of the gimbal suspended gyroscope. Izv. USSR Acad. Sci. MTT, 1967, No. 6.
8. Bruno, A. D. On the motions of a gimbal suspended gyroscope. Ibid., 1972, No. 6.
9. Valeev, K. G. On the precession of a gimbal suspended symmetric gyroscope. In: Rigid body mechanics. Rspubl. Intersect. Proc., Kiev, Naukova Dumka, 1974, No. 7.
10. Bulgakov, V. V. Vibrations. Moscow, Gostekhteorizdat, 1954 (in Russian).
11. Zhuravlev, V. F. Lie series method and the motion separation in nonlinear mechanics. PMM., 1983, Vol. 47, No. 4.

INVERSE PROBLEMS IN DYNAMICS OF CONTROLLED SYSTEMS

M. I. Gusev, A. B. Kurzhanskii

Inverse problems arising in control theory concern the restoration of the inputs of a system from the measured values of its observable coordinates. One possible way to attack these problems is by employing the concept of guaranteed estimation [1, 2], closely related to methods known in the theory of ill-posed problems [3, 4].

1. Formulation. Suppose the motion of a controlled system is described by the equation

$$\dot{x} = f(t, x, u(t))$$
$$x(t_0) = x^0, \quad x \in \mathbf{R}^n, \quad u \in \mathbf{R}^p, \quad t_0 \leq t \leq t_1 \tag{1.1}$$

where $u(t)$ is the input action, or control. If $f(t, x, u)$ is continuous with respect to its variables and satisfies the standard "growth conditions"

$$\|f(t, x, u)\| \leq c_1(1 + \|x\| + \|u\|)$$

($\|\cdot\|$ denoting the Euclidean norm) and the Lipschitz condition

$$\|f(t, x', u) - f(t, x'', u)\| \leq c_2\|x' - x''\|(1 + \|u\|)$$

with respect to x (c_1 and c_2 being known constants), it is known that for any pair $\{x^0, u(\cdot)\}$, where $x^0 \in \mathbf{R}^n$, $u(\cdot) \in L_2^p[t_0, t_1]$, system (1.1) has a unique solution continuable onto the interval $[t_0, t_1]$.

We will assume that the pair $\{x^0, u(\cdot)\}$, or the "input" of the system (1.1), is unknown and that information about this pair is contained in the observation equation

$$y(t) = g(t, x(t)) + \xi(t), \quad y \in \mathbf{R}^m \tag{1.2}$$

where $y(t)$ represents the results of measurements, alias the "output" of the system (1.1), (1.2); $\xi(t)$ represents an error in the observation function and is not assumed to be known *a priori* ($\xi(\cdot) \in L_2^m[t_0, t_1]$). The function $g(t, x)$ will be assumed continuous with respect to the set of its variables. (It is possible also to consider functions $g(t, s, u)$, in which case continuity with respect to t, x, u is assumed.)

In the analysis presented here it will be convenient to suppose that all the information available about the unknowns $\{x^0, u(\cdot), \xi(\cdot)\}$ is expressed by the restriction

$$w = \{x^0, u(\cdot), \xi(\cdot)\}, \quad w \in W \tag{1.3}$$

with a set W known *a priori*.

A typical inverse problem is that of refinement in which the pair $\{x^0, u(\cdot)\}$ is restored from the known function $y(t)$, $t_0 \leqslant t \leqslant t_1$. It should be noted that there is no unique solution to this problem.

Clearly the knowledge of $\{x, u(\cdot)\}$ enables one to construct a corresponding path $x(\cdot, t_0, x^0, u(\cdot)) = x[\cdot]$ for the equation (1.1). There exists an alternative formulation, however, in which instead of seeking for the pair $\{x^0, u(\cdot)\}$, one requires directly the path $x(\cdot, t_0, x^0, u(\cdot))$ compatible with $y(\cdot)$ in view of (1.1)–(1.3). Retrieving $u(\cdot)$ from $x(\cdot, t_0, x^0, u(\cdot))$ then represents a separate problem, which may be treated by standard techniques in some cases (if, for example, the function $f(t, x, u)$ is linear in u).

The above discussion suggests the following formulations for the problem we are considering (we denote by $x(\cdot, t_0, x^0, u(\cdot))$; the solution of (1.1) to which the system evolves from the state x^0 it was in at the initial instant t_0).

Problem 1. Given the function $y(t)$ defined by the system (1.1)–(1.3);

a) to retrieve at least one pair $\{x^0; u(\cdot)\}$ (a path $x(\cdot, t_0, x^0, u(\cdot))$) compatible with equations (1.1)–(1.3) for a given $y(\cdot)$;

b) under the same conditions, and assuming that the set of permissible $y(\cdot)$ obeys the metric $L_2^m[t_0, t_1]$ induced by the class of possible errors $\xi(\cdot)$, to restore a stable (extremal) pair $\{x, u(\cdot)\} \in \mathbf{R}^n \times L_2^p[t_0, t_1]$ (a path $x(\cdot, t_0, x^0, u(\cdot))$) continuous with respect to $y(\cdot)$;

c) to find all the pairs $\{x^0, u(\cdot)\}$ (paths $x(\cdot, t_0, x^0, u(\cdot))$) compatible with the given $y(\cdot)$ and equations (1.1)–(1.3).

A feature of this problem is that the results of measurement are considered to be known all over the interval $[t_0, t_1]$. The inputs to be found are those capable of generating the given $y(\cdot)$. (The reader will note a close similarity of this formulation to the problem of determining a program motion under uncertainty conditions.) In an alternative formulation we are led to the guaranteed filtering problem, that is, to that of evaluating inputs $\{x^0, u(\cdot)\}$ from the measurements known only as of the instantaneous time t. Let $f_t(\cdot) = f(t_0 + \tau)$, $t_0 - t \leqslant \tau \leqslant 0$ denotes the element corresponding to the function $f(\cdot)$ defined on the interval $[t_0, t]$.

Problem 2. Given the function $y_t(\cdot)$ defined by the system (1.1)–(1.3) on the interval $[t_0, t]$;

a) to reconstruct at least one pair $\{x^0, u_t(\cdot)\}$ (a path $x_t(\cdot, t_0, x^0, u_t(\cdot))$) compatible with $y_t(\cdot)$ in view of equations (1.1)–(1.3);

b) under the same conditions to reconstruct a stable (extremal) pair $\{x^0, u_t(\cdot)\}$ (a path $x_t(\cdot, t_0, x^0, u(\cdot))$) depending on $y_t(\cdot)$ in a continuous manner;

c) under the same conditions to find the set $X_t(\cdot, t_0, X_0) = U\{x(\cdot)\}$ of all the paths (all the states $X_t(t, t_0, X_0) = X(t, t_0, X_0)$) compatible with $y_t(\cdot)$ in view of (1.1)–(1.3). From the $X_t(\cdot, t_0, X_0)$ thus obtained to find all the position controls $u_t = u_t(y_t(\cdot), t_0, X_0)$ possessing the property that every one of them, when combined with a certain $x^0 \in X_0$, generates an element $x_t(\cdot) \in X_t(\cdot, t_0, X_0)$.

Note that even when $m = n$ and there is no noise $\xi(\cdot)$ in equation (1.2) (so that $y(t) \equiv x(t)$), the required control u must generally be of the form $u = u(t, x(\cdot))$ rather than $u(t, x)$.

2. Refinement problem. The notation we adopt in this section is as follows. Let $w = (x^0, u(\cdot))$, and let the *a priori* restrictions on w be specified by the condition $w \in W$. Let A denote an input-output operator which associates with the pair $w = (x^0, u(\cdot))$, a vector function $y(\cdot)$ according to

$$y(t) = g(t, x(t), u(t)) \quad pt \in [t_0, t_1] \tag{2.1}$$

where $x(t, x^0, u(\cdot))$ solves the system (1.1). We assume that the operator A is defined for all $w \in X = \mathbf{R}^n \times L_2^p$ and that $y(\cdot) - Aw \in L_2^m$. Given the measurements $y(\cdot)$, it is required to determine the quantity $z = Fw$, where F is a (generally nonlinear) operator acting from X into the Hilbert space Z. Operators F involved in the refinement problem may be of the form $Fw = w$, $Fw = u(\cdot)$. If $Fw = x^0$ and $Fw = x(t_1, x^0, u(\cdot))$, a system status evaluation problem arises.

We thus arrive at the following abstract formulation of the problem. We are given real Hilbert spaces X, Y, Z; operators $A : X \to Y$, $F : X \to Z$: a set $W \subset X$; an element $y \in Y$. It is required to find a set (an information region) $Z(y)$ defined by the equation

$$Z(y) = \{z : z = Fw, Aw = y, w \in W\}$$

An interesting point about this problem is the role of the instability of $Z(y)$ towards errors in the finition of y. For y defined with an error ($\|y - y_\delta\| \leqslant \delta$), the region $Z(y_\delta)$ may differ as much as we please from $Z(y)$ at small δ. To find the solution it is therefore necessary to combine methods known from the theory of ill-posed problems [3, 4] with those developed in the theory of observation under uncertainty conditions [1, 2].

Let W be compact in a weak topology, w^* be an arbitrary element of X, and $\mu = \sup\{\|w - w^*\| : w \in W\}$. We will call a "regularization of $Z(y)$" the extension

$$Z_{\epsilon,\alpha}(y) = \Big\{ z = Fw : Aw + \xi = y \tag{2.2}$$

$$\langle w - w^*, w - w^* \rangle_X + \frac{1}{\epsilon} \langle \xi, \xi \rangle_Y \leqslant \mu^2 + \alpha,\ w \in W \Big\}$$

of the region $Z(y)$. Here $\epsilon > 0$ and $\alpha \geqslant 0$ are the parameters of the regulatization, and $\langle \cdot, \cdot \rangle_X$, $\langle \cdot, \cdot \rangle_Y$ are scalar products in X and Y respectively.

If W is a Hilbert sphere $W = \{w : \|w - w\| \leqslant \mu\}$, we may take the set

$$Z_{\epsilon,\alpha}(y) = \Big\{ z = Fw : Aw + \xi = y \langle w - w^*, w - w^* \rangle_X + \frac{1}{\epsilon} \langle \xi, \xi \rangle_Y \leqslant \mu^2 + \alpha \Big\} \tag{2.3}$$

as an extension $Z_{\epsilon,\alpha}(y)$.

Statement 2.1. a) $Z(y) \subset Z_{\epsilon,\alpha}(y_\delta)$ for $\delta^2/\epsilon \leqslant \alpha$, b) if the operators A, F are weakly closed and F is perfectly continuous, then $d(Z_{\epsilon,\alpha}(y_\delta), Z(y)) \to 0$ for $\epsilon \to 0$, $\alpha \to 0$, $\delta^2/\epsilon \leqslant \alpha$, where d is the Hausdorff metric.

Lemma 2.1. Let the system (1.1), (2.1) be linear in u, that is

$$\dot{x} = A(t, x) + B(t, x)u(t), \quad x(t_0) = x^0 \tag{2.4}$$

$$y(t) = g_1(t, x(t)) + g_2(t_1 x(t))u(t) \tag{2.5}$$

where $A(t, x)$ and $B(t, x)$ satisfy the Lipschitz condition with respect to x, and the conditions

$$\|A(t, x)\| \leqslant c_1(1 + \|x\|), \quad \|B(t, x)\| \leqslant c_2$$

where g_1 and g_2 are continuous. Then the input-output operator $A : \mathbf{R}^n \times L_2^p \to L_2^m$ is weakly closed.

The operator F is weakly closed and perfectly continuous if, for example,

$$Fw = x^0$$

or

$Fw = P_L u(\cdot)$, where the operator P projects onto a finite-dimensional subspace,

or

$Fw = x(t_1, x^0, u(\cdot))$ and system (1.1) has the form of (2.4).

Unfortunately, a typical situation in input evaluation problems ($Fw = u(\cdot)$, $Fw = w$) is that regularized regions obeying Hausdorff metric do not generally converge to $Z(y)$ and only the convergence of finite-dimensional projections, that is, a somewhat weaker statement, can be proved.

Statement 2.2. Let A be weakly closed and F be a linear continuous operator. Then for any $z^* \in Z$

$$\rho(z^*|Z_{\epsilon,\alpha}(y_\delta)) \to \rho(z^*|Z(y))$$

for $\epsilon \to 0$, $\alpha \to 0$, $\delta^2/\epsilon \leq \alpha$ ($\rho(\cdot|Z)$ being the support function for Z).

In particular, if $Fw = w$, $Fw = u(\cdot)$, a strong convergence to the set of exact solutions is only possible for the stable solutions resulting from Tikhonov's regularization method. Let

$$\hat{w}(\epsilon, y_\delta, w^*) = \arg\min\{\|Aw - y_\delta\|^2 + \epsilon\|w - w^*\|^2, w \in W\} \tag{2.6}$$

Theorem 2.1. Let the system (1.1), (2.1) be of the form (2.1), (2.5) and satisfy the conditions of Lemma 1; let W be weakly closed in $\mathbf{R}^n \times L_2^p$. Then a solution $\hat{w}(\epsilon, y_\delta, w^*)$ to the problem (2.6) exists for any $\epsilon > 0$, $w^* \in X$, and if $\epsilon \to 0$, $\delta^2/\epsilon \to 0$, then

$$\hat{w}(\epsilon, y_\delta, w^*) \to \{w : Aw = y, w \in W\}$$

in the metric of the space $\mathbf{R}^n \times L_2^p$.

Assuming that the operator F is uniformly continuous on any bounded set, it may be asserted that regularized evaluations $\hat{z} = F\hat{w}$ converge for $\epsilon \to 0$, $\delta \to 0$, $\delta^2/\epsilon \to 0$ to the region $Z(y)$ in metric Z.

It will be noted that (2.6) is a nonlinear variational problem with constraints. In the case when the phase vector of the system, $x(t)$, is wholly accessible to measurement and when measurement errors are bounded in a uniform metric, it has been shown [5] that the construction of regularized estimations for the system (2.4), (2.5) may be reduced to a linearly-quadratic variational problem. A similar situation is found in the parameter identification problem [6]. Note that a convergence of $Z_{\epsilon,\alpha}(y_\delta)$ to $Z(y)$ in the Hausdorff metric may also be achieved in the refinement problem if certain *a priori* restrictions on the system input are imposed [5].

3. Guaranteed filtering problem. In this section we are concerned with reconstructing the path of a system from the measurement of some of its coordinates [7, 8]. We assume that *a priori* restrictions imposed on x^0 and $u(\cdot)$ are of the form $x^0 \in X$, $u(t) \in P(t)$ for almost all $t \in [t_0, t]$, where X_0, $P(t)$ are compact sets, and the multivalued mapping $P(t)$ is continuous. The measurement equation (1.2) is

$$y(t) = G(t)x(t) + \xi(t) \tag{3.1}$$

where $G(t)$ is a continuous $m \times n$ matrix-function ($m \leq n$), and the restrictions on measurement noise, $\xi(t)$, is specified by the inclusion

$$\xi(t) \in \Xi(t), \quad t \in [t_0, t_1] \tag{3.2}$$

where the continuous multivalued mapping $\Xi(t)$ is a convex compact for any $t \in [t_0, t_1]$.

Setting

$$Y(t) = \{x \in \mathbf{R}^n : G(t)x \in y(t) - \Xi(t)\} \tag{3.3}$$

it is clear that the condition (3.1) is equivalent to the relation

$$x(t) \in Y(t), \quad t \in [t_0, t_1] \tag{3.4}$$

We will assume that the "measured signal" $y(t)$ is continuous and hence so is the multivalued mapping

$$Q(t) = y(t) - \Xi(t)$$

Let us replace (1.1) by the equivalent differential inclusion

$$\dot{x} = F(t, x) = U\{f(t, x, u) : u \in P(t)\} \tag{3.5}$$

Let us fix $\tau \in (t_0, t_1]$ and denote by $X(\cdot, \tau, t_0, X_0)$ the bundle of those paths of the inclusion (3.5) which start at the instant t_0 from the set X_0 and satisfy the condition (3.4) for $t_0 \leqslant t \leqslant \tau$; let $X(\tau, t_0, X_0)$ be the section of the bundle at the time τ.

We denote by $M_{n,m}$ and $M_{n,m}[t_0, \tau]$ respectively the space of $n \times m$ matrices L and that of continuous matrix functions $L(t)$ defined on $[t_0, \tau]$.

Consider the differential inclusion

$$\dot{z} \in F(t, z) - LG(t)z + LQ(t), \quad z \in \mathbf{R}^n, \quad L \in M_{n,m} \tag{3.6}$$

Let $Z(\cdot, \tau, t_0, X_0, L)$ and $Z(\cdot, \tau, t_0, X_0, L(\cdot))$ be the bundles of path of the inclusion (3.6) for $L \in M_{n,m}$, $L(\cdot) \in M_{n,m}[t_0, \tau]$. And let

$$Z(\tau, t_0, X_0, L), \quad (Z(\tau, t_0, X_0, L(\cdot))$$

be the respective sections of these bundles at the time τ.

Theorem 3.1 [8]. The equations

$$X(\cdot, \tau, t_0, X_0) = \cap \{Z(\cdot, \tau, t_0, X_0, L) | L \in M_{n,m}\}$$

$$= \cap \{Z(\cdot, \tau, t_0, X_0, L(\cdot)) | L(\cdot) \in M_{n,m}[t_0, \tau)\}$$

and the inclusions

$$X(\tau, t_0, X_0) \subset \cap \{Z(\tau, t_0, X_0, L(\cdot) | L(\cdot) \in M_{n,m}[t_0, \tau)\}$$

$$\subset \{Z(\tau, t_0, X_0, L) | L \in M_{n,m}\}$$

are valid. If $m = n$ and $G(t) = E$, E being a unitary matrix, we have the following

Theorem 3.2 [8]. Let a mapping $F(t, x)$ be of the form

$$F(t, x) = A(t)x + P(t)$$

where $A(\cdot) \in M_{n,m}[t_0, \tau]$, $P(t)$ is a continuous multivalued mapping with convex compact values.

Then

$$X(\tau, t_0, X_0) = \cap \{Z(\tau, t_0, X_0, L(\cdot)) | L(\cdot) \in M_{r,n}[t_0, \tau)\}$$

for any $\tau \in (t_0, t_1]$.

Since path bundles of the inclusion (3.6) may be constructed independently for different L, it follows from the above theorems that it is possible in principle to treat this problem by parallel computing techniques.

4. Linear systems. We return to the problem of Section 2 and assume that the operators A and F are linear and continuous and that $W = \{w\colon \langle w - w^*, w - w^*\rangle_X \leq \mu^2\}$.

In this case $Z_{\epsilon,\alpha}(y_\delta)$ as defined by (2.3) admits of an analytical description provided by the following

Theorem 4.1 [5]. The region $Z_{\epsilon,\alpha}(y_\delta)$ is an ellipsoidal set with a supporting function ($z^* \in Z$)

$$\rho(z^*|Z_{\epsilon,\alpha}(y_\delta)) = \langle z^*, \hat{z}\rangle + (\mu^2 + \alpha - \langle y - Aw^*, (AA^* + \epsilon E)^{-1}(y - Aw^*)\rangle_Y)^{1/2}(\langle z^*, FTF^*z^*\rangle_Z)^{1/2} \tag{4.1}$$

where

$$\hat{z} = F\hat{w}(\epsilon, \delta, w^*), \quad \hat{w}(\epsilon, y_\delta, w^*) = A^*(AA^* + \epsilon E)_y^{-1} + Tw^*,$$

$$T = \epsilon(A^*A + \epsilon E)^{-1} = E - A^*(AA^* + \epsilon E)^{-1}A$$

E is the identical operator in the corresponding space; and the element $\hat{w}(\epsilon, y_\delta, w^*)$ and the operators involved in (4.1) may be found by solving the following variational problems:

$$\hat{w}(\epsilon, y_\delta, w^*) = Sw^* = \arg\min_w\{\|Aw - y\|^2 + \epsilon\|w - w^*\|^2\} \tag{4.2}$$

$$Tu = \arg\min_w\{\|Aw\|^2 + \epsilon\|w - u\|^2\}(AA^* + \epsilon E)^{-1}x = \frac{1}{\epsilon}(x - A\bar{w}) \tag{4.3}$$

$$\bar{w} = \arg\min\{\|Aw - x\|^2 + \epsilon\|w\|^2\} \tag{4.4}$$

The element $\hat{z}$, or the center of symmetry of the set $Z_{\epsilon,\alpha}(y_\delta)$, is, by the minimax criterion, the best possible estimation of the parameter z [?]. On the other hand, $\hat{z}$ is the image obtained by the action of the mapping F on the Tikhonov regularizator. The problem of describing the set $Z_{\epsilon,\alpha}(y_\delta)$ by theorem 4.1 reduces to the solution of standard linear-quadratic variational problems.

It may be helpful to note that the relations above apply not only to systems governed by ordinary differential equations but to distributed systems as well.

To illustrate the use of the results discussed, let us consider the inverse problem concerned with determining a temperature distribution from measured parameters of a thermal field. Let a temperature field in a region $\Omega \subset \mathbf{R}^n$ be described by the mixed problem

$$\frac{\partial u}{\partial t} = Lu, \quad u(x, t)|_{\partial\Omega} = 0, \quad u(x, 0) = u_0(x) \in L_2(\Omega), Lu \tag{4.5}$$

where Lu is an elliptic operator.

A priori restrictions on the (unknown) initial temperature distribution are given in the form of an integral inequality in $L_2(\Omega)$

$$\int_\Omega u_0(x)M(x)u_0(x)\,dx \leq \mu^2$$

where the function $M(x) > 0$. Given the results of the measurement of the "signal"

$$y(t) = \int_\Omega \varkappa(x, t)u(x, t)\,dx, \quad t_0 \leqslant t \leqslant t_1 \tag{4.6}$$

it is required to estimate the initial temperature distribution from $u_0(x)$ or the temperature distribution $u(x, t_1)$ at the time t_1. (For a more detailed formulation of the problem, see [9].)

The best possible minimax estimation for the initial state $u_0(x)$ is found from the variational problem

$$\int_{t_0}^{t_1} \left| y_\delta(t) - \int_\Omega \int_\Omega \varkappa(x, t)U(t, x, y)u_0(x)\,dx\,dy \right|^2 dt + \epsilon \int_\Omega M(x)u_0^2(x)\,dx \rightarrow \min$$

where U is the Green function of the problem, easily obtainable by familiar techniques [10].

To find the projections of the set $Z_{\epsilon,\alpha}(y_\delta)$ on the direction $z^*(\cdot) \subset L_2(\Omega)$, it is necessary to calculate the quantity Tz^*. This is done by solving the problem

$$\int_{t_0}^{t_1} \left| \int_\Omega \int_\Omega \varkappa(x, t)U(t, x, y)\, u_0(x)\,dx\,dy \right|^2 dt + \epsilon \int_\Omega M(x) \left| u_0(x) - z^*(x) \right|^2 dx \rightarrow \min$$

The minimax estimation for $u(x, t_1)$ is that of the solutions to (4.5) which corresponds to the initial distribution $u_0(x)$. The full description of the region $Z_{\epsilon,\alpha}(y_\delta)$ is given in this case by the minimax filtering equations [9].

5. Example. It is required to evaluate the linear functional $\langle g_0, w\rangle$ in a Hilbert space X from the known values of the n other functionals $\langle g_i, w\rangle = y_i$, $i = 1, \ldots, n$; *a priori* restrictions on w being of the form $\langle w, w\rangle \leqslant l^2$ [11]. This problem is of course obtained as a special case of that discussed in Section 2 if the operators $F : X \rightarrow \mathbf{R}$ and $A : X \rightarrow \mathbf{R}$ are defined by the equation

$$Fw = \langle g_0, w\rangle, \quad Aw = \mathrm{col}\{\langle g_1, w\rangle, \ldots, \langle g_n, w\rangle\}$$

Clearly $Z(y)$ is a closed interval on the number line. The interval $Z_{\epsilon,0}(y)$ includes $Z(y)$ and converges to it as $\epsilon \rightarrow 0$ in Hausdorff's metric in view of statement 2.1 (it being assumed that y is known exactly, so that $\delta = 0$ and hence α may also be set to zero).

Formulas for the middle of the line segment $Z_{\epsilon,0}(y)$ and for its extremities are obtained from theorem 4.1 (at $w^* = 0$). Obviously, $AA^* = K = \|k_{ij}\|$ is the Gram matrix of the vectors g_i, $k_{ij} = \langle g_i g_j\rangle$, $i, j = 1, \ldots, n$. For $x \in \mathbf{R}^n$ we have $FA^*x = k_0^T x$, where $k_0 = \mathrm{col}(\langle g_0, g_1\rangle, \ldots, \langle g_0, g_n\rangle)$. The center of the segment is given by

$$\hat{z}(\epsilon) = k_0^T (K + \epsilon E)^{-1} y \tag{5.1}$$

whereas the extremities of the segment $Z_{\epsilon,0}(y)$ are defined as $z(\epsilon) \pm R(\epsilon)$, where $R(\epsilon) = \rho(1|Z_{\epsilon,0}(y)) - \hat{z}(\epsilon)$. Setting $w^* = 0$ and noting that $F^*1 = g_0$, one finds

$$R(\epsilon) = (l^2 - y^T(K + \epsilon E)_y^{-1})^{1/2} \times (\langle g_0, g_0\rangle - k_0^T(K + \epsilon E)^{-1}k_0)^{1/2} \tag{5.2}$$

from (4.1), showing that $Z(y) = [\bar{z} - \bar{R}, \bar{z} + \bar{R}]$, where $\bar{z} = \lim_{\epsilon \to 0} \hat{z}(\epsilon)$, $R = \lim_{\epsilon \to 0} R(\epsilon)$. This is in full agreement with the result derived by Belov and Kasyanyuk [11]. The additional information we have obtained here is that $Z_{\epsilon,0}(y)$ approximates $Z(y)$ from above.

There has been extensive work on the inverse problems involved in dynamics of mechanical systems (see, for example, the bibliography given in [12], which contains references to all the available literature on the subject as of the time of publication). In the context of controlled systems, the particular features of such problems are errors in parameter measurements and incomplete information about the phase coordinates of the system being studied.

Inverse problems in dynamics of controlled systems are also a subject of considerable current interest. In connection with the input estimation problem, mention should be made of the papers by Kryazhimskii and Osipov [13] and by Kirin [14]. Much attention has been given to the invertibility of the input-output operator and to the related property of perfect observability [15–17].

The principles discussed in this paper offer a wide variety of technical and scientific applications. To take but one example, on-board records of aircraft motion parameters may be used for reconstructing the cause(s) of a flight accident [18].

REFERENCES

1. Krasovskii, N. N. Theory of motion control. Moscow, Nauka, 1968 (in Russian).
2. Kurzhanskii, A. B. Observation and control under uncertainty conditions. Moscow, Nauka, 1977 (in Russian).
3. Tikhonov, A. N. and Arsenin, V. Ya. Method for treating ill-posed problems. Moscow, Nauka, 1974 (in Russian).
4. Ivanov, V. K., Vasin, V. V., and Tanana, V. P. Linear ill-posed problems, theory and its applications. Moscow, Nauka, 1978 (in Russian).
5. Gusev, M. I., On the class of inverse problems in control dynamics. Proc. Intern. Conf. Stochast. Optimization, Springer, 1986.
6. Sonnevend, Gy. A new stable identification scheme and its error estimation for nonlinear systems without inputs. Proceed. IFAC Congr. Budapest, 1984, Vol. 10.
7. Kurzhanskii, A. B. On the analytical description of a bundle of survivable trajectories of a differential system. Dokl. USSR Acad. Sci., 1986, Vol. 287, No. 5, pp. 1047–1050.
8. Kurzhanskii, A. B. and Filippova, T. F. On the description of the set of survivable trajectories of differential inclusions. Ibid., 1988, Vol. 289, No. 1, pp. 38–41.
9. Kurzhanskii, A. B. and Khapalov, A. Yu. On the estimation of distributed fields from observation results. In: Partial differential equations. Novosibirsk, Nauka, 1986 (in Russian).
10. Lions, J. L. Contrôl optimal de systèmes gouverneés par des equations eux deriveés partielles. Paris, Gauthier-Villars, 1968.
11. Belov, Yu. A. and Kasyanyuk, V. S. On the optimal estimation of a linear functional from the values of n other linear functionals in a Hilbert space. Dokl. USSR Acad. Sci., 1986, Vol. 289, No. 5, pp. 1033–1035.
12. Galiullin, A. S. Inverse problems in dynamics. Moscow, Nauka, 1981 (in Russian).
13. Kryazhimskii, A. V. and Osipov, Yu. S. On positional modeling in dynamic system control. Izv. USSR Acad. Sci. TK, 1983, No. 2, pp. 51–60.

14. Kirin, N. E. Successive estimation method in control optimization problems. Leningrad St. Univ., 1975 (in Russian).
15. Nikolskii, M. S. On perfectly observable systems. Dif. Equ., 1971, Vol. 7, No. 4, pp. 631–638.
16. Nijmeijer, H. Invertibility of affine nonlinear control systems: a geometric approach. Syst. Contr. Lett., 1982, Vol. 2, No. 3, pp. 163–186.
17. Hirshorn, R. M. Invertibility of nonlinear control systems. SIAM J., Contr. and Optim., 1979, Vol. 17, pp. 289–297.
18. Ditenberger, M. A., Haines, P. A., and Luert, J. K. Reconstruction of PanAm New Orleans accident. ARS Pap., 1983.

CONTROLLED MODELS AND INVERSE DYNAMICS PROBLEMS

A. V. Kryazhimskii, Yu. S. Osipov

An idea now widely used in the field of control is that an auxiliary controlled model may be introduced into the control loop of a real dynamic system (see, for example, monographs [1, 2] on positional control or [3] on adaptive control). As far as position control problems are concerned, the main attraction of this method is that it is stable towards low-scale measurement noises [1, 2] and permits a certain regularization technique for positional control strategies (in the sense of ill-posed problems [4, 5]).

In this paper the model-assisted control method will be employed for finding the stable solutions of the vast class of inverse problems of controlled system dynamics. To solve such a problem means, given the current state of a dynamic system, restoring the nature of the input influences governing the evolution of the system [6]. The important point to be made is that the approach we suggest here depends on real-time computation for its success and is therefore advantageous from the application point of view in that the required variables can be restored directly in the course of—rather than after—the motion, and may therefore be used either for making decisions as regards the control strategy or for the real-time modeling of the trajectories of unmeasurable variables.

The subject matter of this paper is closely related to that of references [7] to [10].

1. RECONSTRUCTION OF THE INPUT INFLUENCE

1.1. Formulation and the general scheme of solution. Suppose a dynamic system performs a motion of some kind during a given bounded time interval $t \leqslant t \leqslant \vartheta$ and that the state of the system is characterized by an n-dimensional vector x. The motion of the system is determined by a certain time-varying r-dimensional vector v which will be referred to as the input influence and may be thought of as, for example, an external force acting on the system, or a certain characteristic either of the system itself or of the medium in which the system is moving. The motion of the system is assumed to be governed by an ordinary differential equation

$$\dot{x} = f(t, x, v) \tag{1.1}$$

Let $v(t)$ denote the actual input influence (assumed to be *a priori* unknown) and $x(t)$ the corresponding actual motion of the system, that is, the time dependence of its state. The problem is, in synchronism with the motion of the system, to reconstruct the input influence from the measured $x(t)$. All the information available concerning the function $v(t)$ is that it is measurable in the Lebesgue sense and that at any time t its value lies in the set Q assumed to be bounded and closed in an r-dimensional Euclidean space. In practice the states $x(t)$ cannot be generally measured with any degree of accuracy, so that at any moment the actual result of measurement, $\xi(t)$, may only be claimed to satisfy the inequality

$$|\xi(t) - x(t)| \leqslant h \tag{1.2}$$

where the symbol $|\cdot|$ will denote a Euclidean norm in the present discussion (but might denote a different norm in the general case); the magnitude h of measurement error is assumed to be small. Generally speaking, this inaccuracy makes it impossible to restore exactly the function $v(t)$, and the goal we will actually have in mind here is therefore to construct a certain approximation $u(t)$ which may be arbitrarily close to $v(t)$ if h is sufficiently small. Exactly how "close" will be evaluated by means of the functional metric chosen. Although uniform proximity would of course be most preferential, it can be easily shown by simple examples that, because of the possible discontinuity of $v(t)$, this property is generally unachievable under the circumstances being assumed. We therefore limit ourselves to a weaker functional metric and take the mean-square deviation

$$\|u - v\| = \left(\int_{t_0}^{\vartheta} |u(t) - v(t)|^2 dt \right)^{1/2}$$

as a measure of proximity of the functions $u(t)$ and $v(t)$. We will also restrict our attention to those algorithms in obtaining the approximating function $u(t)$ which can be carried out in the real time, which means, in more precise terms, that for any given time t, an algorithm should produce the value of $u(t)$ at the time t at the latest. Let us assume that the initial state of the system

$$x(t_0) = x_0 \tag{1.3}$$

is known beforehand and is fixed.

The essence of our solution procedure is as follows. To the real dynamic system (1.1) is associated an auxiliary controlled model whose equation will be written in the form

$$\dot{w} = \bar{f}(t, \xi(t), w, u) \tag{1.4}$$

The vector w characterizes the state of the system and is assumed, for the sake of simplicity, to have the same dimensionality as the vector x; the function f is specified in the beginning of the procedure and will be corrected later on; the r-dimensional vector u plays the role of control action; the initial state $w(t_0)$ of the system is assumed to be equal to x. During the interval $t_0 \leqslant t \leqslant \vartheta$, we will control the model by employing the feedback principle and taking into account the current measurements of $\xi(t)$ (we will use the discrete control approach, see, e.g., [1, 2]). Let us choose a *a priori* a finite number of instants of time τ_i

$$t_0 = \tau_0 < \cdots < \tau_m = \vartheta, \quad \tau_{i+1} - \tau_i \leqslant \delta \tag{1.5}$$

During the interval $\tau_i < t \pm \tau_{i+1}$, we will apply to the model the control action

$$u_i = U(\tau_i, \xi(\tau_i), \quad w(\tau_i)) \tag{1.6}$$

generated at the time τ_i, the quantities $\xi(\tau_i)$ and $w(\tau_i)$ being respectively the states of the system and of the model at the time τ_i. The motion of the model during this time interval will be described by the equation

$$\dot{w} = \bar{f}(\tau_i, \xi(\tau_i), \quad w(\tau_i), u_i)$$

The above procedure defines in real time, and in synchronism with the motion of the dynamic system, a piecewise continuous function

$$u(t) = u_i, \quad \tau_i < t \leqslant \tau_{i+1} \tag{1.7}$$

which realizes the control action in the model and will be taken as an approximation to the function $v(t)$.

A proper choice of the model (1.1) and of the function U (which we shall call the strategy following the language of the theory of positional differential games) are the two major difficulties one encounters when actually employing the above scheme for obtaining $u(t)$. The point we want to emphasize here is that the model (1.4) is only a certain mental construction and that the above model control process is nothing else than a computer-oriented calculational algorithm to be carried out in real time in the course of the actual motion $x(t)$.

Let us discuss a number of specific examples of how the above scheme can actually be realized. First let us restrict ourselves to the case when the system (1.1) is of the form

$$\dot{x} = f_1(t, x) + f_2(t, x)v \tag{1.8}$$

where the n-vector function $f_1(t, x)$ and the $r \times n$-matrix function $f_2(t, x)$ both satisfy the Lipschitz condition

$$|f_j(t_1, x_1) - f_j(t_2, x_2)| \leqslant c_j^0 (|t_1 - t_2| + |x_1 - x_2|) \quad j = 1, 2 \tag{1.9}$$

the symbol $|\cdot|$ in the left-hand side denoting a matrix norm in the $j = 2$ case. Let us assume for the sake of simplicity that $r \leqslant n$ and that the matrix $f_2(t, x)$ is of rank r. This assumption guarantees up to the Lebesgue measure equivalence the uniqueness of the input $v(t)$ which gives rise to the true motion $x(t)$.

1.2. Extremal shift. Smoothing functional regularization method. As one possible model of (1.4) we will take a copy of the system (1.8), the right-hand side of which has the phase vector x replaced by the result of its measurement and the input action v by the control action u; that is

$$\dot{w} = f_1(t, \xi(t)) + f_2(t, \xi(t))u \tag{1.10}$$

In choosing the strategy U we use the rule of the extremal shift [2] and we find by directly adapting this rule to our situation that the control action u_i is obtained by maximizing the projection of the velocity of the model at the instant τ_i on the direction from the point $w(\tau_i)$ to the point $x(\tau_i)$ or, more precisely, to the point $\xi(\tau_i)$ because the quantity $x(\tau_i)$ is unknown. This method of control is advantageous in that at small h and δ (cf. (1.5)) the motion $w(t)$ of the model follows as closely as we please the motion $x(t)$ of the system. It should be noted, however, that this approach fails to secure the mean-square proximity of the functions $u(t)$ and $v(t)$. It proves possible to remedy this by using the method of a smoothing functional [4, 12] for the regularization of the extremum shift rule. Let the strategy U assign the control action u_i, as defined by (1.6), as a point of minimum (on the set Q) of the function

$$\gamma(u) = (w(\tau_i) - \xi(\tau_i))' f_2(\tau_i, \xi(\tau_i))u + \alpha(h)|u|^2$$

where

$$\alpha(h) > 0; \quad \alpha(h), \quad h/\alpha(h) \to 0 \quad \text{for} \quad h \to 0 \tag{1.11}$$

We also assume that

$$\delta = \delta(h) \leqslant ch, \quad c = \text{const} \tag{1.12}$$

It then follows that $\|u - v\|$ is as small as desired for sufficiently small h ($u(t)$ being defined by (1.17)). (The approximation $u(t)$ will be called stable in the mean-square sense if (1.12) is satisfied).

An interesting question to be discussed is the *a priori* estimation of the deviation $\|u - v\|$, which is the problem of finding a function $\nu(h)$ such that $\nu(h) \to 0$ as $h \to 0$ and $\|u - v\| \leqslant \nu(h)$. If no additional information is available about the input $v(t)$, it can be proved that no estimation can generally be established which would be uniform with respect to all the *a priori* possible realizations of $v(t)$. Let us assume, however, that $v(t)$ has a variation which does not exceed a constant K_0 in its magnitude. In this case it proves possible to obtain the estimation

$$\|u - v\| \leqslant Kh^{1/4}$$

where K may be expressed explicitly in terms of K_0 and certain numerical characteristics of the functions f_1 and f_2.

1.3 Extremum shift. Regularization by the error method. An alternative of extremum shift rule regularization is reminiscent of the method of errors known from the theory of ill-posed problems [1, 12]. Let us use again a model of the form (1.4). In the method we propose that the control assigned by the strategy U depends on the quantities $\xi(\tau_i)$, $\xi(\tau_{i-1})$, and $w(\tau_i)$, viz.,

$$u_1 = U(\tau_i, \xi(\tau_i), \quad \xi(\tau_{i-1}), \quad w(\tau_i)) \tag{1.13}$$

We now assume that at any time $\tau_i (i \geqslant 1)$ we may use the knowledge of the quantities $\xi(\tau_i)$, $\xi(\tau_{i-1})$ and $w(\tau_i)$ to construct a closed set $\Omega_i = \Omega_{h,i}$ contained in a convex shell Q and such that, first, the average value of the function $v(t)$ on the interval $\tau_{i-1} < t \leqslant \tau_i$. i.e., the vector

$$v_i = \frac{1}{\tau_i - \tau_{i-1}} \int_{\tau_{i-1}}^{\tau_i} v(t)dt$$

lies in Ω_i, and, second, if the vector u_i belongs to Ω_i and if h is sufficiently small, the motion $w(t)$ of the model (1.4) is as close as desired to $x(t)$ (in a uniform metric). (As pointed out in Section 1.2, this latter property is guaranteed if the extremum shift method is chosen to control the model). A set Ω_i possessing the above properties will be called a stable local averaging (for the input $v(t)$).

Let the strategy U assign u_i in (1.13) (for $i \geqslant 1$) as a minimal norm vector from Ω_i. Then the approximation $u(t)$ is stable in the mean-square sense.

Two methods for obtaining stable local averagings Ω_i may be considered.

In the first method Ω_i is the intersection of the set Q and the (ellipsoidal) set of all r-dimensional vectors such that

$$|f_{2,i}u - b_i| \leqslant \sigma_i^{(1)}$$

where

$$b_i = \frac{\xi(\tau_i) - \xi(\tau_{i-1})}{\tau_i - \tau_{i-1}} - f_{1,i}$$

$$f_{k,i} = f_k(\tau_i, \xi(\tau_i)) \quad (k = 1, 2)$$

$$\sigma_i^{(1)} = \frac{2h}{\tau_i - \tau_{i-1}} + K_1(\tau_i - \tau_{i-1}) + K_2 h$$

$$K_1 = (c_1^0 + c_2^0)(1 + c^0)$$

$$K_2 = (c_1^0 + c_2^0)c^*$$

c^0, c^* are *a priori* chosen constants such that

$$|f_1(t, x(t)) + f_2(t, x(t))v(t)| \leqslant c^0 |v(t)| \leqslant c^*$$

The above sets will be stable local averagings if there is a certain interelation between the quantities

$$d_h^+ = \max(\tau_{i+1} - \tau_i), \quad d_h^- = \min(\tau_{i+1} - \tau_i)$$

and the quantity h, namely, if

$$h + d_h^+ + \frac{d_h^+}{d_h^-}(d_h^+ + h) \to 0 \quad \text{for} \quad h \to 0 \tag{1.14}$$

This (sufficient) condition is satisfied, for example, if

$$d_h^+ = d_h^- = \delta(h) \quad \text{and} \quad h/\delta(h), \quad \delta(h) \to 0 \quad \text{for} \quad h \to 0 \tag{1.15}$$

With the second method, the set Ω_i is defined as the intersection of Q and the half-space of all u such that

$$l_2' u \leqslant \beta_i + \sigma_i^{(2)}$$

where

$$l_1 = q_i' f_{2,i}; \quad \beta_i = q_i' b_i$$

$$q_i' = w(\tau_i) - \xi(\tau_i)$$

$$\sigma_i^{(2)} = 2(|x_0| + c^0(\vartheta - t_0))(K_1(\tau_i - \tau_{i-1}) + K_2 h + 2h/(\tau_i - \tau_{i-1}))$$

The interrelation between the quantities d_h^+, d_h^- and h must be the same as in the first method, equations (1.14) and (1.15).

1.4. Stable realization of a preset motion. It is instructive to consider the application of the above scheme to a specific control problem. Suppose we have an n-dimensional dynamic system which operates in the time interval $t_0 \leqslant t \leqslant \vartheta$ and is defined by the differential equation

$$\dot{x} = F(t, x, p) + G(t, x, q) \tag{1.16}$$

where p is an r_1-dimensional time-varying control and q an r_2-dimensional unknown time-varying perturbation. All the *a priori* information concerning the actual realization $q(t)$ of the perturbation is limited to the assertion that this is a function measurable in the Lebesgue sense whose values belong to a given bounded and closed set Q_2. The values of p are chosen from an analogous set Q_1. The functions F and G satisfy the Lipschitz condition with respect to the totality of their arguments. The initial state (1.3) of the system is assumed to be known.

Suppose $x_*(t)$ is agreed beforehand to be the motion which the system will have to follow, the realization $q(t)$ of the perturbation being *a priori* unknown. The control $p(t)$ may be constructed in accordance with the measurements $\xi(t)$ of the current state $x(t)$; in the general case these measurements are only accurate in the sense of unequality (1.2). Clearly the measurement errors make it impossible to follow exactly the preset motion $x_*(t)$, and our goal is therefore to approximate it as closely as possible by the true motion $x(t)$. Formally, we state this as the requirement that the quantity

$$v = \max_{t_0 \leq t \leq \vartheta} |x(t) - x_*(x)| + \|\dot{x} - \dot{x}_*\| \tag{1.17}$$

be as small as desired for a sufficiently small h, which implies an arbitrarily close uniform approximation of the motion $x(t)$ to $x_*(t)$ and an arbitrarily close mean-square approximation of its velocity $\dot{x}(t)$ to $\dot{x}_*(t)$.

Before proceeding to the solution proper, let us formulate conditions sufficient from the standpoint of following exactly the motion of the system at a known $q(t)$. Let us denote by $G_*(t)$ the vectogram of the second (perturbing) term in the right-hand side of equation (1.16) at $x = x_*(t)$, so that $G_*(t)$ is in fact the totality of the vectors $v = G(t, x_*(t), q)$ obtained by exhausting all possible q's in the set Q_2. We assume that there exists a function $p_*(t, v)$ (v being an n-dimensional vector) which has its values in Q_1, satisfies the Lipschitz condition, and possesses the property that for almost all t's

$$\dot{x}_*(t) = F(t, x_*(t), p_*(t, v(t))) + v(t) \tag{1.18}$$

for any function $v(t)$ (measurable in the Lebesgue sense) whose value lies in the set $G_*(t)$ for any t.

Clearly, if we knew the value of $q(t)$ at any time t, then setting

$$p(t) = p_*(t, v(t)) \tag{1.19}$$

where

$$v(t) = G_*(t, x_*(t), q(t)) \tag{1.20}$$

it would be possible to make the motion $x(t)$ of the system coincide with $x_*(t)$. Note that to achieve its goal it would in fact be sufficient to know the function $v(t)$ (dependent on $q(t)$) instead of $q(t)$ itself. Since $v(t)$ is unknown, this method of control is infeasible, however. Still, we can employ the above approach to restore $v(t)$ to a certain degree of accuracy, and then replace $v(t)$ in (1.19) by an approximate expression of $u(t)$ obtained by means of a real time procedure. It should be noted that under the assumptions introduced—and disregarding the problem of approximating the derivatives—it is also possible to employ other simple ways to find a uniform approximation of the preset motion $x_*(t)$ by the motion $x(t)$, one example being the method of the extremum aiming at a stable path [1].

A more detailed discussion of the scheme is now in order. To restore $v(t)$, let us introduce a controlled model of the form

$$\dot{w} = F(t, \xi(t), p_*(t, u)) + u \tag{1.21}$$

with the initial state $w(t_0) = x_0$. At any time τ_i (cf. (1.15)) we will again use the feedback principle (1.6) to assign the control of the model. During the time interval $\tau_i < t \leq \tau_{i+1}$, the motion of the model is governed by the equation

$$\dot{w} = F(\tau_i, \xi(\tau_i), p_*(\tau_i, u_i)) + u_i \tag{1.22}$$

The specific form of the function u_i will be borrowed from Section 1.2. Let u_i be the point of the minimum value of the function

$$\gamma(u) = (w(\tau_i) - \xi(\tau_i))'u + \alpha(h)|u|^2 \tag{1.23}$$

on the set $G_*(\tau_i)$. We also assume the conditions (1.11) and (1.12) to be satisfied and we use

$$p(t) = p_i = p_*(\tau_i, u_i) \quad \text{for} \quad \tau_i < t \leqslant \tau_{i+1} \tag{1.24}$$

as the control law for the system (1.16). Note that the stability requirement has been fulfilled in constructing this algorithm because the quantity ν as defined by (1.17) is as small as desired for h sufficiently small.

As remarked earlier, the central point on which the present approach hinges is that the control (1.7) being realized in the model, approximates the function $v(t)$ in the mean-square sense and that it is stable in doing so. It will be understood that this is by no means a mere consequence of the result derived in Section 1.2, because in this case the real motion $x(t)$ of the system responds, through control (1.24), to the results u_i of the approximation to the function $v(t)$.

1.5 Nonlinear input systems. Instead of systems of the type (1.8), with right-hand sides linear in the action $v(t)$, systems of the general type (1.1) will be our concern in this section. We assume that the function $f(t, x, v)$ in the right-hand side of the system is continuous with respect to the totality of its variables and that it satisfies the Lipschitz condition with respect to t and x, with Lipschitz constant being independent of v. The initial state (1.3) of the system is again assumed to be known. We also assume that the function $f(t, x, v)$ is injective with respect to v for any t and x, which ensures (up to the Lebesgue measure zero) the uniqueness of the input $v(t)$ giving rise to the motion $x(t)$ we are for.

Turning now to the problem outlined in Section 1.1, it turns out that it can be solved by the method of Section 1.2 if we apply this latter to a certain auxiliary system and introduce some additional calculations. Let us employ the simplest possible n-dimensional controlling system

$$\dot{x}_* = v_* \tag{1.25}$$

with an initial state $x_*(t_0) = x_0$ and an input $v_* = v_*(t)$. We choose as restrictions on $v_*(t)$ the condition $|v_*(t)| < \lambda$, where λ, in turn, satisfies the condition that for the actual input $v(t)$ and motion $x(t)$ of the system (1.1)

$$\lambda \geqslant |f(t, x(t), v(t))| \tag{1.26}$$

Obviously, $x(t)$ is the motion of the system (1.25) for

$$v_*(t) = f(t, x(t), v(t)) \tag{1.27}$$

so that when the measurements $\xi(t)$, cf. (1.2), are acted upon by the algorithm of Section 1.2 as applied to the system (1.25), we obtain a stable in mean-square approximation to the input $v_*(t)$ in (1.27) (this approximation will be denoted as

$\bar{u}(t)$ and clearly is of the form $\bar{u}(t) = \bar{u}_i$, $\tau_i < t \leqslant \tau_{i+1}$). We are then justified in suggesting that if the function $u(t)$ of the form (1.7) is obtained from the condition that u_i is a point of minimum on Q of the error defined by

$$\beta(u) = |\bar{u}_i - f(\tau_i, \xi(\tau_i), u)| \tag{1.28}$$

then $u(t)$ must be close to $v(t)$ in mean-square (that is, it will represent the required mean-square stable approximation to the input $v(t)$). This supposition proves to be true, thus allowing the following statement.

Let us introduce an n-dimensional model $\dot{w} = u$ with an initial state $w(t_0) = x_0$ and let $\alpha(h)$ and $\delta(h)$ be chosen from the conditions (1.11) and (1.12). At any time τ_i we will construct vectors u_i and u_i from the conditions that (1) u_i is a point of minimum on the sphere $|\bar{u}| \leqslant \lambda$ of the function

$$\bar{\beta}(\bar{u}) = (w(\tau_i) - \xi(\tau_i))'\bar{u} + \alpha(h)|u|^2$$

(2) u_i is the point of minimum on the set Q of the function (1.28), and (3) on the interval $\tau_i < t \leqslant \tau_{i+1}$ the motion of the model is determined by the equation $\dot{w} = u_i$. The approximation (1.7) resulting from the algorithm is stable in mean-square.

1.6. Remarks. 1) Examples of methods by which the input of the system (1.8) can be restored (Secs. 1.2, 1.3) could be multiplied. In particular, we may abandon the extremum shift rule and use the rule of aiming at a concomitant point [2] instead of deriving a corresponding control law for the model.

2) As in the case of the method of Sec. 1.2, any of the methods may be applied to the problem of finding a stable following procedure for the prescribed motion (Sec. 1.4). The scheme we illustrated in Sec. 1.4 can be modified to conform to other positional control problems under perturbations, the stable following of a variety for example.

3) It was assumed at the end of Sec. 1.1 that the rank of the matrix $f_2(t, x)$ equals r, the dimensionality of the input action. If this assumption is removed, it turns out that the input generating the measured motion $x(t)$ is not generally unique (up to Lebesgue measure equivalence). The formulations which naturally suggest themselves in this case are, for example, as follows:

a) to restore an arbitrary input generating $x(t)$;
b) to restore an input generating $x(t)$ and satisfying a certain additional conditions;
c) to restore the totality of all inputs generating $x(t)$.

4) The methods of Secs. 1.2, 1.3 prove to be applicable to the formulation b) provided an additional condition separates out the minimal norm input. Let the input $v_*(t)$ generate $x(t)$, such that for any other such input $v(t)$

$$|v_*(t)| \leqslant |\bar{v}(t)| \tag{1.29}$$

for almost all t (the existence of $v_*(t)$ is easy to demonstrate). Suppose the set Q

is convex (and hence $v_*(t)$ is unique). Then the control $u(t)$ obtained by the algorithms of Secs. 1.2, 1.3 is such that the deviation $\|u - v_*\|$ may be made as small as desired for h sufficiently small.

5) The above algorithms can be easily modified for the case when, instead of condition (1.29), $v_*(t)$ is specified by the condition

$$|v_*(t) - v_0| \leqslant |\bar{v}(t) - v_0|$$

for almost all t, v_0 being a fixed r-dimensional vector. All that is needed is:

in the algorithm of Sec. 1.2, to replace $|u|$ by $|u - v_0|$ in $\gamma(u)$;

in the algorithm of Sec. 1.3, to choose as the vector u_i the one closest to v_0 in Ω_i.

6) The method presented in Sec. 1.6 for systems of the general form (1.1) provides the solution for problem a). The control $u(t)$ obtained by this method has the property that the quantity $\inf \|u - \bar{v}\|$ is as small as desired for sufficiently small h (the operator inf being assumed to act on all inputs $v(t)$ generating $x(t)$).

7) Problem a) for the system (1.8) and a convex set Q can also be solved by algorithms similar to those outlined in secs. 1.2, 1.3. To illustrate the procedure, let us consider the simplest possible case of a scalar input. The set Q is then represented by a line segment $a \leqslant v \leqslant b$, and the restored set (i.e., the totality of all inputs resulting in the motion $x(t)$) can be easily shown to include all functions $v(t)$ such that

$$v_-(t) \leqslant \bar{v}(t) \leqslant v_+(t)$$

for almost all $t(a \leqslant v_-(t) \leqslant v_+(t) \leqslant b)$. The given set is thus completely specified by the functions $v_-(t)$ and $v_+(t)$, and the problem is reduced to that of restoring these functions. This can be done by applying in parallel the algorithms described in Remark 5) and setting $v_0 = a$ in the first of them and $v_0 = b$ in the second. The resulting functions $u_-(t)$ and $u_+(t)$ will of course approximate $v_-(t)$ and $v_+(t)$, respectively, as closely as desired in the mean-square sense if h is sufficiently small.

2. PARTIALLY MEASURABLE COORDINATES

2.1. Unique input restoration. It turns out that the above method can be incorporated into algorithms for more general problems, such as the restoration of the input in situations when only a certain part of the coordinates of the vector $x(t)$ are measurable. Let us concentrate on situations in which the unmeasurable coordinates admit of reasonably accurate restoration in the course of motion.

We thus assume that the vector x is composed of the n_1-dimensional vector y and n_2-dimensional vector z, and that equation (1.8) may be written (under the assumptions of Sec. 1.1) in the form

$$\dot{y} = \varphi_1(t, y, z) + \varphi_2(t, y, z)v \quad (2.1)$$

$$\dot{z} = \psi_1(t, y, z) + \psi_2(t, y, z)v \quad (2.2)$$

with an initial state

$$y(t_0) = y_0, \quad z(t_0) = z_0 \tag{2.3}$$

We designate by $y(t)$, $z(t)$ the true motion of the system due to the true input $v(t)$ and we assume that only the vector $y(t)$ is capable of being measured, and that the result of measurement, $\eta(t)$, is generally inaccurate in the sense

$$|\eta(t) - y(t)| \leqslant h \tag{2.4}$$

The problem is obtaining mean-square stable approximations to the input $v(t)$ from real-time measurements $\eta(t)$.

The solution will be sought for under the condition that the rank of the matrix $\varphi_2(t, x)$ is equal to r. Equation (2.1) may then be formally solved with respect to v giving

$$v = s(t, y, z, \dot{y}) = \overset{+}{\varphi}_2(t, y, z)(\dot{y} - \varphi_1(t, y, z))$$

where the matrix $\overset{+}{\varphi}_2$ is pseudo-inverse to φ_2. We will use this solution to construct a controlled model capable of approximately restoring the unmeasurable vector $z(t)$, the state of the model being determined by the n_1-dimensional vector w_y and n_2-dimensional vector w_z. We assume that

$$w_y(t_0) = y_0, \quad w_z(t_0) = z_0 \tag{2.5}$$

initially.

The equation of motion of the model is of the form

$$\dot{w}_y = \bar{u} \tag{2.6}$$

$$\dot{w}_z = \psi(t, \eta(t), w_z, \bar{u}) = \psi_1(t, \eta(t), w_z) + \psi_2(t, \eta(t), w_z)s(t, \eta(t), w_z, \bar{u}) \tag{2.7}$$

where u is a controlling action supposed to represent a suitable approximation to the derivative $\dot{y}(t)$; the motion of the component $w_z(t)$ is supposed to follow the motion $z(t)$. The model control (2.6), (2.7) will be carried out by the scheme of Sec. 1, that is, by assigning for any time τ_i a control u_i by the feedback principle

$$\bar{u}_i = \bar{U}(\tau_i, \eta(\tau_i), w_y(\tau_i), w_z(\tau_i))$$

and defining by the equations

$$\dot{w}_y = \bar{u}_i \tag{2.8}$$

$$\dot{w}_z = \psi(\tau_i, \eta(\tau_i), w_z(\tau_i), \bar{u}_i) \tag{2.9}$$

the motion of the model over the interval $\tau_i < t \leqslant \tau_{i+1}$.

We will employ the following modification of the extremum shift rule to form the action u_i. For any v from Q, let no one of the coordinates of the right-hand side of (2.1) exceed in its absolute magnitude the number λ along the motion $y(t)$; the k-th coordinate $\bar{u}_{k,i}$ of the vector $\bar{u}_i$ will be defined by putting

$$\bar{u}_{k,i} = -\lambda \,\text{sign}(w_{k,y}(\tau_i) - \eta_k(\tau_i)) \tag{2.10}$$

If we also assume the condition (1.12), it turns out (see [10]) that

$$|w_z(t) - z(t)| \leqslant Kh$$

where the constant K can, but need not, be written down explicitly. Noting (2.4), we may thus consider the pair $(\eta(t), w_z(t))$ as an inaccurate measurement of the whole vector $x(t) = (y(t), z(t))$ (the error of this "pseudomeasurement" not exceeding $(K + 1)h$) and proceeding by employing one of the algorithms from Secs. 1.2 or 1.3. Alternatively, we may define $u(t)$ by using equation (2.1) alone. For this, we define by

$$\dot{w} = \varphi(t, \eta(t), w_z(t), u) = \varphi_1(t, \eta(t), w_z(t)) + \varphi_2(t, \eta(t), w_z(t))u \tag{2.11}$$

the model corresponding to this equation and assume

$$w(t_0) = y_0 \tag{2.12}$$

as the initial state.

The motion of the model over the interval $\tau_i < t \leqslant \tau_{i+1}$ will be defined by the equation

$$\dot{w} = \varphi(\tau_i, \eta(\tau_i), w_z(\tau_i), u_i) \tag{2.13}$$

The control law u_i may be defined, for example, in a similar way to that of Sec. 1.2, that is, as a point of minimum (on the set Q) of the function

$$\gamma(u) = (w_y(\tau_i) - \eta(\tau_i))'\varphi_2(\tau_i, \eta(\tau_i), w_z(\tau_i))u + \alpha(h)|u|^2 \tag{2.14}$$

for $\alpha(h)$ satisfying (2.12).

The entire algorithm for restoring $v(t)$ is thus as follows. We introduce a joint $(2n_1 + n_2)$-dimensional model having a state (w_y, w_z, w) and described by equations (2.6), (2.7), (2.11). The initial state of the model is determined by equations (2.5), (2.12). At the time τ_i the control $(\bar{u}_i, u_i)$ is formed, $\bar{u}_i$ being determined from (2.10) and u_i as the point of minimum (on Q) of the function (2.14). On the interval $\tau_i < t \leqslant \tau_{i+1}$, the motion of the model is specified by equations (2.8), (2.9), (2.13).

The resulting approximation $u(t)$ (cf. (1.7)) is stable in the mean-square sense.

One further application of the above method is for the real-time modeling of the additional unmeasurable coordinates. Suppose the system (2.1), (2.2) is augmented by the equation

$$\dot{p} = \chi(t, y, z, p, v) \tag{2.15}$$

with an initial condition $p(t_0) = p_0$, the function χ satisfying the Lipschitz condition with respect to the totality of its variables. The problem is, given the measurements $\eta(t)$, following in real time regime the motion $p(t)$ corresponding to the input $v(t)$.

To solve the problem it proves sufficient to introduce one more component, w_p, into the above algorithm, under the assumption that the initial state of this new component is $w_p(t_0) = p_0$ and that on the time interval $\tau_i < t \leqslant \tau_{i+1}$ it is governed by the equation

$$\dot{w} = \chi(\tau_i, \eta(\tau_i), w_z(\tau_i), w_P(\tau_i), u_i)$$

Then the motion $w_p(t)$ will approximate $p(t)$ as closely as desired in a uniform metric, on condition that a large enough value of h is chosen.

2.2. Piecewise-continuous system, not all coordinates measurable. Let a dynamic (noncontrolled) system be governed by the equations

$$\dot{y} = \Phi(t, y, z) \tag{2.16}$$

$$\dot{z} = \Psi(t, y, z) \tag{2.17}$$

with an initial state of the form (2.3). We are assuming as before that y and z are respectively n_1- and n_2-dimensional vectors. Let the functions Φ and Ψ undergo discontinuities with respect to the phase variables y and z on a certain "surface." The phase space is divided into three sets G_0, G_1, and G_2 (that is, the sets do not intersect and combine to represent the entire phase space of the problem). The sets G_1 and G_2 are open, and G_0, of Lebesgue measure zero, is closed. The functions Φ and Ψ are of the form

$$\Phi(t, y, z) = \begin{cases} \Phi_1(t, y, z) & \text{if} \quad (y, z) \quad \text{is in } G_1 \\ \Phi_2(t, y, z) & \text{if} \quad (y, z) \quad \text{is in } G_2 \end{cases}$$

$$\Psi(t, y, z) = \begin{cases} \Psi_1(t, y, z) & \text{if} \quad (y, z) \quad \text{is in } G_1 \\ \Psi_2(t, y, z) & \text{if} \quad (y, z) \quad \text{is in } G_2 \end{cases}$$

where the functions Φ_1, Φ_2, Ψ_1, Ψ_2 satisfy the Lipschitz condition.

Equations (2.16), (2.17) thus represent a system with a discontinuous right-hand side and by its motion we mean, following [13], the solution of the differential equation

$$\dot{y} = v(t)\Phi_1(t, y, z) + (1 - v(t))\Phi_2(t, y, z) \tag{2.18}$$

$$\dot{z} = v(t)\Psi_1(t, y, z) + (1 - v(t))\Psi_2(t, y, z) \tag{2.19}$$

with an initial condition (2.3), $v(t)$ being an arbitrary measurable function satisfying the conditions

$$\begin{aligned} v(t) &= 1 \text{ for } (y(t), z(t)) \text{ from } G_1 \\ v(t) &= 0 \text{ for } (y(t), z(t)) \text{ from } G_2 \end{aligned} \tag{2.20}$$

(If $y(t)$, $z(t)$ is in G_0, $v(t)$ may take any value from 0 to 1).

We thus conclude that the motion of the system is not unique and depends on the function $v(t)$ satisfying (2.20). We assume as before that we measure $y(t)$ in the course of motion and that the results $\eta(t)$ of the measurements obey the inequality (2.4). Under such circumstances it is required to approximate in real time the functions $y(t)$ and $z(t)$ by functions $w_y(t)$ and $w_{1,z}(t)$ with small deviations

$$v_y = \max_{t_0 \leq t \leq \vartheta} |w_y(t) - y(t)| + \|\dot{w}_y - \dot{y}\| \tag{2.21}$$

$$v_z = \max_{t_0 \leq t \leq \vartheta} |w_{1,z}(t) - z(t)| + \|\dot{w}_{1,z} - \dot{z}\| \tag{2.22}$$

It required, in other words, that the true motion of the system, together with its derivatives, should be approximated on the basis of the inaccurate measurement of part of the system's coordinates.

The system (2.18), (2.19) is easily shown to be of the form (2.1), (2.2) with

$$\varphi_1 = \Phi_2, \quad \varphi_2 = \Phi_1 - \Phi_2$$

$$\Psi_1 = \Psi_2, \quad \Psi_2 = \Psi_1 - \Psi_2$$

Now if $\Phi_1(t, y, z) \neq \Phi_2(t, y, z)$, we clearly satisfy the condition on the rank of φ_2 introduced in Sec. 2.1, and hence the approximation $u(t)$ obtained by the algorithm of Sec. 2.1 is stable in mean-square. This suggests the following approach to the problem. Let the model we adopted, equations (2.6), (2.7), (2.11), be augmented by the component

$$\dot{w}_{1,z} = \psi_1(t, \eta(t), w_z) + \psi_2(t, \eta(t), w_z)u$$

with an initial state $w_{1,z}(t_0) = z_0$. The motion of the component on the interval $\tau_i < t < \tau_{i+1}$ will thus be determined by the equation

$$\dot{w}_{1,z} = \psi_1(\tau_i, \eta(\tau_i), w_z(\tau_i)) + \psi_2(\tau_i, \eta(\tau_i), w_z(\tau_i))u_i$$

That the deviations (2.21), (2.22) may be made as small as desired for h small enough is easily seen from the stability in mean-square of the approximation $u(t)$, (1.7).

2.3. Remarks. 1) We have thus discussed one special example of a situation in which only part of the coordinates of the system is measured. Another example of this kind was investigated by the authors elsewhere [10]. It should be observed that with the aid of the above approach, problems of much more general nature can be treated, including those concerned with the restoration of the entire family of inputs capable of generating a given measurement (such a family usually contains more than one function $v(t)$).

2) The scheme discussed in Sec. 2.1 may also be applied—in a manner similar to that illustrated in Sec. 1.4—to the problem of the stable following of a prescribed motion on the basis of the (inaccurate) measurements of a part of the coordinates of the system.

3. INPUT RESTORATION IN OTHER CLASSES OF DYNAMIC SYSTEMS

The restoration method outlined in the present discussion is also applicable to dynamic systems governed by other types of differential equations, such as

1. System (1.1) with a discontinuous right-hand side: the function $f(t, x, v)$ undergoes a discontinuity with respect to x on a certain surface (cf. Sec. 2.4); the motion $x(t)$ is understood in the sense of [13].

2. System described by an equation with an aftereffect:

$$\dot{x} = f(t, x_t, v)$$

$x_t = x(t + s), \quad -\tau \leqslant s \leqslant 0, \quad \tau = \text{const} > 0$

3. System in which the role of an input action is played by a time-varying lag:

$\dot{x} = f(t, x(t - v))$

$0 \leqslant v \leqslant \tau, \quad \tau = \text{const} > 0$

4. System with a lag in both coordinates and input:

$\dot{x} = f(t, x(t), x(t - \tau), v(t), v(t - \tau))$

$\tau = \text{const} > 0$

5. Systems described by the Cauchy problem or by standard mixed problems expressed by equations of parabolic and hyperbolic types and by other types of partial differential equations.

REFERENCES

1. Krasovskii, N. N. and Subbotin, A. I. Positional differential games. Moscow, Nauka, 1974 (in Russian).
2. Krasovskii, N. N. Dynamic system control: the problem of a minimized guaranteed result. Moscow, Nauka, 1985 (in Russian).
3. Fomin, V. V., Fradkov, A. L., and Yakubovich, V. A. Adaptive control of dynamic objects. Moscow, Nauka, 1981 (in Russian).
4. Tikhonov, A. N. and Arsenin, V. Ya. Ill-posed problems and their solution. Moscow, Nauka, 1979 (in Russian).
5. Ivanov, V. K., Vasin, V. V., and Tanana, V. P. Linear ill-posed problems. Theory and its applications. Moscow, Nauka, 1978 (in Russian).
6. Petrov, B. N. and Krut'ko, P. D. Inverse problems of the dynamics of controlled systems. Izv. USSR Acad. Sci. TK, 1980, No. 5, pp. 149–155.
7. Osipov, Yu. S. and Kryazhimskii, A. V. Lyapunov function method in the motion modeling problem. In: Stability of motion. Novosibirsk, Nauka, 1985, pp. 53–56.
8. Kryazhimskii, A. V. and Osipov, Yu. S. On the control modelling in a dynamic system. Izv. USSR Acad. Sci. TK, 1983, No. 2, pp. 51–60.
9. Kryazhimskii, A. V. and Osipov, Yu. S. On the optimally approximated differentiation operator in the class of anticipatory operators. Mat. Zametki, 1985, Vol. 37, No. 2, pp. 192–199.
10. Kryazhimskii A. V. and Osipov, Yu. S. On positional calculation of Ω-normal controls in dynamical systems. Probl. contr. inform. theory, 1984, Vol. 13, No. 6, pp. 427–436.
11. Kryazhimskii, A. V., Maksimov, V. I., and Osipov, Yu. S. On positional modeling in dynamic systems. PMM, 1983, Vol. 47, No. 6, pp. 883–889.
12. Vasilyev, F. V. Extreme problems and their solution. Moscow, Nauka, 1981 (in Russian).
13. Filippov, A. F. Differential equations with a discontinuous right-hand side. Moscow, Nauka, 1885 (in Russian).

MOTION ESTIMATION GAME PROBLEMS WITH UNMODELABLE ACCELERATIONS

M. L. Lidov

The basic assumption underlying classical statistical estimation methods is that the probability distribution of the measurement errors—or at least the mathematical expectation and the covariation matrix of this distribution—are known for the problem. This is not always the case, however. In space experimentation, for example, or in other fields where one-of-a-kind measurement systems are employed, it is often impossible to obtain experimental data in a volume sufficiently large that the parameters of the error distribution could be determined unambiguously and with a sufficient degree of accuracy. In such cases other estimation schemes should be used, such as minimax methods or methods of guaranteed or robust estimation.

Robust Estimation Problems

The idea of robust estimation had been advanced by Peter Huber back in 1964 [1], and the state of the art of the subject—as of 1984—is well described in Huber's recent monograph [2]. As is the case with classical methods, Huber assumes that measurement errors are independent quantities all having the same probability distribution laws, but the novelty of his approach is that the only *a priori* information about these laws is the *class* of distributions to which the actual distribution belongs. Given a class of allowable estimation methods, one is then required to solve a minimax problem of choosing such an estimation method from the given class which would minimize the asymptotic error variance of the estimated controlled parameter relative to its true value; the minimization is carried out for a distribution assumed to be—in a certain known sense—the "worst" in the given class. The asymptotic behavior is investigated for a sufficiently large number n of measurements of the same kind. A corresponding variational problem then yields an (essentially nonlinear) estimation method with a certain degree of stability. The following is the most typical problem for this approach.

Shift parameter estimation for the class of ϵ-contaminated normal distributions [1].

1. Given n measurements $z_k = m + \xi_k$, $k = \overline{1, n}$, to obtain an estimation $\hat{m}$ of the parameter m.
2. ξ_k, $k = \overline{1, n}$ are independent and equally distributed measurement errors.
3. The distribution density $p(x)$ of the random variables ξ_k assumed to be the form

$$p(x) = (1 - \epsilon)g(x) + \epsilon h(x) \tag{1.1}$$

where $g(x)$ is the density of the standard normal distribution, and $h(x)$ an arbitrary distribution density. The real situation corresponding to the density (1) is one in which $(1 - \epsilon)$ is the probability of the measuring equipment functioning adequately and ϵ is the probability of failure at which any error is possible.

4. With the aid of functions $\rho(x)$ the class

$$\hat{m} = \arg\min \sum_{k=1}^{n} \rho(z_k - m)$$

of M-estimation is defined. If the distribution density $p(x)$ were known, it would be easy to see how, by putting $\rho = -ln\, p(x)$, the M-estimations are the same as those given by the maximum likelihood method, which coincides with the mean square method in this particular case (at $\epsilon = 0$).

5. If an additional assumption of the symmetry of the function $p(x)$ is made, it is possible to show by a proper choice of an (also symmetrical) function $\rho(x)$ that the method of M-estimations leads to consistent

$$\hat{m} - \underset{n \to \infty}{m} \to 0$$

asymptotically normal estimations with errors characterized by the quantity

$$A(p, \rho) = \lim_{n \to \infty} M[\sqrt{n}(\hat{m} - m)]^2$$

proportional to the asymptotic variation (here and henceforth M is the mathematical expectation).

6. The solution of the game problem

$$\min_{\rho} \max_{p} A(p, \rho)$$

leads in this case to the function

$$\rho(x) = \begin{cases} x^2/2 & \text{for } |x| \leq a(\epsilon) \\ a(\epsilon)|x| - a^2(\epsilon)/2 & \text{for } |x| > a(\epsilon) \end{cases}$$

where $a(\epsilon) \to \infty$ as $\epsilon \to 0$, and $a(\epsilon) \to 0$ as $\epsilon \to 1$. This function is represented schematically in Fig. 1. We thus conclude that the minimax algorithm leads to the least square method at $\epsilon \to 0$ and to the least modulus method at $\epsilon \to 1$.

7. For a nonsymmetric $p(x)$, the estimations are no longer consistent, viz.,

$$\max_{p} |\hat{m} - m| \xrightarrow[n\to\infty]{p} \Delta(\rho) > 0$$

and a method of estimation (of the function $\rho(x)$) is determined from the solution of the problem min $\Delta(\rho)$. In the above example we see that $\rho(x) = |x|$, showing that we have the method of the least moduli as an estimation method.

8. This approach is also well suited for other distribution classes, and for applied problems in which the state space is represented by an m-dimensional vector $q \in \mathbf{R}^m$, and the measurements z_k, $k = \overline{1, n}$ and the vector q are related, for example, by linear equations of the form $z_k = H_k^T q + \xi_k$, where the $H_k \in \mathbf{R}^m$ are known vectors.

It should be remarked, however, that the errors ξ_k are always considered to be independent and equally distributed quantities, and that the solutions are obtained for an asymptotically large number of measurements [2–5].

Minimax Problems with Correlated Measurement Errors

"Worst correlation" problem. The simplest limiting problem of this kind has been considered in [6] and is formulated as follows.

1. A system is defined by an m-dimensional state vector $q \in \mathbf{R}^m$.

2. The state vector and the measurements z_k are related by the equation $z_k = H_k^T q + \xi_k$, $k = \overline{1, n}$, where the $H_k \in \mathbf{R}^m$ are known vectors.

3. Given the measurements z_k, $k = \overline{1, n}$, one is required to estimate the parameter $l = a_q^T$, where $a \in \mathbf{R}^m$ is a given vector.

4. The estimation l is determined in the class of linear estimations

$$\hat{l} = \sum_{k=1}^{n} \Phi_k z_k$$

where the estimator $\Phi^T = (\Phi_1, \ldots \Phi_n)$ is defined below.

5. Let $\xi^T = (\xi_1, \ldots, \xi_n)$. The probability hypotheses are as follows:

ρ

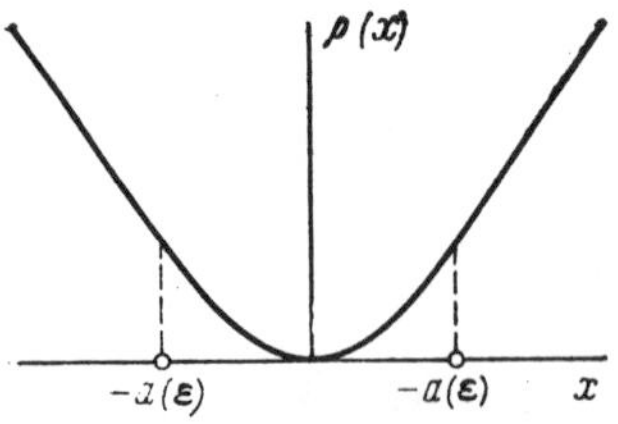

Figure 1

a) the mathematical expectation of ξ is zero, $M\xi = 0$;
b) measurement error variations are known, $M\xi_k^2 = \sigma_k$;
c) correlation coefficients $k_{ij} = M\xi_i\xi_j/\sigma_i\sigma_j$ are unknown and may assume any values provided that $|k_{ij}| < 1$ and that the covariation matrix $P = M\xi\xi^T$ is nonnegative definite. These two conditions specify the class $\{P\}$ of permissible covariation matrices.

6. The variation of estimation error is defined by

$$D(P, \Phi) = M(\hat{l} - l)^2$$

and depend on the actual form of P and the choice of the estimator Φ.

7. The estimator is determined from the game problem

$$\min_{\Phi} \max_{\rho \in \{P\}} \sqrt{D(\rho, \Phi)} \tag{2}$$

under the nondisplacement condition,

$$M(\hat{l} - l) = 0$$

The above formulation leads to the following linear programming problem: Given the conditions

$$\begin{gathered} \sum_{k=1}^{n} H_k \Phi_k = a \\ \sum_{k=1}^{n} \sigma_k |\Phi_k| - \min_{\Phi} \end{gathered} \tag{3}$$

it is required to determine Φ_k, $k = \overline{1, n}$.

Remarks. The problem (3) can also be arrived at in a somewhat different formulation known as the "board scheme," in which all the *a priori* knowledge about the errors ξ is limited to the fact that they belong to the papallelepiped

$$\Pi = \{\xi \in \mathbf{R}^m, |\xi_k| \leqslant \sigma_k, k = \overline{1, n}\}$$

We are led exactly to the problem (3) if we denote $S(\xi, \Phi) = |\hat{l} - l|$ and consider the game problem

$$S(0, \Phi) = 0 \tag{4}$$

$$\min_{\Phi} \max_{\xi \in \Pi} S(\xi, \Phi)$$

The most important conclusion to be drawn from (3) is that only m components of the optimum estimator $\hat{\Phi}$ will be different from zero (by linear programming theory), and that for the calculation of the optimum estimation $\hat{l}$ only m measurements are actually needed. Since $n \gg m$ in most cases, it follows that the overwhelming majority of measurements turn out to be of no use for the procedure. It will be noticed, however, that the optimum composition of measurements will in general be different for different parameters l (for different vectors a).

The ability to provide a "guaranteed" estimation; the relative simplicity of the required measurement program, which may be determined as early as at the preliminary analysis stage; and the reliability of numerical computer calculations for even very large dimensionalities of the state vector q (linear programming) are the factors which have made the above approach quite attractive for a number of applications.

Minimax estimation problem: the continuous formulation. Since the author's 1964 paper [6] a number of attempts have been undertaken at the generalization of the proposed formulation, emphasis usually being placed on a more detailed description of the initial set $\{P\}$ of all possible covariation matrices. A common—and only very natural—assumption in these attempts is that it is possible to determine experimentally the corresponding numerical coefficients of the more complex model, which of course is in itself a rather difficult task to perform.

The most accurate description of the results achieved in this area may be found in the monograph by Bakhshiyan et al. [7]. The approach the authors employ in their discussion is first to calculate max $D(P, \Phi)$ and then to reduce the problem to a linear programming formulation. It should be kept in mind, however, that no computational algorithms are as yet available for the effective determination of the minimizing estimator $\tilde{\Phi}$ in nonlinear problems.

The continuous formulation of the minimax estimation problem [8] will serve to illustrate a more detailed approach to the problem under consideration. We will discuss a somewhat simplified version of the problem as compared with [8] and will stress only those points in our argument where the formulation differs from the discrete case outlined above.

1. The equation of measurement is taken to be of the form

$$z(t) = H^T(t)q + \xi(t), \quad t \in [0, T] \tag{5}$$

where $H(t) \in \mathbf{R}^m$ is a known vector-function.

2. The unknown estimation $\hat{l}$ is written as a linear-functional

$$\hat{l} = \int_0^T \Phi(t)z(t)dt$$

3. The measurement errors $\xi(t)$ are assumed to be a sum of two random and mutually independent processes $\xi_i(t)$, $i = 1, 2$

$$\xi(t) = \xi_1(t) + \xi_2(t)$$

of which $\xi_1(t)$ is a "white noise" of intensity $c(t) \geqslant 0$, and $\xi_2(t)$ is the correlated process, in the sense

$$M\xi_2(t) = 0, M\xi_2(t)\xi_2(s) = \sigma(t)\sigma(s)r(t, s)$$

where the variances $\sigma^2(t)$ are considered to be given, and as regards the autocorrelation function $r(t, s)$, it is only known that, given $\theta \geqslant 0$ and $\nu \geqslant 0$ ($|\theta \pm \nu| \leqslant 1$), we have

$$|r(t, s) - \nu| \leqslant \theta \tag{6}$$

(The case $\nu = 0$, $\theta = 1$ corresponds to the worst correlation hypothesis for the process $\xi_2(t)$; a noncorrelated process takes place when $\nu = 0$, $\theta = 0$).

We consider a game problem of determining the estimator from the conditions

$$M(\hat{l} - l) = 0$$
$$\min_{\Phi(t)} \max_{r(t,s)} M(\hat{l} - l)^2 \tag{7}$$

The formulation (7) leads to the following variational problem: it is required to determine $\Phi(t)$ from conditions

$$\int_0^T H(t)\Phi(t)\,dt = a$$
$$D(\Phi) = \int_0^T c(t)\Phi^2(t)\,dt + \nu\left(\int_0^T \sigma(t)\Phi(t)\,dt\right)^2 + \theta\left(\int_0^T \sigma(t)|\Phi(t)|\,dt\right)^2 - \min_{\Phi(t)} \tag{8}$$

If $c(t) \equiv 0$ (no white noise) and $\nu = 0$, the system (8) represents a continuous analogue of the worst correlation problem; the value of θ ($0 \leqslant \theta \leqslant 1$) affects only the variance of the estimation error $D(\Phi)$ and has no relevance to the shape of the minimizing estimator. In the $\nu = \theta = 0$ case we are led to the classical version of the problem, in which $\xi(t)$ is a white noise process, and the solution to (8) is given by

$$\Phi(t) = H^T(t)K^{-1}a, \quad K = \int_0^T \frac{1}{c(t)} H(t)H^T(t)\,dt \tag{9}$$

Using Lagrange's indeterminate multipliers, the minimizing estimator for the general case of the problem (8) is found to be

$$\Phi(t) = \begin{cases} -f^+(t)/c(t), & \text{if} \quad f^+(t) < 0 \\ -f^-(t)/c(t), & \text{if} \quad f^-(t) > 0 \\ 0 & \text{otherwise} \end{cases} \tag{10}$$

where

$$f^{\pm}(t) = \frac{1}{2}\lambda^T H(t) + \sigma(t)(\nu\gamma \pm \theta\beta) \tag{11}$$

$$\gamma = \int_0^T \sigma(t)\Phi(t)\,dt, \quad \beta = \int_0^T \sigma(t)|\Phi(t)|\,dt \tag{12}$$

$\lambda \in \mathbf{R}^m$ being the vector of Lagrange's multipliers. The problem thus is reduced to that of finding the constants λ, γ, β from the system (10)–(12) and can be solved only numerically except in some very special cases.

The approach we have outlined makes it possible to gain some useful insights into the qualitative aspects of the problem. Consider the case $\nu = 0$. If $\theta = 0$ is also zero, then the estimator is found from (9). The function $\Phi(t)$ is almost everywhere different from zero and hence all available measurements will be necessary for the optimum calculation of the estimation $\hat{l}$. Even at arbitrarily small positive θ, the estimator vanishes in certain intervals $J_s = [t_s, t_{s+1}]$ showing that the measurements $z(t)$ for $t \in J_s$ are of no use for the optimum estimation $\hat{l}$. Note that these intervals become greater in length as θ is increased.

To illustrate the use of these results we consider the application of the above minimax problem (in the continuous formulation) to a simple model example of determining a linear combination of the parameters $q = (q_1, q_2)$ for an almost circular orbit from the measurements of the radial velocity

$$z(\tau) = \dot{r}(\tau) = q_1 \cos \tau + q_2 \sin \tau + \xi(\tau), \quad \tau \in [0, 2\pi]$$

If we assume that $\nu = 0$, $\sigma(\tau) \equiv \sigma =$ const., and $c(\tau) \equiv c =$ const., it turns out that there are two isolated intervals where the optimum estimator is different from zero, each having an angular length $2y$ shown in Fig. 2. The value of y is determined from the relation

$$\mu = 4(\tan y - y) \tag{13}$$

where $\mu = c/\theta\sigma^2$ is the ratio of the white noise intensity to the characteristic uncertainty of the correlation function. If $\mu \to 0$ (low intensity white noise), it follows from (13) that also $y \to 0$, leaving us with only two measurements, at points τ^* and $\tau^* + \pi$. The estimator $\Phi(\tau)$, cf. (10), is the sum of two delta functions centered at $\tau = \tau^*$ and $\tau = \tau^* + \pi$.

The other extreme occurs when $\theta\sigma^2 \to 0$. In this (classical) case we find that $\mu \to \infty$ and $y \to \pi/2$, and are led to the well-known result that in order to obtain the optimum estimation $\hat{l}$ all measurements on the interval $\tau \in [0, 2\pi]$ must be used.

Game Estimation Problem with Unmodelable Accelerations. The Solution Algorithm

An estimation with unmodelable accelerations was first formulated by Arkhangelskii and Belousov in 1979 [19] as a natural extension of the board scheme [6]

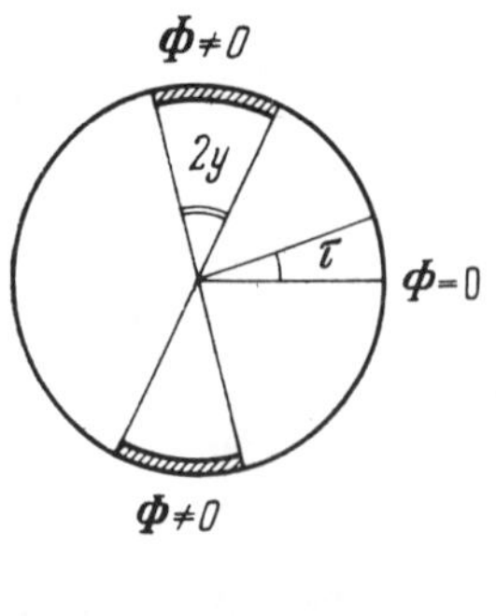

Figure 2

and received considerable attention in the last few years because of a serious error that had found its way into the original analysis and in connection with major difficulties arising in developing computation algorithms for high dimensional applications. The formulation we discuss below is a simplified version of that given in [10].

1. The evolution $x(t) \in \mathbf{R}^m$ of a system over the interval $t \in [0, T]$ is described by the set of linear differential equations

$$x' = A(t)x + B(t)u \tag{14}$$

where $u(t) \in \mathbf{R}^r$ is the vector of the unmodelable accelerations, and the matrices $A(t)$ and $B(t)$ have dimensionalities $m \times m$ and $m \times r$, respectively.

2. The measurements $z(t) \in \mathbf{R}$ carried out in the interval $[0, T]$ are of the form

$$Z(t) = H^T(t)x(t) + \xi(t) \tag{15}$$

where $H(t) \in \mathbf{R}^m$ is a known vector function; $\xi(t) \in \mathbf{R}$ represents the measurement errors.

3. The restrictions on the errors and unmodelable accelerations are given by

$$|\xi(t)| \leqslant \sigma(t), \quad \|u(t)\| \leqslant \gamma(t) \tag{16}$$

where $\|\cdot\|$ is the Euclidean norm of the vector; $\sigma(t)$ and $\gamma(t)$ are known non-negative functions.

4. The basic problem is to find an estimation for the parameter $l = a^T x(0)$.

5. The allowable estimations are of the linear class

$$\hat{l} = \int_0^T \Phi(t) z(t)\, dt \tag{17}$$

By analogy with the board approach, see above, we take the estimation error

$$S(\xi(t), u(t), \Phi(t)) \equiv |\hat{l} - l| \tag{18}$$

as the quantity criterion. The estimator $\Phi(t)$ is determined from the minimax problem

$$\begin{gathered} S(0, 0, \Phi(t)) = 0 \\ \min_{\Phi} \max_{\xi, u} S(\xi, u, \Phi) \end{gathered} \tag{19}$$

where the first equation is analogous to the estimation nondisplacement condition.

Let us define

$$h(t) = X^T(t)H(t) \in \mathbf{R}^m, \quad b(t) = X^{-1}(t)B(t) \tag{20}$$

conditions both for the continuous set of measurements and for the discrete formulation, in which it is assumed at the outset that it is only in certain discrete instants of time ($0 \leqslant t_1 \leqslant t_2 \leqslant \ldots \leqslant t_n \leqslant T$) that the system is accessible to measurement.

Theorem. For the pair $\tilde{\Phi}(t)$, $\tilde{u}(t)$ to satisfy the saddle point inequalities

$$J(\tilde{\Phi}, u) \leqslant J(\tilde{\Phi}, \tilde{u}) \leqslant J(\Phi, \tilde{u})$$

it is necessary that there be a constant vector $\beta \in \mathbf{R}^m$ for which the following four conditions are fulfilled.

Condition 1. For the m-dimensional vector function

$$p(t) = a - \int_0^t h(t)\Phi(t)\,dt$$

it is necessary that $p(T) = 0$. It is possible in the general case that there exists $0 < T^* < T$ such that $\tilde{\Phi}(t) \equiv 0$ on $(T^*, T]$ and, by condition 1, $p(t) \equiv 0$ on $(T, T]$.

Condition 2. On the interval $[0, T^*)$

$$\tilde{u}(t) = \gamma(t)\frac{b^T(t)p(t)}{\|b^T(t)p(t)\|}$$

Condition 3. Let us introduce the m-dimensional vector function

$$g(t) = \int_0^t b(\tau)\tilde{u}(\tau)d\tau$$

and denote by $X(t)$ the matrix of the fundamental system of solutions

$$dX/dt = A(t)X, \quad X(0) = I_m$$

where I_m is an identity matrix of size $m \times m$.

After performing some manipulations on the equations above, it is now an easy task to determine $\max_\xi S(\xi, u, \Phi)$ and to rewrite the problem (19) in an explicit form. It is necessary to determine $\Phi(t)$ and $u(t)$ from the nonlinear game problem

$$\int_0^T h(t)\Phi(t)\,dt = a$$

$$J = \int_0^T |\Phi(t)|\sigma(t)dt + \int_0^T \Phi(t)h^T(t)\int_0^t b(\tau)u(\tau)d\tau - \min_\Phi \max_u \tag{21}$$

with $u(t)$ subjected to the restriction (16).

The general theory of differential games [11] makes it possible to formulate the necessary minimax conditions, and additional analysis [10] determines the set of conditions (including all the necessary ones) which are sufficient for ensuring the saddle point inequalities. It can be shown without difficulty that the minimizing estimator of the problem is realized in the form of a sum of delta functions $\Phi_i\delta(t_i)$ centered in the discrete set of the points $t = t_i$ and having Φ_i as their weight factors. We will list further below the set of necessary conditions under the additional assumption that $|\Phi(t)| \leqslant M$, where M is a sufficiently large positive number. Letting M tend to infinity makes it possible [10] to lift this restriction and to derive the corresponding the function $f(t) = h^T(\beta + g(t))$ and the following subsets of the interval $[0, T^*)$:

$$F^+ = \{t \in [0, T^*), \quad \sigma(t) + f(t) < 0\}$$

$$F^- = \{t \in [0, T^*), \quad -\sigma(t) + f(t) > 0\}$$

$$F^0 = \{t \in [0, T^*), \quad |f(t)| \leq \sigma(t)\}$$

It is then necessary that

$$\tilde{\Phi}(t) \equiv \begin{cases} M, & \text{if} \quad t \in F^+ \\ -M, & \text{if} \quad t \in F^- \\ 0, & \text{if} \quad t \in F^0 \end{cases}$$

Condition 4. On the interval $t \in (T^*, T]$ there must exist a function $u(t)$ obeying the restriction (16) and securing the inequalities

$$-\sigma(t) \leq h^T(t)(\beta + g(T^*)) + h^T(t) \int_{T^*}^{t} b(\tau)\tilde{u}(\tau)\, d\tau \leq \sigma(t) \quad t \in [T^*, T]$$

Two simple two-dimensional problems will serve to illustrate the use of the sufficient conditions above. In both cases we consider the two-dimensional system

$$dx_1/dt = u, \quad dx_2/dt = x_1$$

and measure only the coordinate $x_2(t)$, that is, $z(t) = x_2(t) + \xi(t)$. We have

$$h(t) = \begin{pmatrix} t \\ 1 \end{pmatrix}, \quad b(t) = \begin{pmatrix} 1 \\ -t \end{pmatrix}$$

Example 1. Measurements $z(t)$ are allowed only at $t \in [0, 1]$ for $\sigma(t) \equiv \sigma$ and $\gamma(t) \equiv \gamma$ constant (the ratio $\alpha = \gamma/\sigma$ is the key parameter of the problem). One is required to estimate $x_2(t)$, (that is, $a^T \equiv (1, 0)$). It follows from the analysis of the sufficient conditions that two special cases are possible, namely

$$A) \quad 0 \leq \alpha \leq 4, \quad B) \quad \alpha > 4$$

In the case *A* the optimum estimator does not depend on α and is the same as in the absence of unmodelable accelerations ($\alpha = 0$). The only two measurements involved in the minimax program are those at the extremities of the segment, $t = 0$ and $t = 1$. We have

$$\Phi(0) = -\delta(0), \quad \Phi(1) = \delta(1), \quad \Phi(t) \equiv 0 \quad \text{for} \quad t \in (0, 1)$$

In the case *B*, again only two measurements are necessary, but the second of them is this time displaced into the point $t_1 = 2/\sqrt{\alpha}$, and the optimum estimator is of the form

$$\Phi(0) = -\frac{\sqrt{\alpha}}{2}\delta(0), \quad \Phi(t_1) = \frac{\sqrt{\alpha}}{2}\delta(t_1), \quad \Phi(t) \equiv 0 \quad \text{for} \quad t \in (0, t_1) \cup (t_1, 1]$$

whereas the required functional is

$$J/\sigma = \begin{cases} 2 + \alpha/2, & \text{for} \quad \alpha \leqslant 4 \\ 2\sqrt{\alpha}, & \text{for} \quad \alpha > 4 \end{cases}$$

Example 2. The measurement time interval is [0, 3] and measurements are possible only at $t = 1, 2, 3$. The function $\gamma(t)$ is defined in the form

$$\gamma(t) = \begin{cases} 0, & t \in [0, 1) \\ \gamma_1, & t \in (1, 2) \\ \gamma_2, & t \in [2, 3] \end{cases}$$

where γ_1 and γ_2 are known constants, and $\sigma(t) = \sigma = \text{const}$. The problem is to estimate the linear combination $l = t_0 x_1(0) + x_2(0)$, $0 \leqslant t_0 \leqslant 1$.

It follows from the analysis of the sufficient conditions that depending on the values of γ_1, γ_2, σ three situations may arise, which are:

A. $0 \leqslant \gamma_1 \leqslant -\gamma_2 + 4\sigma$

B. $(\gamma_1 + \gamma_2)/2 - 2\sigma \geqslant (\gamma_2 - \gamma_1)/2 + 4\sigma \geqslant 0$

C. $(\gamma_2 - \gamma_1)/2 + 4\sigma \geqslant (\gamma_1 + \gamma_2)/2 - 2\sigma \geqslant 0$

These inequalities represent the whole of the quadrant $\gamma_1 \geqslant 0$, $\gamma_2 \geqslant 0$, see Fig. 3, and describe therefore all the possible versions of the problem.

In the case *A* the optimum estimator turns out to be the same as with no unmodelable accelerations

$$\Phi(1) = \frac{1}{2}(3 - t_0)\Phi(1), \quad \Phi(2) = 0, \quad \Phi(3) = -\frac{1}{2}(1 - t_0)\delta(3)$$

In the case *B* the estimator is independent of the specific values of γ_1, γ_2 in the region *B*,

$$\Phi(1) = (2 - t_0)\,\delta(1), \quad \Phi(2) = -(1 - t_0)\,\delta(2), \quad \Phi(3) = 0$$

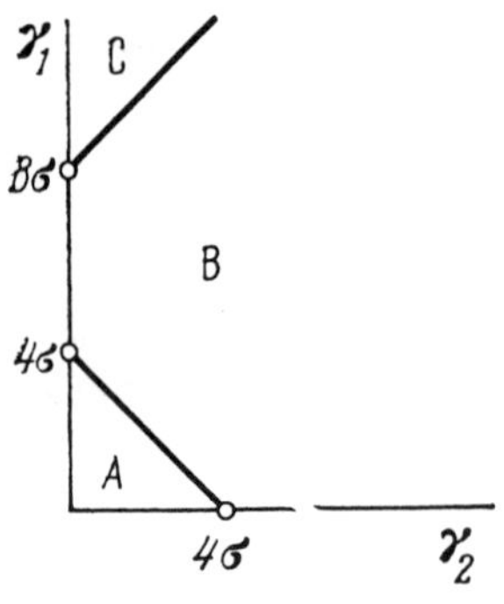

Figure 3

In the case C the parameters of the estimator are continuous functions of γ_1 and γ_2

$$\Phi(1) = (2 - t_0 + \Phi_3)\,\delta(1), \quad \Phi(2) = -(1 - t_0 + 2\Phi_3)\,\delta(2), \quad \Phi(3) = \Phi_3\,\delta(3)$$

$$\Phi_3 = \frac{z(1 - t_0)}{1 + z(1 - t_0)}, \quad z = 1 - \sqrt{\frac{\gamma_1 + \gamma_2 + 8\sigma}{2\gamma_1}}$$

In the minimax program of the case C is it is necessary that all three measurements should be used, which makes an essential difference between the unmodelable acceleration problem and the ordinary board scheme, where at most two measurements are needed in two-dimensional formulations. There may be situations in real problems with large m and very large n in which the minimax program depends on a very large number of measurements for its success. This, together with nonlinearity, is the major complicating factor in developing practical computation procedures for such problems.

Numerical Algorithms for the Problem (21)

Thus far only one version of numerical algorithm has been successfully applied to the problem (21). The method itself is described in [10]; in [12], a detailed account of the numerical work that has been performed by this method on a number of relatively simple models will be found.

The basic assumption in the discrete version of the problem is that measurements may be carried out in certain fixed instants of time

$$0 \leqslant t_1 \leqslant t_2 \leqslant \ldots \leqslant t_N \leqslant T$$

In this case the estimator can be represented as a sum of delta functions

$$\Phi(t) = \sum_{k=1}^{N} \Phi_k\,\delta(t_k)$$

where $\delta(t_k)$ is the delta function centered about the instant t_k. The problem (21) may then be rewritten as

$$\sum_{k=1}^{N} h_k \Phi_k = a$$

$$J = \sum_{k=1}^{N} |\Phi_k|\,\sigma_k + \sum_{k=1}^{N} \Phi_k h_k^T \sum_{s=1}^{K} \int_{t_{s-1}}^{t_s} b(\tau) u(\tau)\, d\tau - \min_{\Phi} \max_{u} \tag{22}$$

where $h_k = h(t_k)$, $\sigma_k = \sigma(t_k)$.

The principal idea of the algorithm is to submerge the problem (22) into a one-parameter family of such problems, each differing from the original one in having the restriction $\|u(t)\| \leqslant \gamma(t)$ replaced by $\|u(t)\| \leqslant \epsilon\gamma(t)$, where $0 \leqslant \epsilon \leqslant 1$. The first step in the algorithm is to find a solution satisfying the sufficient minimax conditions for a certain $\epsilon = \epsilon^0$. This done, one linearizes these conditions to build a solution for $\epsilon = \epsilon^0 + \Delta$, where Δ is a sufficiently small positive number. The solution of the original problem is completed when the value $\epsilon = 1$ is achieved. The important point to be made is that at $\epsilon = 0$ the problem (22) coincides with the problem (3), the latter admitting of effective solution by the simplex method known in linear programming, thus supplying initial data for the procedure.

The algorithm showed itself quite workable when applied to problems of the kind discussed in Example 2 (the number of measurement points was actually greater in [12]) and is suitable—in principle—for the numerical solution of discrete problems of greater complexity. In practice, however, the algorithm was found rather difficult to work with and as far as high-dimensional problems are concerned, clearly more effective numerical methods are needed. Bakhshiyan [13] reduces the problem (22) to a certain high-dimensionality problem of generalized linear programming (with mN equality-type conditions). A GLP problem can then be treated by the simplex method, modified so as to be capable of repeatedly solving at each iteration a certain "internal problem" of (generally) nonlinear programming.

The work by Bakhshiyan suggested to us the idea of reducing the problem (22) to the well-known problem of generalized linear impulse correction or GLIC [14]. The general method of proceeding is as follows. The second of equations (22) is transformed to

$$J = \sum_{u=1}^{N} |\Phi_k| \sigma_k + \sum_{s=1}^{N} \int_{t_{s-1}}^{t_s} \sum_{k=s}^{N} \Phi_k h_k^T b(\tau) u(\tau)\, d\tau$$

(where $t_0 = 0$), which very much simplifies the determination of $\tilde{u} = \arg\max_u J$. For $t_{s-1} \leqslant \tau \leqslant t_s$, $s = \overline{1, N}$ we have

$$\tilde{u}(\tau) = \gamma(\tau) \sum_{k=s}^{N} \Phi_k h_k^T b(\tau) \Bigg/ \left\| \sum_{k=s}^{N} \Phi_k h_k^T b(\tau) \right\|$$

$$\bar{J} = \max_u J = \sum_{k=1}^{N} |\Phi_k| \sigma_k + \sum_{s=1}^{N} \int_{t_{s-1}}^{t_s} \gamma(\tau) \left\| \sum_{k=s}^{N} \Phi_k b^T(\tau) h_k \right\| d\tau$$

If the quantity $\max_{s=\overline{1,N}} (t_s - t_{s-1})$ is sufficiently small in a particular subdivision of the segment $[0, T)$, then the approximation

$$\int_{t_{s-1}}^{t_s} \gamma(\tau) \left\| \sum_{k=s}^{N} \Phi_k b^T(\tau) h_k \right\| d\tau \approx \left\| \sum_{k=s}^{N} \Phi_k b_s^T h_k \right\|$$

is in principle possible (where $b_s = (t_s^*) b(t_s^*)(t_s - t_{s-1})$ is an $m \times m$ matrix and $t_s^* = (t_s + t_{s-1})/2$), and we are lead to the functional

$$\bar{J} = \sum_{k=1}^{N} |\Phi_k|\, \sigma_k + \sum_{s=1}^{N} \left\| \sum_{k=s}^{N} \Phi_k b_s^T h_k \right\|$$

Using the representation

$$\Phi_k = v_k\alpha_k, \quad |\alpha_k| = 1, \quad v_k \geqslant 0$$

$$\sum_{k=s}^{N} \Phi_k b_s^T h_k = w_s\beta_s, \quad \beta_s \in \mathbf{R}^r, \quad \|\beta_s\| = 1, \quad w_s \geqslant 0$$

the problem (22) is now reduced to the following GLIC problem.

It is required to determine v_k, w_k, α_k, β_k from the conditions

$$\begin{gathered}\sum_{k=1}^{N} h_k v_k \alpha_k = a \\ \sum_{k=s}^{N} (b_s^T h_k) v_k \alpha_k = w_s\beta_s, \quad s = \overline{1, N} \\ v_k \geqslant 0, \quad w_s \geqslant 0, \quad \alpha_k = \pm 1, \quad \beta_s \in \mathbf{R}^r, \quad \|\beta_s\| = 1 \\ \sum_{k=1}^{N} v_k\sigma_k + \sum_{s=1}^{N} w_s - \min_{\substack{v_k,\alpha_k \\ w_s,\beta_s}}\end{gathered} \tag{23}$$

The GLIC problem is also a problem of generalized linear programming, but is by far much simpler in comparison with the problem discussed in [13]. First, the "inner problem" is elementary to solve; second, the GLIC approach requires less computer time because the dimensionality of the equality type conditions in (23) is $rN + m$ against mN in [13], and $m \gg r$ in most applications.

REFERENCES

1. Huber, P. J. Robust estimations of a location parameter. Ann. Math. Statist., 1964, Vol. 35, No. 1, p. 73.
2. Huber, P. J. Robust statistics. Wiley, N.Y., 1981.
3. Smolyak, S. A. and Totorenko, B. P. Robust estimation methods. Moscow, Statistika, 1980 (in Russian).
4. Tsypkin, Ya. Z. and Polyak, B. T. Robust version of the maximum likelyhood method. In: Dynamics of systems. Gorkii, 1977 (in Russian).
5. Tsypkin, Ya. Z. Adaptation and instruction in automatic systems. Moscow, Nauka, 1968 (in Russian).
6. Lidov, M. L. On *a priori* accuracy in least-square method parameter determination. Kosm. Issled., 1964 Vol. 2, No. 5, (in Russian).
7. Bakhshiyan, B. Ts., Nazirov, R. R., and Elyasberg, P. E. Determination and correction of motion. Moscow, Nauka, 1980 (in Russian).
8. Lidov, M. L. Continuous formulation for path parameter minimax estimation. Kosm. Issled., 1984, Vol. 22, No. 4 (in Russian).
9. Arkhangelskii, B. A. and Belousov, L. Yu. Minimax accuracy in the spacecraft orbit determination problem with unmodelable accelerations. Ibid., 1979, Vol. 17, No. 3, p. 342 (in Russian).
10. Lidov, M. L. Game theory estimation problem for unmodelable accelerations. Solution algorithm. Ibid., 1986, Vol. 24, No. 2, p. 246.
11. Isaacs, R. Differential games. Wiley, N.Y., 1965.

12. Lidov, M. L. and Lyakhova, V. A. Motion parameter minimax estimation problem: numerical solution in the presence of unmodelable accelerations. Kosm. Issled., 1987, Vol. 25, No. 1.
13. Bakhshiyan, B. Ts. Simplex algorithm for the guaranteed estimation optimum problem with unmodelable perturbations. Ibid., in print.
14. Lidov, M. L. Path correction and measurement composition problems. Analogy and solution algorithms. Ibid., 1971, Vol. 9, No. 5.

DYNAMICS OF SPACE TETHERED SYSTEMS

V. V. Beletskii

1. Introduction. Dynamics of tethered systems is a subject of long-standing tradition in this country and may be associated with such well-known names in Russian mechanical science as Krylov [1], Kochin [2], and Minakov [3], among others. Advances in space science technology have given a new dimension to the subject in recent years, when it has become increasingly evident that tethers can be as effective in space as they have ever been on the Earth in towing, load hoisting, connecting construction parts, and similar applications. It was found appropriate during the preparation of the 25th Congress of the International Astronautical Federation (Lausanne, October, 1984) to organize a special section, "Space Tether Applications", whose topic was the new possibilities offered by static, dynamic, and electric properties of tether in the study of space.

The first step in this direction was made in 1965 when the cosmonaut Leonov remained attached by a so-called umbilical cord to the Vostok-2 spacecraft during his (man's first) fly-away. The idea of a spacecraft being tethered to the last stage of the carrier-rocket was tested in the 1967 Gemini-11 and Gemini-12 missions. In 1984, NASA and ESA (European Space Agency) joined their efforts to develop a program of the first three flights of the Shuttle-Tethered-Subsatellite System, scheduled originally for the years 1987–1990. Two of the flights are intended for testing a 20 km-long electromagnetic tether system; in the third an athmospheric probe will be deployed from an altitude of 120 km by means of a 100 km-long tether.

Not unnaturally, the promises of tethered space systems have stimulated extensive theoretical examination of their dynamical properties in general, and of their stability in particular. The first works in this field were published about a quarter century ago. Today, we are fully justified in calling the subject of space tethered systems a quite independent and full-fledged direction in the vast field of space flight mechanics, with its own investigation methods, well-tested models, and quite a number of reliable results (see [4]–[30]). The present paper is in fact a systematized review of these results and is based largely on the works of the author in cooperation with E. M. Levin and other colleagues.

2. Equations of motion of an orbital tethered system (OTS). The system is idealized as two point masses m_A and m_B (shown by points A and B in Fig. 1) connected by a ponderable, flexible, extensible string. The string may in general be reeled in or out by either of the bodies so that its length will vary, as also will the masses m_A and m_B. Let $0XYZ$ be an inertial coordinate system, the origin 0 being the center of masses of the attracting body (the Earth). An inherent variable of the problem is the length measured along the unstressed string. We will assume this length to be measured from point A and will denote it by the letter s; clearly, $0 \leqslant s \leqslant l$, where l is the total length of the unstressed string. There are three regions which may be identified at any time t, namely, $0 \leqslant s \leqslant s_A(t)$ (string within the body A), $s_A(t) \leqslant s \leqslant s_B(t)$ (string between the bodies A and B), and $s_B(t) \leqslant s \leqslant l$ (string within the body B). If the string is constant in length, $s_A = 0$, and $s_B = l$. Let $R_A = R(s_A, t)$ and $R_B = R(s_B, t)$ be the radius vectors of the end points and $R = R(s, t)$ the radius vector of a current point. The equations of motion are of the form

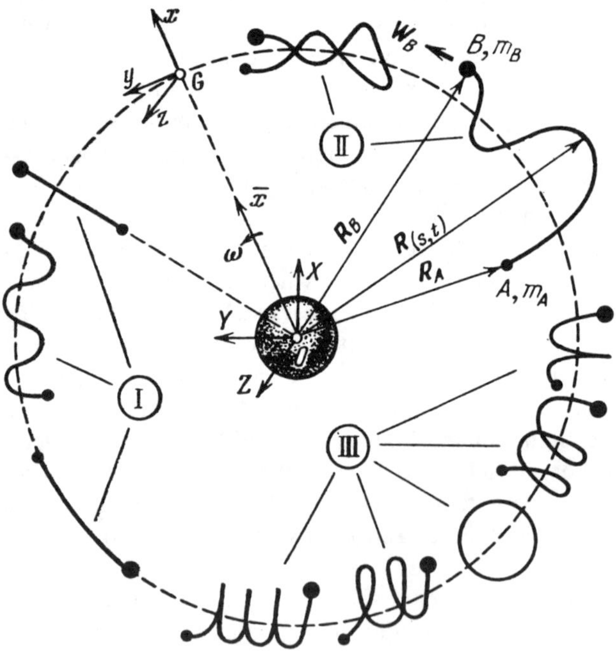

Figure 1

$$\rho \frac{\partial^2 R}{\partial t^2} = \frac{\partial T}{\partial s} - \rho\mu \frac{R}{R^3} + F \tag{2.1}$$

$$T = T\tau, \quad \tau = \frac{\partial R}{\partial s} : \left| \frac{\partial R}{\partial s} \right| \tag{2.2}$$

$$T = E(\gamma - 1), \quad \gamma = \left| \frac{\partial R}{\partial s} \right| \tag{2.3}$$

$$m_A \frac{d^2 R_A}{dt^2} = \tau_A \left[T_A - \rho\gamma \left(\frac{dS_A}{dt} \right)^2 \right] - m_A \mu \frac{R_A}{R_A^3} + F_A + W_A \tag{2.4}$$

$$m_B \frac{d^2 R_B}{dt^2} = -\tau_B \left[T_B - \rho\gamma_B \left(\frac{dS_B}{dt} \right)^2 \right] \quad m_B \mu \frac{R_B}{R_B^3} + F_B + W_B \tag{2.5}$$

$$\tau_{A,B} = \left(\frac{\partial R}{\partial s} \right)_{A,B} \Big/ \gamma_{A,B}; \quad \gamma_{A,B} = \left| \frac{\partial R}{\partial s} \right|_{A,B}; \quad T_{A,B} = E(\gamma_{A,B} - 1) \tag{2.6}$$

The system (2.1)–(2.6) contains both ordinary and partial differential equations: equation (2.1) describes the dynamics of the string of linear density $\rho = \rho(s)$; equation (2.2) shows that the vector $\mathbf{T}$ of string tension is tangent to the string, $\boldsymbol{\tau}$ being the unit vector of the tangent; (2.3) expresses the Hook's law assumption adopted in the problem, with γ denoting elongation and $E = E(s) = E_M f$, where E is Young's modulus and f the sectional area of the tether. In equations (2.1), (2.4), (2.5), the terms containing the gravitational constant μ describe the attraction on the string element and on the bodies A and B; $\mathbf{F}$, $\mathbf{F}_A$ and $\mathbf{F}_B$ are the remaining external (perturbing) forces acting on these objects (in the same order); W_A and W_B are the thrusts of the propulsive motors of the bodies A and B (if there are any). Equations (2.4)–(2.6) are the equations of motion (or Meshcherskii's equations) for the bodies A and B; they contain reactive forces (the terms proportional to ρ) due to the release or the retrieval of the string by the bodies A and B. The reactive forces, thrusts, and some other forces (of electromagnetic nature) may serve as controls in the system (2.1)–(2.6).

There are situations in which the above equations admit of some simplifications.

3. Equation of stationary motions of a tethered system [18, 19]. We introduce a rotating coordinate system $Oxyz$ whose origin coincides with that of the system $0XYZ$, see above; the vector of angular velocity $\boldsymbol{\omega}$ is constant and directed along the axis $0Z$, which coincides with the $0z$ axis. It turns out that under certain non-restrictive conditions the tethered system may be in equilibrium relative to the $0xyz$ coordinate system, and, when referred to the absolute space, will travel as a solid body along the (circular) orbit with an orbital angular velocity $\boldsymbol{\omega}$. One of the necessary conditions for such a *stationary* motion is the constancy of the tether length. The equations of stationary motions are derived from the general equations (2.1)–(2.6). Referring the tethered system to $0xyz$ axes and assuming that $\mathbf{F} =$

$\mathbf{F}(\boldsymbol{\omega}, s, \mathbf{R}, d\mathbf{R}/ds)$ and that $\mathbf{F}_A$ and $\mathbf{F}_B$ are constant (stationary motion!) we obtain

$$\frac{dT}{ds} = \rho\bar{\omega} \times (\omega \times R) + \rho\mu \frac{R}{R^3} - \bar{F} \tag{3.1}$$

$$\frac{dR}{ds} = T\left(\frac{1}{T} + \frac{1}{E}\right) \tag{3.2}$$

$$T_A - m_A\, \omega\, (\omega \times R_A) - m_A\mu \frac{R_A}{R_A^3} + F_A + w_A = 0 \tag{3.3}$$

$$-T_B - m_B\omega \times (\omega \times R_B) - m_B\mu \frac{R_B}{R_B^3} + F_B + w_B = 0 \tag{3.4}$$

We thus conclude that the stationary configuration $R(s)$ and the stationary tension $T(s)$ of the tether are found by integration of the system (3.1)–(3.2) subjected to the boundary conditions (3.3)–(3.4). It proves more convenient, however, to reduce the boundary value problem (3.1)–(3.4) to a certain Cauchy problem using the presence of the arbitrary thrust forces $\mathbf{W}_A$ and $\mathbf{W}_B$ in the system [18, 19].

4. Stationary configurations in a gravitational field [18, 19]. If $\mathbf{F} = 0$, the system (3.1)–(3.2) admits of two first integrals, the tension integral

$$T\left(1 + \frac{T}{2E}\right) + \rho\frac{\omega^2 r^2}{2} + \rho\frac{\mu}{R} = c_1 \tag{4.1}$$

for a uniform tether (ρ = const., E = const.), and the integral of the moment of tension forces relative the $0z$ axis

$$rT_\varphi = c_2 \tag{4.2}$$

for the general case of a non-uniform tether. Here we have denoted by r the distance of the tether element s from the axis of rotation $0z$, and by the T_φ the transverse component of the tension force (lying in the plane $0xy$ and normal to $\mathbf{r}$). In the plane case we have $r = R$ and the tether configuration can be determined in quadratures using the integrals (4.1) and (4.2). The stationary configuration of an infinitely long tether is of the nature of a wave; owing to the presence of the thrust forces $\mathbf{W}_A$ and $\mathbf{W}_B$, each finite portion of this wave configuration is realized as a stationary configuration of a finite-length tether; two special cases are the radial configuration (the tether may be folded) and the orbital configuration (the tether coincides with the circular orbit).

These special cases are also realizable when $\mathbf{W}_A = \mathbf{W}_B = 0$ and represent therefore the natural equilibrium configurations of the tether in a gravitational field (see Fig. 1*I*).

It is convenient to introduce the characteristic length

$$L_* = \sqrt{\frac{2\sigma_*}{3\rho_M\omega^2}} \tag{4.3}$$

where $\sigma_* = T_*/f$ is the tether tension at a certain reference altitude $R_* \approx R_0$; $\rho_M = \rho/f$ is the volume density of the cable material; f is the tether cross-sectional area. The terrestrial analogue of (4.3) is $l_* = \sigma_*/\rho_M g$, the maximum length still permissible—in terms of strength—for a vertical tether suspended above the Earth's surface; here g is the free fall acceleration and σ_* the maximum allowable stress for the given material (not exceeding rupture stress value). We have listed in Table 1 the values of L_* calculated from σ_* for a number of tether materials for an altitude of 220 km. The values of l_* are also given for comparison.

When in the natural vertical configuration, the parameters of the tether system and the orbital angular velocity are related by ($E = \infty$, $R_B > R_A$)

$$\omega^2\left[m_A R_A + m_B R_B + \frac{\rho}{2}(R_B^2 - R_A^2)\right] = \mu\left[\frac{m_A}{R_A^2} + \frac{m_B}{R_B^2} + \rho\left(\frac{1}{R_A} - \frac{1}{R_B}\right)\right] \tag{4.4}$$

showing that the mass center of the system follows a non-Keplerian orbit in its motion. With no end masses, it follows from (4.4) that the half-length ($l_0/2$) of the "Space elevator" (a tether reaching the Earth's surface) must be about 62,000 km in length, which by far exceeds allowable L_* values (if $R_c = 42{,}000$ km, for example, L_* is about 10,000 km for a Kevlar string).

5. Weightless tether. Radial configuration stability. In the tether system model we will consider in this section two point masses are connected by a weightless inextensible tether; gravity is the only force present; the system is moving in a Keplerian orbit; gravity gradient effects are considered to be linear in displacements relative to the system mass center.

Referring to Fig. 1, we associate with the system mass center G an orbital set of axes $Gxyz$ such that the x axis is directed along the radius vector of the mass center; the y axis is tangent to the orbit and is considered positive in the direction of motion; the z axis is normal to the orbit plane. It will be convenient to introduce the assumption, of minor consequence for the problem as a whole, that of the bodies of the system, one is by far heavier than the other (to the degree, in fact, that it represents the center of gravity of the system). We use the dimensionless time $\tau = \omega t$ and normalize with respect to the tether length l the coordinate of the lighter body. With this as background, the equations of motion of the lighter body are

$$\ddot{x} - 2\dot{y} - 3x = -\lambda x \tag{5.1}$$

Table 1

Tether	$\sigma_*\left(\frac{kg}{mm^2}\right)$	$\rho_M\left(\frac{g}{cm^3}\right)$	l_* (km)	L_* (km)
Tungsten string	140	19.3	7.2	185
Steel string	100	7.8	12.8	245
Titan string	70	4.55	15.4	268
Glass fiber	150	2.5	60.0	530
Boron fiber	350	2.4	146	840
Kevlar fiber	370	1.45	255	1080

$$\ddot{y} + 2\dot{x} = -\lambda y$$

$$\ddot{z} + z = -\lambda z \qquad (5.2)$$

$$x^2 + y^2 + z^2 \leqslant 1$$

These equations admit of a stationary solution representing a radial arrangement of the system

$$x = \pm 1, \quad y = z = 0, \quad \dot{x} = \dot{y} = \dot{z} = 0, \quad \lambda = 3 \qquad (5.3)$$

The perturbed motion generally contains free motion phases ($\lambda = 0$; strict inequality in (5.2)) and coupled motion phases ($\lambda > 0$; equal sign in (5.2)), both possessing the energy integral

$$\dot{x}^2 + \dot{y}^2 + \dot{z}^2 + z^2 + 3(1 - x^2) = h \qquad (5.4)$$

to secure the stability of the stationary motion (5.3). Fig. 2 shows schematically the surfaces bounding the allowable motion regions [6].

6. Weightless tether. The evolution of the system. In the (x, y)-plane the coupled motion is described by the integral

$$\dot{\varphi}^2 - 3\sin^2\varphi = h \qquad (6.1)$$

and the decoupling condition may be written as

$$\dot{\varphi}^2 - 2\dot{\varphi} + 3\sin^2\varphi < 0 \qquad (6.2)$$

where φ is the angle measured from the orbit tangent to the direction of the stretched tether; the region $\dot{\varphi} > 0$ corresponds to rotations opposite in direction to the orbital motion. Fig. 3 shows the phase paths (6.2) and the decoupling zones of the motion (dotted regions). The free motion is bound to get coupled, the subsequent evolution of the system depending on the nature of the impact accompanying the coupling formation. For the case of a perfectly inelastic impact, Fig. 4 illustrates the mapping $h_1 \rightarrow h_2$ of the pre-impact into the post-impact energy. This depen-

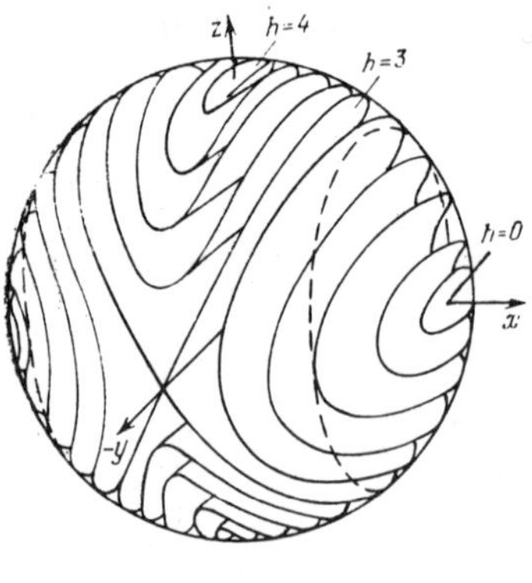

Figure 2

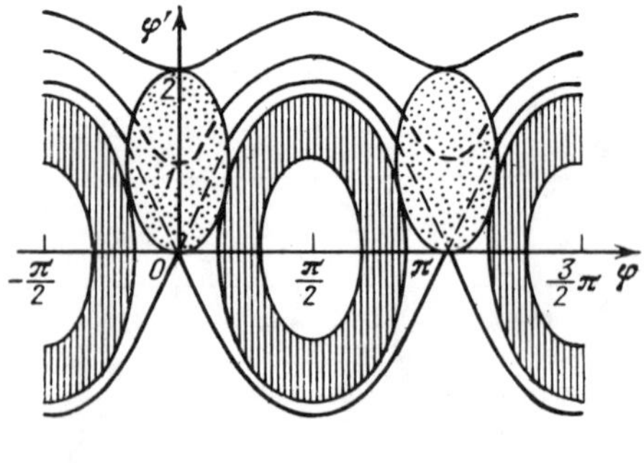

Figure 3

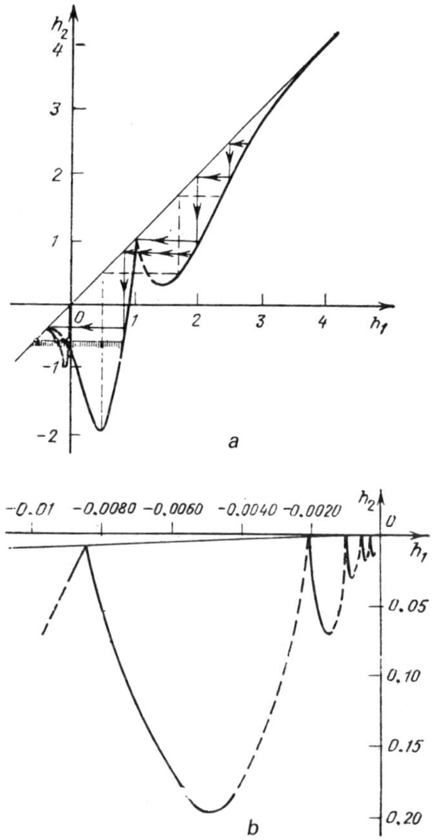

Figure 4

dence determines the evolution of the system to limit regimes. Calculations predict that in a finite time the system will assume with probability 0.8 one of the possible forms of coupled oscillations about the local vertical, the energy constant ranging from -0.5 to -0.85 (48 to 66 deg. amplitude range). With probability of about 0.2, the motion will tend for an infinitely long time to the limit cycle, or $h = -0.5$ oscillations (amplitude 66 deg.). Finally, there is zero probability (from the zero measure initial data set) that in a finite time the motion will take an unstable mixed (coupled-free) form with $h = 1$ or will assume one of the values of the infinite, discrete, and converging to zero set $h = -0.00825$; $h = -0.00215$; $h = 0.00095$; ... $h = -0$.

7. Tether system deployment model [21]. The operational stationary arrangement of the tether system is obtained from an initial folded state. The deployment of the tethered system is a controlled process, in which the tether tension may be taken as control. The simplest model description of such a process is given by equation (5.1) with λ, the tether tension, considered as control. The $\lambda = 0$ case corresponds to the free deployment process in which the end mass drifts along the orbit tangential ($x = c$, $y = -(3/2)c\tau + c_0$ being the partial solution of (5.1)). This type of deployment is of course a far cry from the required local-vertical tether orientation. If the tension control is such that $\lambda = 3$ at all times, the system does drift along the required vertical because the equations (5.1) have $x = c$, $x = -(3/2c\tau + c_0)$ as their partial solution.

8. Heavy tether. Radial configuration stability. The wave-like stationary configurations of a tethered system are known to be unstable [22]. Stability may be achieved for monotonic configurations, particularly for a radial configuration in a gravitational field. This statement admits of a strict proof [14].

Assuming the tether to be ponderable and making the same assumptions used in Sec. 5, equations (2.1)–(2.6) referred to the orbital coordinate system $Gxyz$ become

$$\left.\begin{aligned} &\ddot{x} - 2\omega\dot{y} - 3\omega^2 x = \frac{1}{\rho}(Tx')' \\ &\ddot{y} + 2\omega\dot{x} = \frac{1}{\rho}(Ty')' \\ &\ddot{z} + \omega^2 z = \frac{1}{\rho}(Tz')' \end{aligned}\right| = \frac{1}{m_A}\begin{pmatrix} Tx' \\ Ty' \\ Tz' \end{pmatrix}_A \left| = -\frac{1}{m_B}\begin{pmatrix} Tx' \\ Ty' \\ Tz' \end{pmatrix}_B \right. \tag{8.1}$$

$$x'^2 + y'^2 + z'^2 = 1 \tag{8.2}$$

where dots denote differentiation with respect to be dimensionless time, and primes with respect to s; equations (8.1) describe the oscillations of the system; (8.2) represents the inextensibility condition; the tension T plays the role of a Lagrange's multiplier in the constraint problem (8.2). The left-hand side of the system (8.1) has the same form for all points of the tethered system, whereas in the right-hand side the first column refers to the points of the tether and the second and the third to the end bodies A and B respectively.

There exists a stationary radial configuration

$$\begin{aligned} &y = 0, \quad z = 0, \quad x = x_0(s) = s - s_c \\ &T = T_c - \int_0^x 3\rho\omega^2 \times dx, \quad T_A = -3m_A\omega^2 x_A, \quad T_B = 3m_B\omega^2 \times B \end{aligned} \tag{8.3}$$

where the index C refers (in the stationary configuration) to the system mass center (moving in the unperturbed circular orbit under the assumptions adopted).

The stability of the relative equilibrium (8.3) follows from the fact that the relative motion of the system (8.1), (8.2) has an energy integral which serves as Lyapunov's functional.

9. Tether system pendular motions. Equations (8.1), (8.2) have an exact partial solution

$$x = (s - s_c)\cos\psi\cos\Theta, \quad y = (s - s_c)\cos\psi\sin\Theta, \quad z = (s - s_c)\sin\psi \tag{9.1}$$

corresponding to situations in which the tethered system behaves as a rigid rectilinear rod (pendular motion). Here ψ is the angle between the tether and the mass center orbit plane; the angle Θ is made by the tether with the radial direction in the orbit plane and is measured positively from Gx to Gy. The quantities $\Theta(t)$ and $\psi(t)$ are determined by the equations

$$\begin{aligned} &\ddot{\psi} + \sin\psi\cos\psi[(\omega + \dot{\Theta})^2 + 3\omega^2\cos^2\Theta] = 0 \\ &\ddot{\Theta} + \dot{\omega} - 2\dot{\psi}tg\psi(\omega + \dot{\Theta}) + 3\omega^2\sin\Theta\cos\Theta = 0 \end{aligned} \tag{9.2}$$

and the tether tension by

$$T = T_A + \lambda\int_{s_A}^{s}(s_c - s)\rho ds = T_B + \lambda\int_{s}^{s_B}(s - s_c)\,\rho ds \tag{9.3}$$

$$\begin{aligned} &T_A = m_A(s_c - s_A)\lambda; \quad T_B = m_B(s_B - s_c)\lambda \\ &\lambda = \dot{\psi}^2 + \cos^2\psi(\omega + \dot{\Theta})^2 - \omega^2 + 3\omega^2\cos^2\psi\cos^2\Theta \end{aligned} \tag{9.4}$$

The condition for the stretched-tether motion is $\lambda > 0$. In the plane motion we have $\psi = 0$, and equations (9.2) together with the tether slackening condition $\lambda < 0$ for a circular orbit ($\dot{\omega} = 0$) transform into (6.1) and (6.2) provided necessary notation modifications.

For a uniform tether $\rho = \text{const.}$, simplifying (9.3) to

$$T = T_A + \frac{1}{2}\rho\lambda(s - s_A)(2s_c - s - s_A) \tag{9.5}$$

Since the tether tension is proportional to the same factor $\lambda(t)$ at all points of the tether length, it follows that the tether tension distribution is parabolic in s, regardless of the particular pendular motion mode. This tension increases from the value T_A at point A to its maximum value $T_c = T_A + (\frac{1}{2})\rho\lambda(s_c - s_A)^2$ and then decreases to the value T_B at point B.

10. Small oscillations of a radial tethered system: shape and frequencies. Taking variations with respect to the stationary motion (8.3), it is seen from (8.1) and (8.2) that $\delta x' \equiv 0$, $(\delta x \equiv 0)$, and

$$\left.\begin{aligned} &\delta\ddot{y} = \frac{1}{\rho}(T\delta y')' \\ &\delta\ddot{z} + \omega^2\delta z = \frac{1}{\rho}(T\delta z')' \end{aligned}\right| = \frac{1}{m_A}\begin{pmatrix} T\delta y' \\ T\delta z' \end{pmatrix}_A \left| = -\frac{1}{m_B}\begin{pmatrix} T\delta y' \\ T\delta z' \end{pmatrix}_B \right. \tag{10.1}$$

where the tension T is given by (8.3). Assuming a partial solution of the form

$$\delta y = U_y(s)\cos\beta_y t, \quad \delta z = U_z(s)\cos\beta_z t \tag{10.2}$$

we are led to the same Sturm problem

$$(TU')' + \beta^2\rho U = 0; \quad (TU')_A + m_A\beta^2 U_A = 0; \quad (TU')_B - m_B\beta^2 U_B = 0 \tag{10.3}$$

$$\beta_z^2 = \beta_y^2 + \omega^2 \tag{10.4}$$

for both transverse and axial oscillations. We conclude that both types of oscillations coincide in shape and that the frequencies β_{nz} and β_{ny} are related by (10.4).

For any $\rho(s)$, equations (10.3), (10.4) have a strict solution $\beta_{1y} = \omega\sqrt{3}$, $\beta_{1z} = 2\omega$, $U_{1y} = U_{1z} = s - s_c = x_0$ representing dumb-bell pendular oscillations of the system. The remaining possible oscillatory modes depend on the specific mass distribution law over the tether length. For a uniform tether, (8.3) yields

$$T = \frac{3}{2}\rho\omega^2(L^2 - X_0^2), \quad L = \sqrt{\frac{2T_c}{3\rho\omega^2}} \tag{10.5}$$

and the first of equations (10.3) becomes

$$\frac{d}{d\xi}\left[(1 - \xi^2)\frac{dU}{d\xi}\right] + \nu(\nu + 1)U = 0, \quad \xi = \frac{x_0(s)}{L}, \quad \nu(\nu + 1) = \frac{2\beta^2}{3\omega^2} \tag{10.6}$$

The general solution to (10.6) is expressed in terms of Legendre functions of the first and second kind,

$$U = C_1P_\nu(\xi) + C_2Q_\nu(\xi)$$

which when substituted into the second and third of equations (10.3) yields a homogeneous linear system for C_1, C_2. The characteristic equation for determining the frequency spectrum is obtained in the usual manner by equating the determinant of the system to zero

$$\left[\frac{dP_\nu}{d\xi} - \frac{\nu(\nu + 1)}{2\xi}P_\nu\right]_A\left[\frac{dQ_\nu}{d\xi} - \frac{\nu(\nu + 1)}{2\xi}Q_\nu\right]_B$$
$$= \left[\frac{dQ_\nu}{d\xi} - \frac{\nu(\nu + 1)}{2\xi}Q_\nu\right]_A\left[\frac{dP_\nu}{d\xi} - \frac{\nu(\nu + 1)}{2\xi}P_\nu\right]_B \tag{10.7}$$

The sequence of the roots ν_n of this equation always contain the root $\nu_1 = 1$ corresponding to the form $U_1 = \xi$ (dumb-bell oscillations). In the general case, equation (10.7) must be solved numerically. The following are the special cases which lend themselves to analytical solution:

1) No end bodies are present [8], so that $m_A = m_B = 0$, $x_A = -L$, $x_B = L$. The roots of (10.7) are all the natural numbers ($\nu_n = n = 1, 2, \ldots$) the eigenfunctions are the Legendre polynomials of the first kind, and the eigenfrequencies are given by

$$U_n = P_n(\xi)$$

$$\beta_{ny} = \omega\sqrt{\frac{3}{2}n(n + 1)}; \quad \beta_{nz} = \omega\sqrt{\frac{3}{2}n(n + 1) + 1} \tag{10.8}$$

2) A satellite with a string: $m_A = 0$, $m_B = \infty$, $x_A = -L$. The roots of (10.7) are the odd numbers ($\nu_n = 2n + 1$, $n = 0, 1, 2, \ldots$), the eigenfunctions are the odd-power Legendre polynomials, and the eigenfrequencies are given by (10.8) with n odd.

For a uniformly stressed tether, (8.3) gives

$$T = T_c \exp(-x_0^2/L^2), \quad L = \sqrt{2\sigma/(3\rho_M\omega^2)} \tag{10.9}$$

where σ is a constant stress; ρ_M is the mass density of the material. The first of equations (10.3) reduces to the form

$$\frac{d^2U}{d\xi^2} - 2\xi\frac{dU}{d\xi} + 2\nu U = 0, \quad \xi = \frac{x_0}{L}, \quad 2\nu = \frac{2\beta^2}{3\omega^2} \tag{10.10}$$

to be solved in terms of parabolic cylinder functions.

11. Atmospheric probe. Stationary motion [18, 19]. For situations in which the resistance of the atmosphere is of relevance, the stationary motion of the tethered system is determined from equations (3.1)–(3.4), where $\mathbf{F}$, $\mathbf{F}_A$, and $\mathbf{F}_B$ are the aerodynamic forces on a unit length of the tether, on the probe A, and on the satellite-carrier B, respectively. Let

$$\mathbf{F}_{A,B} = -\frac{1}{2}C_{A,B}\cdot S_{A,B}(\rho_a)_{A,B}\left|\boldsymbol{v}_{A,B}\right|\boldsymbol{v}_{A,B}$$
$$\mathbf{F} = -\frac{1}{2}c\rho_a d\gamma\left|\sin\alpha\right|\left|\boldsymbol{v}\right|\boldsymbol{v} = -\frac{1}{2}cd\rho_a\left|\frac{\partial\mathbf{R}}{\partial s}\times\boldsymbol{v}\right|\boldsymbol{v} \tag{11.1}$$

where $C_{A,B}$, C are the coefficients of aerodynamic resistance; $S_{A,B}$—the midship section areas of the bodies A and B; $\boldsymbol{v}_{A,B}$ and $\boldsymbol{v}$—the velocities of the bodies A and B and of a tether element relative to the ongoing flow; α—the angle of attack of a tether element; d—the tether diameter; γ—$= |\partial\mathbf{R}/\partial s|$—the elongation of d (it being assumed that the area of a tether element is $S = dl_1$, where an element length of the stressed tether, l_1, and that of the unstressed tether, l_0, are related by $l_1 = \gamma l_0$).

In the general case the stationary configuration is of a wave nature and may contain a "sail" loop in it as shown in Fig. 1*II*. Of practical importance are only monotonic configurations. The results illustrated in Figs. 5–8 were calculated for the following probe parameters:

1) $m_A = 100$ kg, $D = 1.0$ m

2) $m_A = 10$ kg, $D = 0.3$ m

The tether material was taken to be either glass fiber or steel (in the figures, dashed and solid lines, respectively), the major satellite mass $m_B = 10{,}000$ kg, and the altitude $H_B = 220$ km. Figure 5 shows the effect of the probe altitude H_A on d_* and m_t^*, which are, respectively, the tether thickness and tether mass least allowable with respect to the strength criterion. Figure 6 shows the equilibrium configurations; Figure 7, the required motor thrust; Figure 8, the probe surface temperature versus flight altitude. The surface temperature was calculated from

$$T_P(\rho_a v^3/(8\sigma))^{1/4} \tag{11.2}$$

where σ is the Stefan-Boltzmann constant.

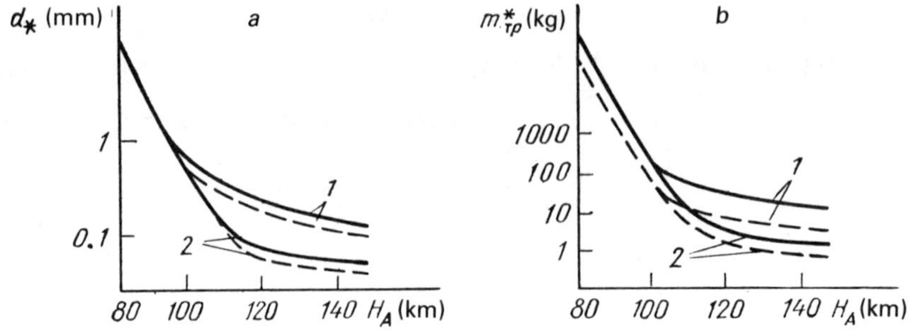

Figure 5

The results obtained suggest atmosphere probing at altitudes of about 100 to 110 km and above, by a probe connected with a satellite by an approximately 100 km tether of small thickness (of the order of 0.1 mm) and moderate mass (of about 100 kg). The required compensating motor thrust is only about 1 kg, and the friction-induced surface temperatures are reasonably moderate, of about 500 to 800 K, both on the tether and the probe under these circumstances.

At the lower altitudes of about 90 km, atmosphere probing presents too much difficulty to be practical (due to critically high mass and temperature parameters) and it seems to be totally infeasible below 80 km.

12. Atmospheric probe. Stability [22]. An approximate stability analysis of probe motion is conveniently carried out in the framework of a model similar to that used for treating a tethered system in a gravitational field, see Sec. 5. Using the orbital coordinate system, the equations of motion of the probe may be written in the following dimensionless form

$$\ddot{x} - 2\dot{y} - 3x = -c\rho\dot{x} - E(r-1)\frac{x}{r}$$

$$\ddot{y} + 2\dot{x} = -c\rho v(R_0 + \dot{y}) - E(r-1)\frac{y}{r} \tag{12.1}$$

$$r = \sqrt{x^2 + y^2}; \quad v = \sqrt{(R_0 + \dot{y})^2 + \dot{x}^2}; \quad \rho = \rho(R)$$

$$R = \sqrt{(R_0 + x)^2 + y^2}$$

where dots denote differentiation with respect to the dimensionless time $\tau = \omega t$; all the (dimensionless) coordinates are referred to the tether length l; velocity, to the quantity ωl; the atmosphere density ρ varies with the distance from the center of attraction. Account is taken of the elastic extensibility of the tether, $E = E_p f / m_A \omega^2 l$ being the (dimensionless) Young's modulus.

The stationary solution to (12.1) is

$$r_0 = \frac{E}{E-3}\,(E > 3); \quad y_0 = -\frac{c\rho_0}{3}R_0^2 \tag{12.2}$$

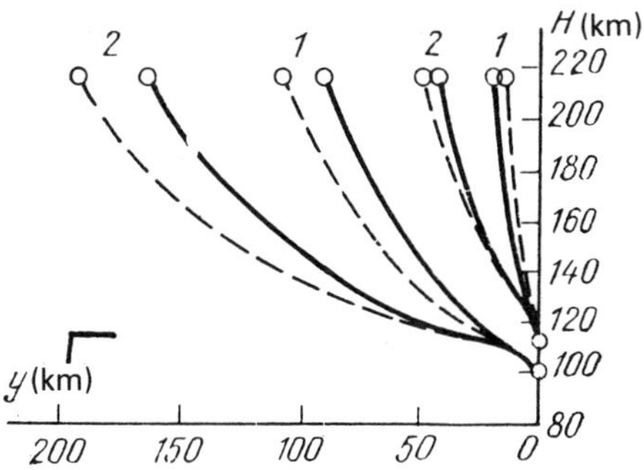

Figure 6

equations in variations are

$$\begin{pmatrix}\delta\ddot{x}\\ \delta\ddot{y}\end{pmatrix} + 2\begin{pmatrix}-\delta\dot{y}\\ \delta\dot{x}\end{pmatrix} + \begin{pmatrix}a_{11}a_{12}\\ a_{21}a_{22}\end{pmatrix}\begin{pmatrix}\delta x\\ \delta y\end{pmatrix} + \beta\begin{pmatrix}\delta\dot{x}\\ 2\delta\dot{y}\end{pmatrix} + \varkappa\begin{pmatrix}0\\ \delta x\end{pmatrix} = 0$$

$$\varkappa = c\left.\frac{\partial\rho}{\partial R}\right|_0 R_0^2; \quad \beta = c\rho(R_0)R_0; \quad a_{12} = a_{21} = E\frac{x_0 y_0}{r_0^3}; \tag{12.3}$$

$$a_{11} = E\frac{x_0^2}{r_0^3}; \quad a_{22} = 3 + E\frac{y_0^2}{r_0^3}$$

and the characteristic equation is

$$\lambda^4 + 3\beta\lambda^3 + [4 + 2\beta^2 + E]\lambda^2 + \left[2\varkappa + \beta E\left(1 + \frac{x_0^2}{r_0^3}\right)\right]\lambda + 3E\frac{x_0^2}{r_0^3} = 0 \tag{12.4}$$

The requirement that the stationary solution (12.2) be asymptotically stable is that all the roots of (12.4) must have their real parts negative; this is secured by the Routh Gurwitz criterion, which may be represented reasonably accurately in the form of single condition in our case, namely

$$E > -(\partial\rho/\partial R)R/\rho \tag{12.5}$$

The system loses its stability if this condition is invalidated. It is seen from

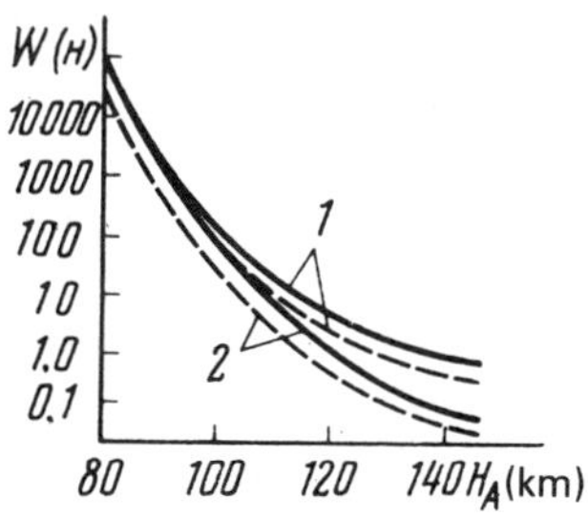

Figure 7

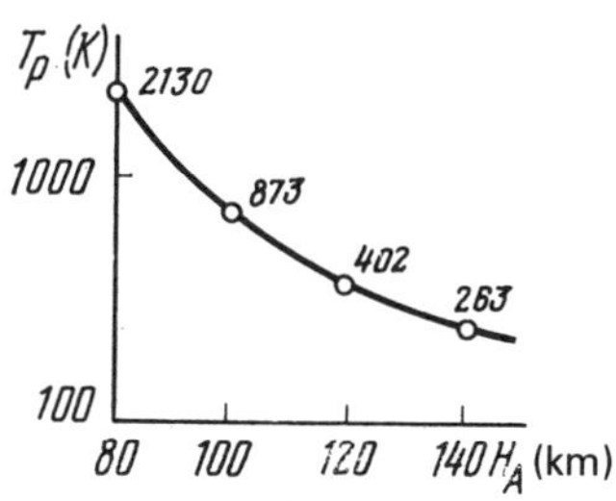

Figure 8

(12.5) that the mechanism of instability depends on the interplay of two factors, the atmosphere density gradient and the elastic extensibility of the tether. Analysis shows that stability requirements turn out to be more restrictive than those on strength, so that tether thickness may prove insufficient in terms of stability even when securing enough strength.

This conclusion has been verified by a more rigorous treatment of the problem [22] using a numerical algorithm based on the variation of the general system of equations of motion, (2.1) to (2.6). Referring to Fig. 9, the hatched region in the (d, H)-plane represents the region of stability for one possible flight regime. For comparison, a curve corresponding to the strength requirement and a curve bounding the region (12.5) are shown (by dashes and dots, respectively); for $H_A > 110$ km, the latter curve coincides with the exact boundary of the stability region.

Asymptotic stability is secured by the friction-related damping effect of the atmosphere. It turns out that the high-frequency tether oscillations decay over a few orbital revolutions of the system, whereas the pendular oscillations (of the system considered as a solid body) may continue for a relatively long period of time.

13. Electromagnetic tethered orbital system. Let an electroconducting tether be attached to a massive satellite B and have a device on its end A functioning as an absorber of or a drain for the electrons of the ionospheric plasma. It is then possible in principle to make an electric current flow through the tether. This will interact with the Earth's magnetic field thus giving rise to a mechanical force $\mathbf{F}$ which is distributed along the tether and may be utilized for correcting the orbital motion. We have

$$\mathbf{F} = I\boldsymbol{\tau} \times \mathbf{B} \tag{13.1}$$

where I is the strength of the current; $\boldsymbol{\tau}$—the unit vector of the tether tangent; $\mathbf{B}$—the vector of the magnetic flux density of the Earth's magnetic field.

If we assume that $\mathbf{B}$ is constant and normal to the satellite's orbit (which is very nearly the case for equatorial orbits), then the equations of the stationary motion of the tether will have the same tension integral as given by (4.1), whereas the integral (4.2) becomes $RT_\varphi - BR^{-1} = c_2$ (for plane motion). As in the purely gravitational case, the stationary forms are waves located along the orbit, but the tether tangent may now change its sign thus giving rise to loops in the crests and troughs of the wave, see Fig. 1*III*. According to Levin, the tether may form a closed loop in its stationary motion under certain circumstances.

The case of a weightless tether lying in the orbit plane is conveniently discussed in terms of the argument developed by Levin. The force $F = IB$ is constant in magnitude and is everywhere normal to the tether tangent, so that the tension T of the weightless tether is the same at any time at any point of the tether. So too is the elongation $\gamma = 1 + T/E$. Since unit length tether portions are subjected to equal forces throughout the tether, it follows that the tether has the form of a circular arc at any instant of time (as shown in Fig. 10), the radius of the circle being $r_T = T/IB$. The geometrical conditions of the problem may be written

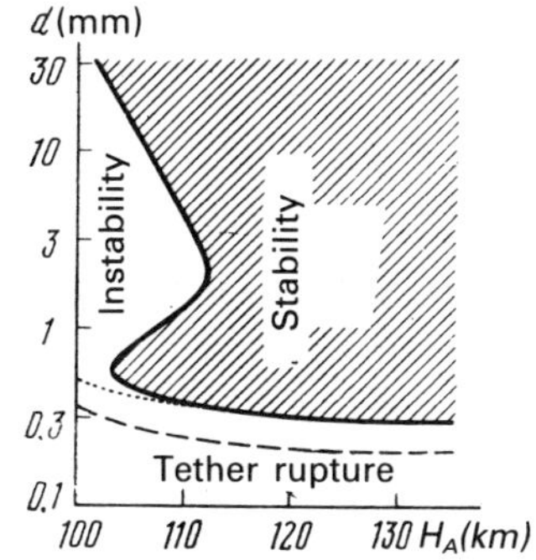

Figure 9

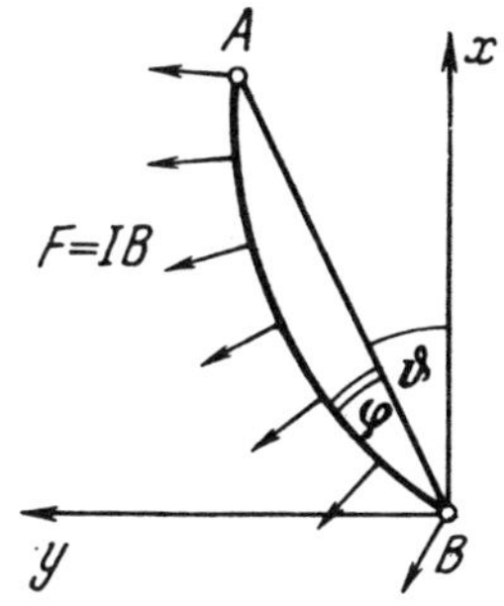

Figure 10

$$l\gamma = 2r_T\,\psi, \quad r = 2r_T \sin\psi$$

giving

$$r = \frac{l \sin\psi}{\psi(1-\epsilon)}, \quad T \sin\psi = \frac{1}{2} IBr, \quad \epsilon = \frac{T}{T+E} \tag{13.2}$$

in terms of the polar coordinates r, ϑ of point $A(x = r\cos\vartheta,\ y = r\sin\vartheta)$. Note that the quantities r, T, ψ, and r_T are always related by the equations above even though all of them are functions of time. The equations of motion of point A are

$$\frac{\ddot{r}}{r} - \dot{\vartheta}^2 - 2\dot{\vartheta} - 3\cos^2\vartheta = -h \operatorname{ctg}\psi \tag{13.3}$$

$$\ddot{\vartheta} + 2(\dot{\vartheta} + 1)\frac{\dot{r}}{r} + 3\sin\vartheta\cos\vartheta = h, \quad h = IB/2m_A\omega^2 \tag{13.4}$$

where the angle ψ is formed by the tether line and AB at points A and B. The stationary motion is

$$h = 3\sin\vartheta\cos\vartheta; \quad \vartheta = \psi \tag{13.5}$$

If $2h < 3$, the stationary motion may have a nonmonotonous configuration, and if $2h > 3$, it does not exist at all. To the monotonous configuration there corresponds $\vartheta \leqslant \pi/4$. In variations

$$\begin{aligned} &\ddot{u} - 2\dot{v} + 3\sin 2\vartheta v - 3a_2 u = -(1-3a_1)\operatorname{ctg}\vartheta r\delta h \\ &\ddot{v} = 2\dot{u} + 3\cos 2\vartheta v = r\delta h \end{aligned} \tag{13.6}$$

where ϑ, r are the stationary values: $u = \delta r$, $v = r\delta\vartheta$. Variation has also been taken from δh (through the variation δI of the current strength). Also

$$a_1 = \epsilon\vartheta/(3\vartheta\cos\vartheta), \quad a_2 = \vartheta\cos\vartheta(1-\epsilon)/\vartheta$$

$$\vartheta = \sin\vartheta - \vartheta\cos\vartheta + \epsilon\vartheta\cos\vartheta$$

If the current strength is constant ($\delta h = 0$), the characteristic equation of the system (13.6) takes the form

$$\lambda^4 + \lambda^2(3\cos 2\vartheta + 3a_2 + 4) - 6\lambda \sin 2\vartheta + 9a_2 \cos 2\vartheta = 0 \tag{13.7}$$

Verifying the Routh-Gurwitz criterion (real parts of λ positive) we find that one of the inequalities involved in the criterion is contradictory, $(6 \sin 2\vartheta)^2 < 0$, showing that the stationary motion is always unstable at a constant current.

We must therefore control the current strength if we wish to achieve stability. Since there is no damping in (13.6) when $\delta h = 0$, let us introduce it by assuming control of the form

$$\delta I/I = \delta h/h = -b\delta\dot{\vartheta} \tag{13.8}$$

The characteristic equation is then modified and if the condition

$$b > b_*(\vartheta) \tag{13.9}$$

is fulfilled, asymptotic stability is attained. For ϑ and ϵ small we have $b_* = (\frac{4}{3})(\epsilon + \vartheta^2/3)$. When ϑ assumes its maximum value, $\pi/4$, we have $b_* \approx 0.4$. There exist, of course, other control strategies capable of stabilizing the relative equilibrium of the system.

An electromagnetic tethered system can provide a thrust of the order of 2.5 N for a tether length $l = 10$ km and an operating current of 10 A. In the power generation regime the system will produce about 20 kW of electrical energy (emf of inductor being 200 V per km, or 2 kV over 10 km). For aluminum wire of 6.5 mm in diameter having 0.5 mm-thick insulation coating, the weight of the tether will be somewhere about 1 ton. The system has a 90% efficiency in transforming mechanical energy into electrical energy or vice versa.

REFERENCES

1. Krylov, A. N. On the equilibrium of a spheric mine in a current. Izv. Mining, 1909, No. 44.
2. Kochin, N. E. On the wind-induced bending of a shake-like aerostat. PMM, 1946, Vol. 10, No. 1.
3. Minakov, A. P. Fundamentals of thread mechanics. Moscow, 1941 (in Russian).
4. Beletskii, V. V. and Novikova, E. T. On the relative motions of two tethered orbiting bodies (I), Kosm. Issled., 1969, Vol. 7, No. 13, pp. 377–384.
5. Beletskii, V. V. On the relative motions of two tethered orbiting bodies (II), Ibid., No. 6, pp. 827–840.
6. Beletskii, V. V. and Novikova, E. T. On the three-dimensional motions of two tethered orbiting bodies. Izv. USSR Acad. Sci. MTT, No. 5, pp. 23–28.
7. Dokuchayev, L. V. and Efimenko, G. G. Effects of the atmosphere on the relative motions of two tethered orbiting bodies. Kosm. Issled., 1972, Vol. 10, No. 1, pp. 57–65.
8. Komarov, V. I. Liberations of a heavy string in a central field. Ibid., pp. 46–56.
9. Polyakov, G. G. Generalized space-elevator problems. Izv. USSR Acad. Sci. MTT, 1972, No. 6, pp. 54–59.
10. Singkh, R. B. Three-dimensional motions of two tethered bodies in an elliptic orbit. Vestnik MGU, 1973, No. 4, pp. 59–64.

11. Komarov, V. I. On the gravitational twisting of a two-component satellite. Kosm. Issled., 1974, Vol. 12, No. 6, p. 856.
12. Efimenko, G. G. On the motion of a loaded elastic string in a Newtonian force field. Izv. USSR Acad. Sci. MTT, 1975, No. 2, pp. 24–31.
13. Vereshchgin, I. F. and Ivanov, A. P. On the relative motions of a point-solid body system on a circular orbit. In: Analytical mechanics, stability theory and control theory. Moscow, Nauka, 1975, pp. 105–109.
14. Bolotina, N. E. and Wilke, V. G. On the equilibrium stability of a heavy flexible string attached to a satellite moving in a circular orbit. Kosm. Issled., 1978, Vol. 16, No. 4, pp. 621–626.
15. Sarychev, V. A. Man-made satellite orientation. In: Science and technology achievements. Space Studies. Moscow, VINITI, 1978, Vol. 2, p. 224.
16. Merkin, D. R. Introduction into the mechanics of the flexible string. Moscow, Nauka, 1980 (in Russian).
17. Polyakov, G. G. Radial system of tethered satellites. Kosm. Issled., 1981, Vol. 19, No. 3, pp. 467–470.
18. Beletskii, V. V. and Levin, E. M. Mechanics of the orbital tethered system. Ibid., 1980, Vol. 18, No. 5, pp. 678–688.
19. Beletskii, V. V. and Levin, E. M. Orbital tethered systems. Prepr. Keldysh Inst. Math. ISSR Acad. Sci., Moscow, 1981, No. 13.
20. Beletskii, V. V. and Levin, E. M. Mechanics of a moon tethered system. Kosm. Issled., 1982, Vol. 20, No. 5, pp. 760–765.
21. Levin, E. M. Deployment of two tethered orbiting bodies. Long tether case. Ibid., 1983, Vol. 21, No. 1, pp. 121–124.
22. Levin, E. M. On the stability of the stationary motions of two tethered bodies orbiting under gravitational and aerodynamic forces. Ibid., 1984, Vol. 22, No. 5, pp. 675–682.
23. Beletskii, V. V. and Levin, E. M. Satellite ring rotation stability. PMM, 1986, Vol. 50, No. 2, pp. 179–186.
24. Komarov, V. I. Motions of two string-connected material bodies under the action of gravitational and ampère forces. Kosm. Issled., 1986, Vol. 24, No. 3, pp. 380–387.
25. Ivanov, V. A. and Sitarskii, Yu. S. Flight dynamics of a flexible system of tethered spacecrafts. Moscow, Mashinostroyeniye, 1986 (in Russian).
26. Chobotov, V. A. Gravity-gradient excitation of a rotating cable-connected space station in orbit. Appl. Mech., 1963, Vol. 30, No. 4, pp. 547–554.
27. Weiffenbach, G. G., Colombo, G., Gaposhkin, E. M., and Grossi, M. D. Gravity gradient measurements in the vicinity of 100 km height by long-tethered satellites. Proc. 27th Intern. Astronautic. Congr., Anaheim, 1976.
28. Bainum, P. M. and Kumar, V. K. Optimal control of the shuttle-tethered subsatellite system. Acta Astronaut, 1980, Vol. 7, No. 12, p. 1333.
29. Modi, V. J. and Misra, A. K. On the deployment dynamics of tether connected two-body systems. Ibid., 1979, Vol. 6, No. 9, pp. 1183–1197.
30. Breakwell, V. V. Stability of an orbit rings. J. Guid. Contr., 1981, Vol. 4, No. 2, pp. 197–200.

SPACECRAFT ENTRY PATHS AND THEIR CONTROL

V. A. Yaroshevskii

1. Although spacecraft entry (or reentry) paths are perfectly amenable to numerical calculation and can be determined with reasonable accuracy on modern high-speed computers, approximate analytical solutions retain their significance for a number of reasons.

First, an analytical solution has the advantage of providing insights into the manner in which the initial conditions of the problem affect the flight regime of the spacecraft. One example to illustrate this point is a rather paradoxical analytical result, obtained by a number of workers [1–5], that the maximum overload on a spacecraft traveling in an isothermal atmosphere is independent of the ballistic parameter $\sigma_x = c_x S/m$, c_x being the resistance coefficient, S—the characteristic area, and m—the mass of the vehicle.

It is perhaps worth remembering, in this connection, the situation in classical celestial mechanics where, to cite but one example, effective semi-analytical methods have been developed for calculating mooncraft trajectories using the concept of a sphere of action [6] even though the problem lends itself fairly readily to numerical treatment.

Second, the important practical problem of optimizing the spacecraft flight regime becomes considerably more tractable if approximate equations of motion are employed [7–9].

Third, the advent of onboard computer systems have permitted—and indeed necessitated—the inflight calculation of the future path of the spacecraft. The way it is usually done (by the onboard computer) is either by numerical integration of approximate motion equations [10] or by employing approximate solutions of (exact) motion equations.

2. The development of approximate equations for the motion of a spacecraft (re)entering the atmosphere of a planet has been the subject of several investigations [2–5, 11], and, not unnaturally, the equations obtained differ in their form on account of different variables used and approximations made. We can reduce these equations to one general form, however, if the following basic assumptions are introduced

$$\begin{cases} \rho \approx \rho_0 e^{-\lambda H}, & c_x = \text{const}, \quad H \ll R \\ \Theta^2 \ll 1, & |\sin \Theta| \ll n_x \end{cases} \tag{1}$$

where R is the radius of the planet; ρ—the density of the atmosphere; H—the altitude; λ—the mean logarithmic density gradient; Θ—the inclination of the path; n_x—the longitudinal overload, i.e., the ratio of the resisting force to the spacecraft weight. With these approximations, it can be shown that all the variants of equations are reduced to

$$\frac{d^2y}{dx^2} = -\eta + \frac{e^{2x} - 1}{y} \tag{2}$$

where $y = (c_x S/2m)(R/\lambda)^{1/2} \rho$ is the dimensionless density; $x = ln(\sqrt{Rg}/V) = ln\ 1/\bar{V}$ is an independent variable determined by the velocity V; $\eta = (R\lambda)^{1/2} K \cos \gamma$ is the parameter of the planet (for the Earth, $\eta \approx 30K \cos \gamma$); $K = c_y/c_x$—the aerodynamic efficiency; c_y—the lift coefficient; g—the free fall acceleration; γ—the roll angle; $\bar{V}$—the velocity/angular velocity ratio.

The remaining parameters of the path can be found from the relations

$$\sin \Theta \approx \Theta - -\frac{1}{\sqrt{R\lambda}} \frac{dy}{dx}$$

$$\frac{ds}{dx} = \frac{1}{\sqrt{R\lambda}\, y}$$

$$\frac{d\bar{t}}{dx} = \frac{e^x}{2\pi\sqrt{R\lambda}\, y}$$

$$n_x = \sqrt{R\lambda}\, y e^{-2x}$$

where s is the dimensionless slant range and the dimensionless time $\bar{t}$ is measured in fractions of the period of revolution in a near-circular near-surface orbit.

For a spacecraft descending from a near-circular orbit, the initial conditions (2) may be written as

$$x = 0, \quad y = 0, \quad \frac{dy}{dx} = -\sqrt{R\lambda} \sin \Theta_0 = c_1 \tag{3}$$

Θ_0 denoting the entrance angle. This problem may be treated either numerically or analytically.

The following approximate solutions are valid for a limited range of velocity variations [5]:

(1) Ballistic spacecraft ($\eta = 0$)

$$\text{a)}\quad c_1 = 0, \quad y = \sqrt{\frac{8}{3}}\, x^{3/2}\left(1 + \frac{x}{6} + \frac{x^2}{24} + \dots\right)$$

$$n_{x\,\max} = 0.28\sqrt{R\lambda} \approx 0.83 \text{ (for the Earth)}$$

$$\text{b)}\quad c_1 = 1 \div 5, \quad y = c_1 x + \frac{x^2}{c_1} + \frac{(c_1^2 - 1)\, x^3}{3\, c_1^3} + \dots$$

$$\text{c)}\quad c_1 > 5, \quad y \approx c_1 x, \quad n_{x\,\max} \approx \frac{R\lambda}{2e}|\sin\Theta_0| \approx 170|\sin\Theta_0|$$

(Allen-Eggers helical path).

(2) Lifting spacecraft

$$\text{a)}\quad c_1 = 0, \quad |\eta| \leqslant 3$$

$$y \approx \sqrt{\frac{8}{3}}\, x^{3/2} - \frac{4\eta}{11} x^2 + \sqrt{\frac{8}{3}}\left(\frac{1}{6} + \frac{\eta^2}{121}\right) x^{5/2} + \dots \tag{4}$$

$$\text{b)}\quad c_1 \approx 0, \quad \eta \geqslant 15, \quad y \approx (e^{2x} - 1)/\eta$$

$$n_{n\max} \approx 1/(k\cos\gamma) \quad \text{(equilibrium glide)}$$

$$\text{c)}\quad c_1 > 1, \quad \eta \gg 1/c_1 \quad - \text{ a path}$$

with reflections,

$$x_1 \leqslant 2c_1/\eta, \quad y \approx c_1 x - \eta x^2/2$$

$$x \geqslant 2\, c_1/\eta, \quad y_m \approx (c_1^2/2 + e^{2x} - 1)/\eta$$

where y_m envelopes the maximum (dimensionless) density values assumed in the motion along the oscillating path.

The accuracy of the formulas obtained is illustrated in Figs. 1, 2, and 3 for $K = 0$, $K = 0.25$ and $K = 2$, respectively. The approximate solutions were found by expanding in a series of powers of x or $x^{1/2}$ using recurrence formulas. The envelope $y_m(x)$ (see Fig. 3) may be found using the adiabate invariance of the action integral at large η.

Of most practical importance is an equilibrium glide path with a slowly varying inclination angle. The simplest possible formula, $y_1 \approx (e^{2x} - 1)/\eta + e^x$ is adequate only for "high" lift to drag ratios $\eta \geqslant 15$ (for the Earth, $K \geqslant 0.5$) in a limited velocity range $0.99 \geqslant \bar{V} \geqslant 10/(K\sqrt{R\lambda})$.

One way of improving this formula is by using the method of round-off stationary solutions, in which successive changes of the dependent variable $y_{k+1} =$

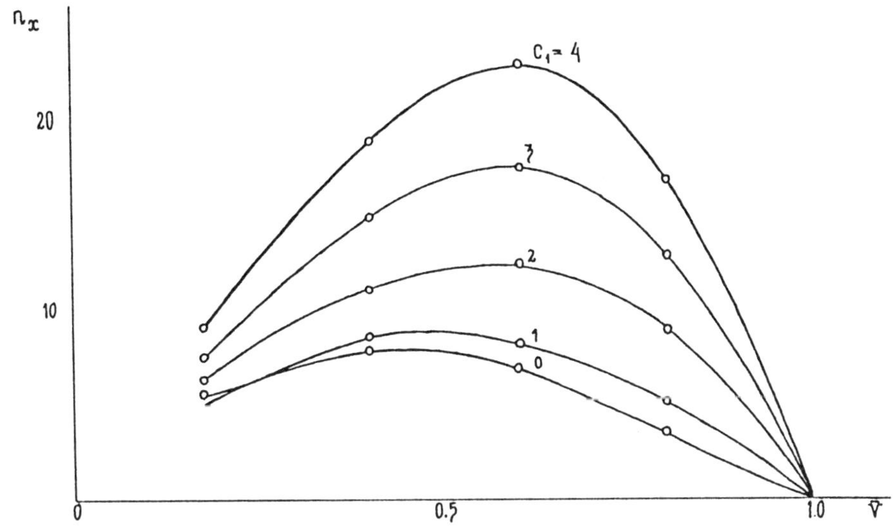

Figure 1

$\varphi(x)y_k$ are used to transform equation (2) to a form $d^2y_n/dx^2 = F(x, dy_n/dx, y_n)$ such that the stationary solution given by $F(x, 0, y_n) = 0$ satisfy the equation with good accuracy.

The more accurate solutions found in this way for the equilibrium glide path

$$y_{(1)} \approx \frac{e^{2x} - 1}{\eta + e^x}$$

$$y_{(2)} \approx \frac{2(e^{2x} - 1)}{\eta\left\{1 + \sqrt{1 + \frac{4e^x}{\eta}\left(1 + \frac{e^x}{\eta}\right)\left[1 - \frac{(\eta^2 - 1)(\eta - e^x)}{(\eta + e^x)^3}\right]}\right\}}$$

make it possible to determine more reliably the parameters of the path (see Fig. 2, where open circles correspond to the numerical calculation).

Equation (2) describes the entry process for a limited velocity range, $0.99 \geqslant \bar{V} \geqslant 0.1$ on the so-called main portion of the entry path.

Equation 2 becomes somewhat more complicated if the conditions $H \ll R$ and $|\sin\Theta| \ll n_x$ are lifted in (1). We have

$$\frac{d^2y}{dx^2} = -\sqrt{R_*\lambda}\, K\cos\gamma + \frac{e^{2x} - \frac{R_*}{R + H}(y)}{y}$$

$$\approx -\sqrt{R_*\lambda}\, K\cos\gamma + \frac{2x - \frac{1}{R_*\lambda}\ln\left(\frac{y}{y}\right)}{y}$$

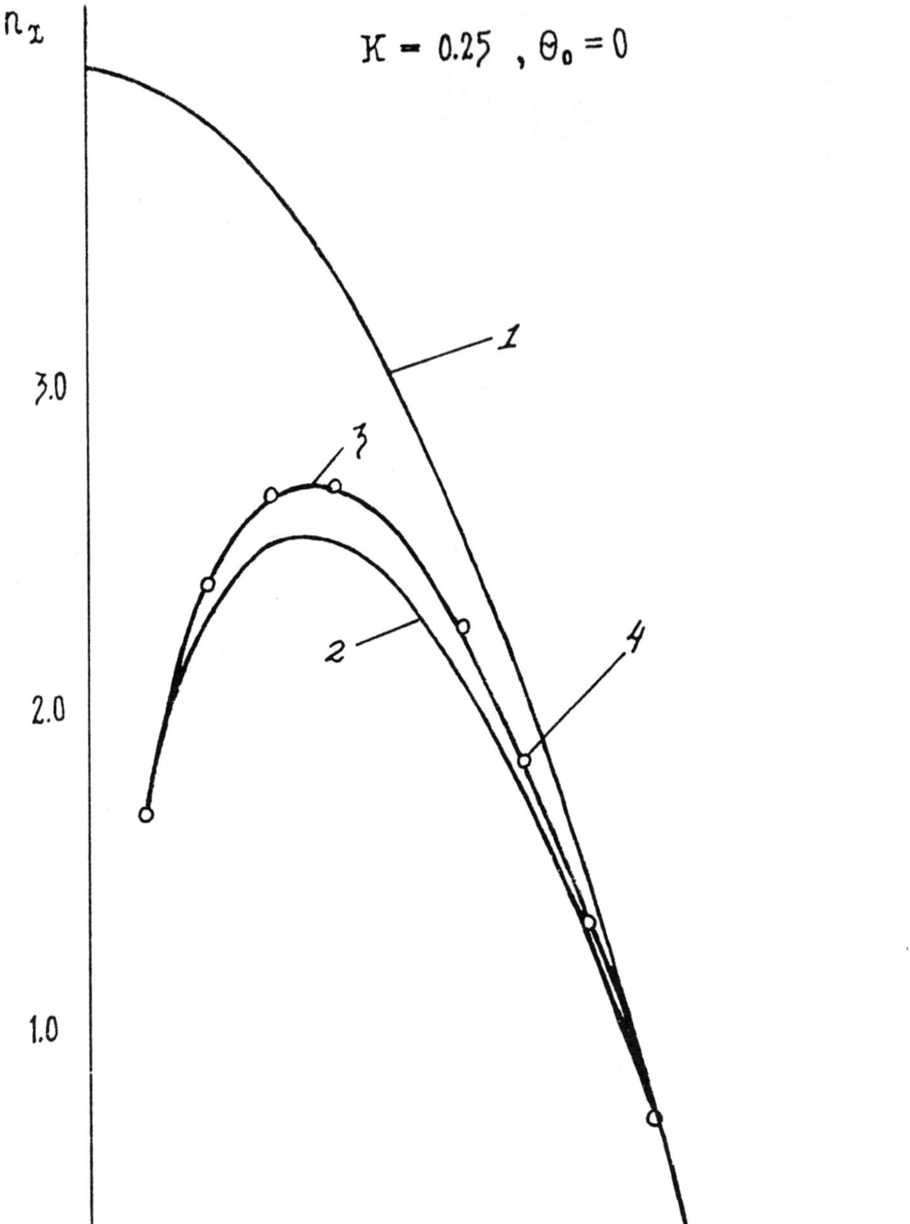

Figure 2

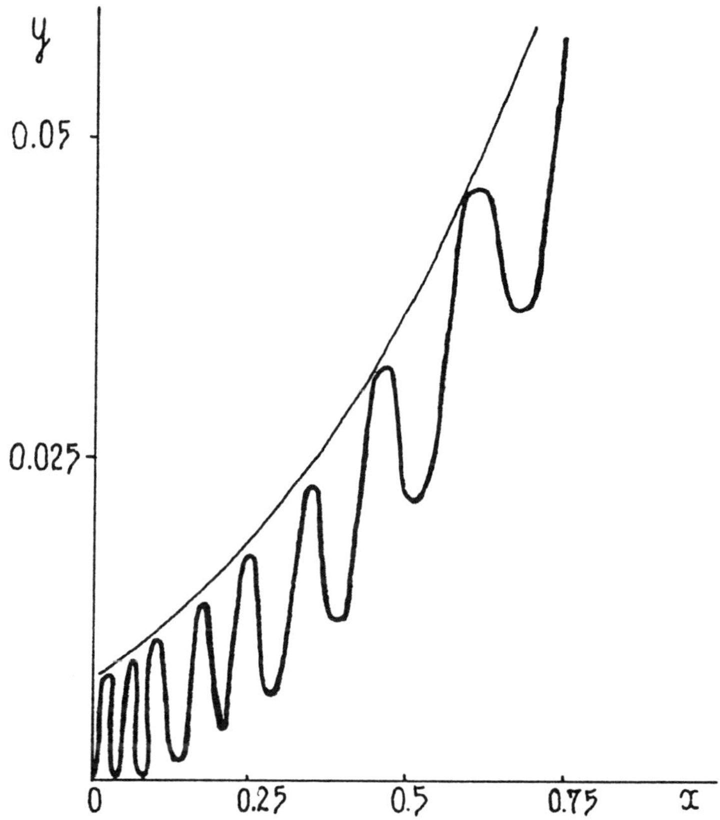

Figure 3

where

$$x = \ln \frac{\sqrt{R_* g(R_*)}}{V \cos \Theta}, \quad y = \frac{C_x S}{2m} \sqrt{\frac{R_*}{\lambda}}\, \rho$$

$$\frac{R_*}{R + H}(y) \approx 1 + \frac{1}{R_* \lambda} \ln\left(\frac{y}{y_*}\right), \quad y_* = \frac{C_x S}{2m} \sqrt{\frac{R_*}{\lambda}}\, \rho\,(R_*)$$

R_* denoting a certain characteristic distance from the planet's center. This equation applies both to rarified atmosphere layers, and to dense low-x layers.

The quasi-stationary solution of this equation controls the evolution of the orbit under the action of aerodynamic forces. It is easy to verify that if $K \geqslant 0.5$, the solution may be approximated by the formula

$$y \approx \frac{2x - \dfrac{1}{R_* \lambda} \ln \dfrac{y}{y_*}}{\sqrt{R_* \lambda}\, K \cos \gamma}$$

At small x's these solutions smoothly go over into quasi-stationary glide mode solutions applicable at the main portion of the descent path.

The equations will be extended to a finite interval of the path if the conditions $\Theta^2 \ll 1$ and $|\sin\Theta| \ll n_x$ are lifted in (1). If we limit ourselves to a quasi-stationary descent path and join low-V to moderate-V solution, the result

$$n_x \approx \sqrt{\frac{1 + 2K\bar{u} + \bar{u}^2}{(1 + K^2) + 2K(1 + K)\bar{u} + K^2\bar{u}^2}}\left(1 - \frac{\bar{u}^2}{R\lambda}\right)$$

is obtained [12], where $\bar{u} = (R\lambda)^{1/2}$ and $\bar{V} = V(\lambda/g)^{1/2}$. The accuracy of this formula is illustrated in Fig. 4.

The above solutions are simple enough to be effectively used by the onboard computer for the spacecraft trajectory prediction.

Equation (2), valid for a limited velocity range, can be used for analyzing the entry situation in which the spacecraft velocity is greater than the circular velocity ($y_0 \approx 0$, $\bar{V}_0 > 1$).

According to [4], the initial conditions for this problem are conveniently written as

$$x_0 = -\ln \bar{V}_0 < 0, \quad y(x_0) = y_0 > 0$$

$$\frac{dy}{dx}(x_0) = \sqrt{2(1 - e^{2x_0}) \ln \frac{y_p}{y_0}} \tag{5}$$

where $y_p = (C_x S/2m)\rho(R_p)(R\lambda)^{1/2}$, R_p being the perigee distance calculated neglecting aerodynamic forces (Fig. 5a).

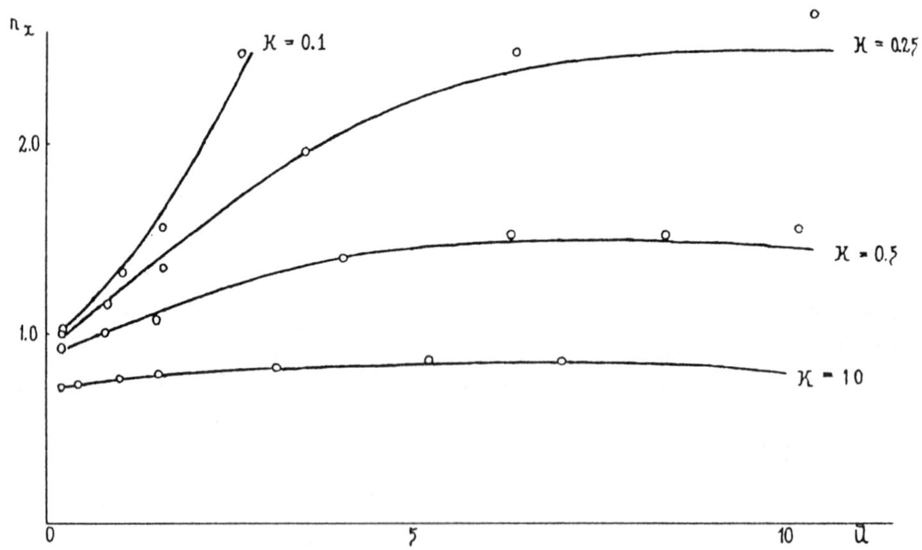

Figure 4

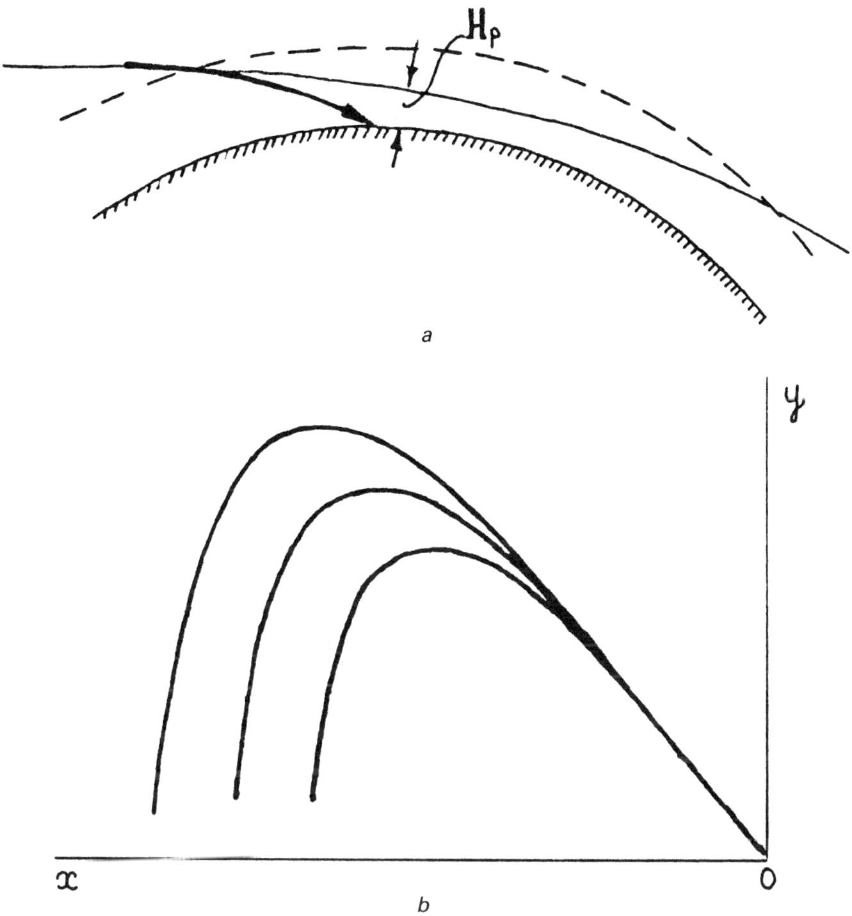

Figure 5

For a given value of $\bar{V}_0$, the limiting values of y_p may be found. One of these, $y_{p\max}$, determines the path in which the maximum overload value coincides with the maximum allowable overload for the maximum effective lift regime ($\gamma = 0$). The second value, $y_{p\min}$, refers to the capture trajectory, in which given the minimum effective lift ($\gamma = 180$ degrees), the spacecraft has $\bar{V} = 1$ when it reaches the exit point $y \approx 0$ (Fig. 5*b*). It will be noted that the "capture trajectory" is rather a difficult term to define in the exact formulation of the problem; we may require, for example, that the spacecraft should not exceed a certain altitude when leaving the atmosphere, but this is by no means the only possible criterion. In the approximate approach, equation (2), a perfectly unambiguous definition becomes possible.

The solution of equation (2) under the initial conditions (5) cannot be expanded in a series form in terms of a set of elementary functions.

Instead, useful approximate relations between entrance and minimum-altitude parameters can be obtained for moderate longitudinal overloads on the descending spacecraft ($n_x \sim 10$). Using these relations it proves possible to calculate $y_{p\max}$ and $y_{p\min}$ and then the entry corridor width [5]

$$\Delta H_p = \frac{1}{\lambda} \ln \frac{y_{p\max}}{y_{p\min}} \approx \frac{1}{\lambda} \left[\mu + \ln \mu + 1 + \sqrt{\frac{2\pi}{1 + \frac{\pi\mu}{4}}} (1 + \mu)\, \nu \right]$$

where

$$\mu = \frac{K n_{\max}}{(\bar{V}_0^2 - 1)\sqrt{1 + K^2}}$$

$$\nu = \frac{n_{\max}}{\bar{V}_0 \sqrt{\bar{V}_0^2 - 1}\sqrt{R\lambda}\sqrt{1 + K^2}}$$

The accuracy of formula (6) is illustrated in Fig. 6.

It should be noticed that the accuracy of the approximate solutions in the above discussion has always been estimated by comparison with the numerical solutions either of equation (2) or of more rigorous equations subjected to the $\rho = \rho_0 \exp(-\lambda H)$ assumption. This assumption—equivalent, in fact, to the isothermal atmosphere condition—turns out to be most unsatisfactory, however, as far as the computatioral accuracy is concerned. Luckily, there exists a simple enough method by which the nonisothermicity of the atmosphere can be corrected for. It turns out, for example, that for large entrance angles $|\Theta_0|$ the maximum

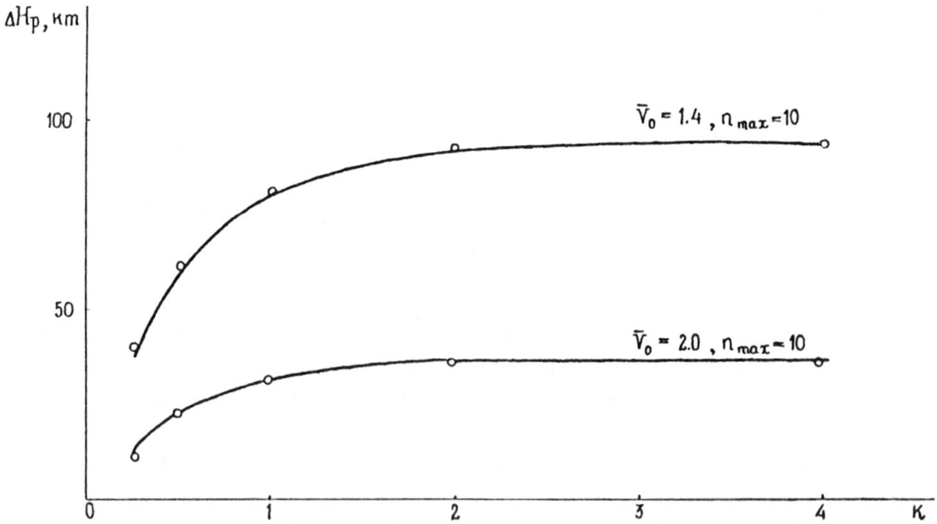

Figure 6

overload is proportional to the value of λ. To correct the result it is necessary—using values of the parameter $\sigma_x = c_x S/m$—to calculate from $y|n_{x\,\max}$ the altitude H_n at which the overload is a maximum; to determine the "local" value of $\lambda(H_n)$, linear in the atmosphere temperature in the first approximation; and to utilize this value for calculating the maximum overload. In this way it proves possible, for example, to derive the nonmonotonous maximum overload versus σ_x curve (Fig. 7) from a nonmonotonous temperature-altitude dependence.

3. The approximate equations of motion prove to be a powerful tool in solving certain variation optimization problems arising in spacecraft entry path control. In the simplest example, the problem is to optimize the longitudinal profile of the path for the case of roll-angle control (assuming the spatial shape of the path to be of no importance). In the majority of cases, time is not a significant parameter of a problem, and a relation between the density (altitude) and velocity variations, equation (2), proves to be the most suitable form of equation of motion.

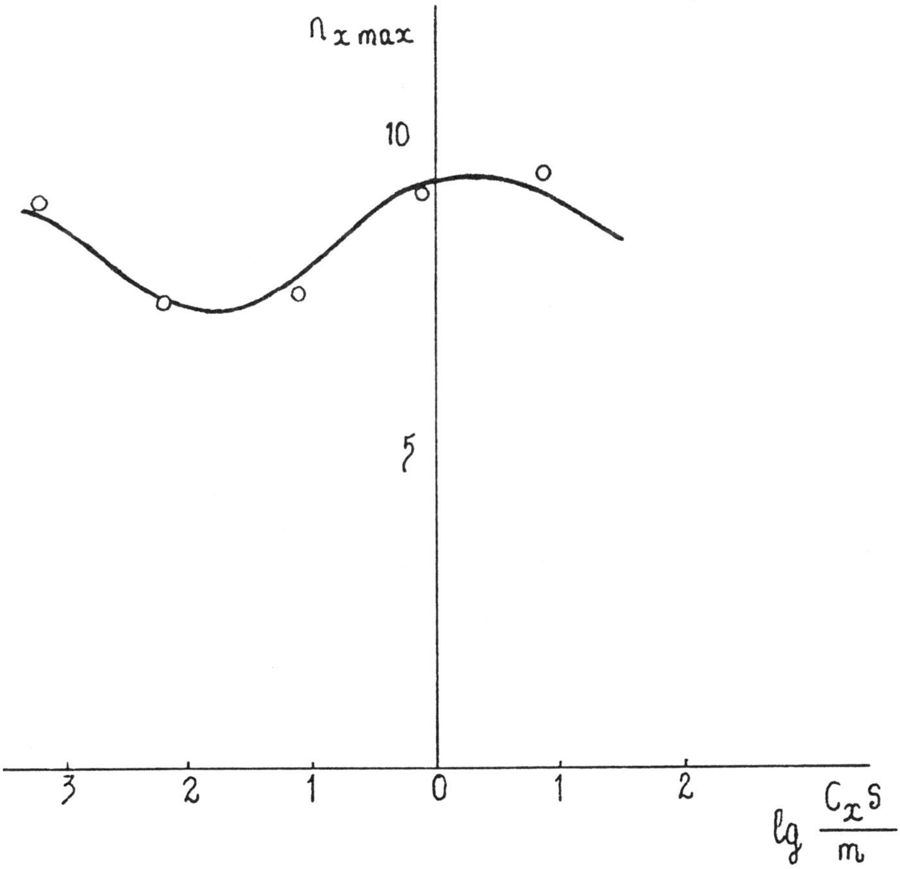

Figure 7

If the optimization criterion is taken in the form of the functional

$$I = F(x_k, y_k, y_k') + \int_{x_0}^{x_k} f(x, y, y')\, dx$$

where the subscript k refers to the final point, it follows from the maximum principle that the equation for conjugate variables may be written as [7]

$$\frac{d^2 P_y'}{dx^2} + \frac{e^{2x} - 1}{y^2} P_y' = \frac{d}{dx}\left(\frac{\partial f}{\partial y'}\right) - \frac{\partial f}{\partial y} \tag{7}$$

and the Hamiltonian takes the form

$$H = P_y y' + P_y' \left(-\eta + \frac{e^{2x} - 1}{y}\right)$$

with a resultant optimum control law

$$\eta = -\eta_{\max} \operatorname{sign} P_y'$$

$$\text{if} \quad \cos\gamma = -\operatorname{sign} P_y'$$

provided the restriction $|\eta| \leqslant \eta_{\max}$.

From the above equations, a number of properties of optimum paths can be deduced. If, for example, a non-integral criterion is taken and the case $\bar{V} > 1$ is considered, no more than one control change-over proves to be possible, since there is no more than one zero in the homogeneous equation (7). If $\bar{V} < 1$, several change-overs may occur. If, say, the velocity at a given altitude is to be minimized (or the value of x_k for a given y_k maximized), the optimum paths include both constant-altitude and zero-γ phases as shown in Fig. 8. Figure 9 illustrates certain results obtained by the present author in cooperation with Karyagin and Pichhadze for the case of a spacecraft entering the Martian atmosphere. Note the marked decrease in the optimum path final velocity relative to the $\gamma = 0$ case.

Of great importance are the problems concerned with the determination of the so-called accessibility region [8] bounded by the locus of the final points (for $x = x_k$) of the optimal paths obeying the $|\gamma| \leqslant \gamma_{\max}$ condition (Figs. 10, 11). It is convenient for practical control purposes to have available an explicit expression for the roll angle variation along an optimal path [8]. Ivanchikhina and the present author suggested the following approximately optimum law for the roll angle variation

$$\frac{\sin\gamma}{\cos^2\gamma} \approx K \frac{1 - \bar{V}^2}{2\bar{V}^2} \ln\left(\frac{1}{1 - \bar{V}^2}\right) tg\, \Delta\psi \tag{8}$$

where, in analogy with [8], $\Delta\psi = \psi_* - \psi$, ψ_* being a parameter and ψ the instantaneous relative bearing calculated using the rules of spheric trigonometry. To different values of ψ_* there correspond approximate optimum paths terminating in different points of the accessibility region.

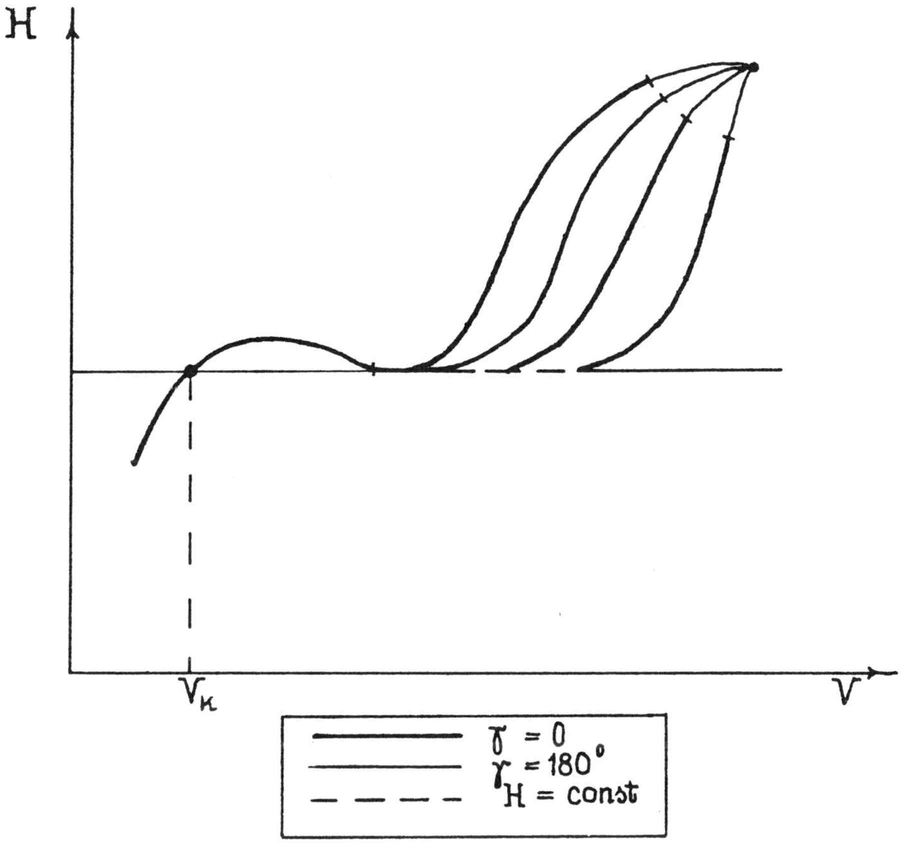

Figure 8

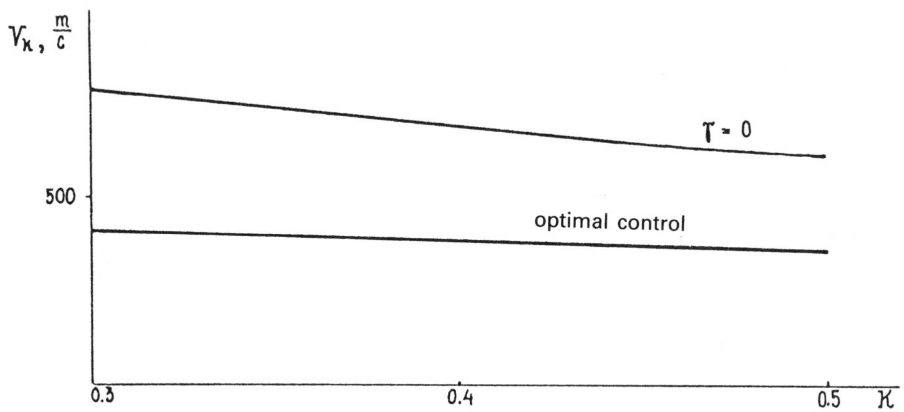

Figure 9

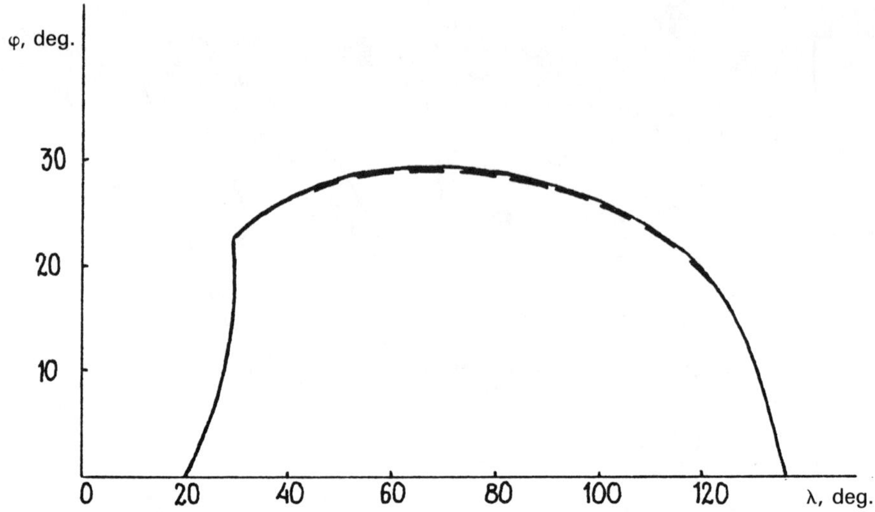

Figure 10

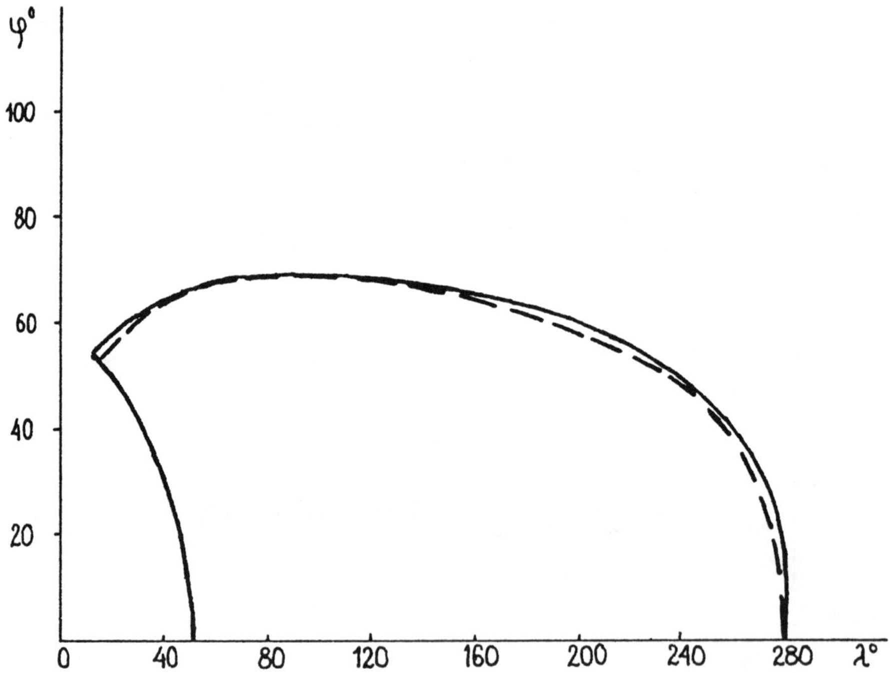

Figure 11

Fig. 10 ($K = 1.5$) and Fig. 11 ($K = 3$) show the regions found by the exact numerical solution (solid curves) and those calculated from (8). The conclusion to be drawn from the comparison is that the inflight prediction of the accessibility region is quite a feasible task for an on-board computer.

It is sometimes necessary that the accessibility region (for a finite portion of the path) be determined for a given value of the final point relative bearing. This problem is considered in [13] for the case when the limiting radius of path curvature, r, decreases exponentially as a function of the "commanded" angle range. Some of the results obtained are illustrated in Fig. 12.

4. Although it is quite a natural requirement in entry path control that the guidance algorithm be optimized (or rather stochastically optimized) in some respect(s), difficulties already arise at the formulation stage of the problem. First, a control optimization problem is hard to formalize because of the lack of criteria by which the algorithm should be judged acceptable. Second, the statistical characteristics of disturbances and of measurement errors are only poorly known. Finally, the structure of an optimum algorithm is often too complicated to be practical.

A stochastical optimum control problem is reduced to that of solving the corresponding nonlinear Bellman equation of the parabolic type in the space of phase coordinates [14, 15]. A closed form solution is only possible in on-dimensional formulation [16] when a control law for the longitudinal range is sought for. If a problem of controlling the spatial path by means of roll angle variation is con-

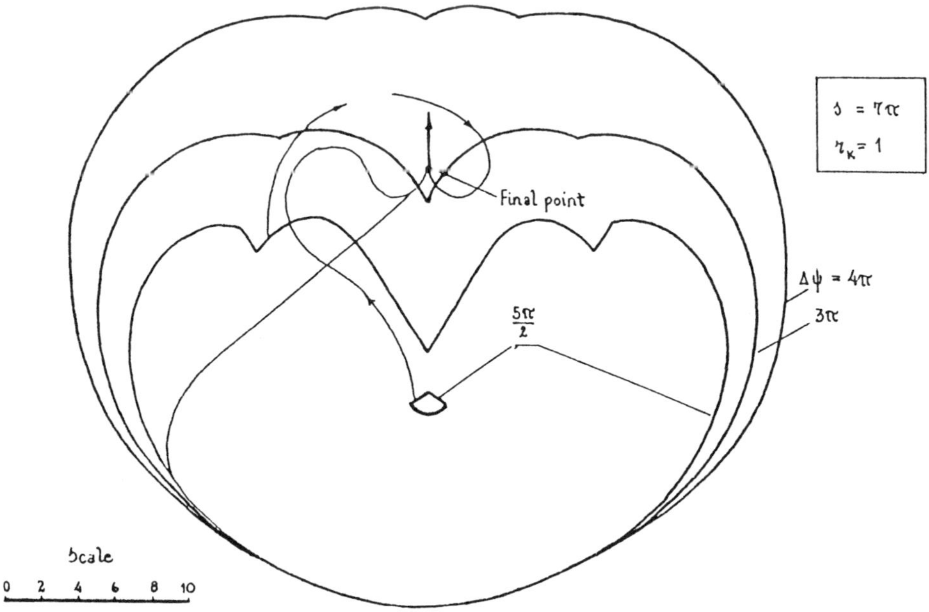

Figure 12

sidered, the following rule may be recommended for determining the first approximation optimum control strategy. For any instant of time find the accessibility region; calculate the position of the required target point in the accessibility region; determine the manner in which the "commanded" roll angle must be varied so as the target point coincides as rapidly as possible with the center of the accessibility region (Fig. 13*a*). In this way the maximum "allowance" is maintained to compensate for random disturbances (of either sign) which may affect the spacecraft motion on the remaining part of its path. It should be noted, however, that as the target point gets closer to the center of the accessibility region, too rapid variations of the commanded roll angle arise, and the resulting spacecraft control regime becomes close to the glide mode and unsatisfactory in a number of respects. This discredits the approach in which the roll angle orientation is taken as "control" instead of being considered as a phase coordinate. The above optimum control concept thus turns out to be impractical.

The way out is to introduce the hierarchial principle into the picture; that is, to control the longitudinal range by continuously varying the modulus of the role angle, and the lateral range by discretely changing the sign of the roll angle and using the relay-hysteresis law [9, 17, 18]. This program is conveniently carried out by first determining the boundaries of the accessibility region for any instant of time, by proportionally narrowing down these boundaries and, finally, by changing the sign of the roll each time when the representative point (the target point) crosses these new boundaries when going from the inside (Fig. 14*b*).

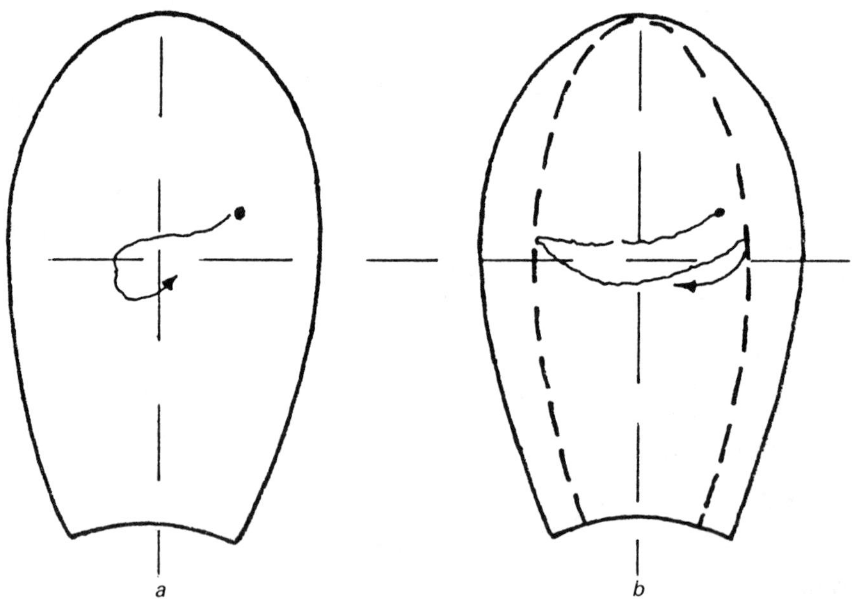

Figure 13

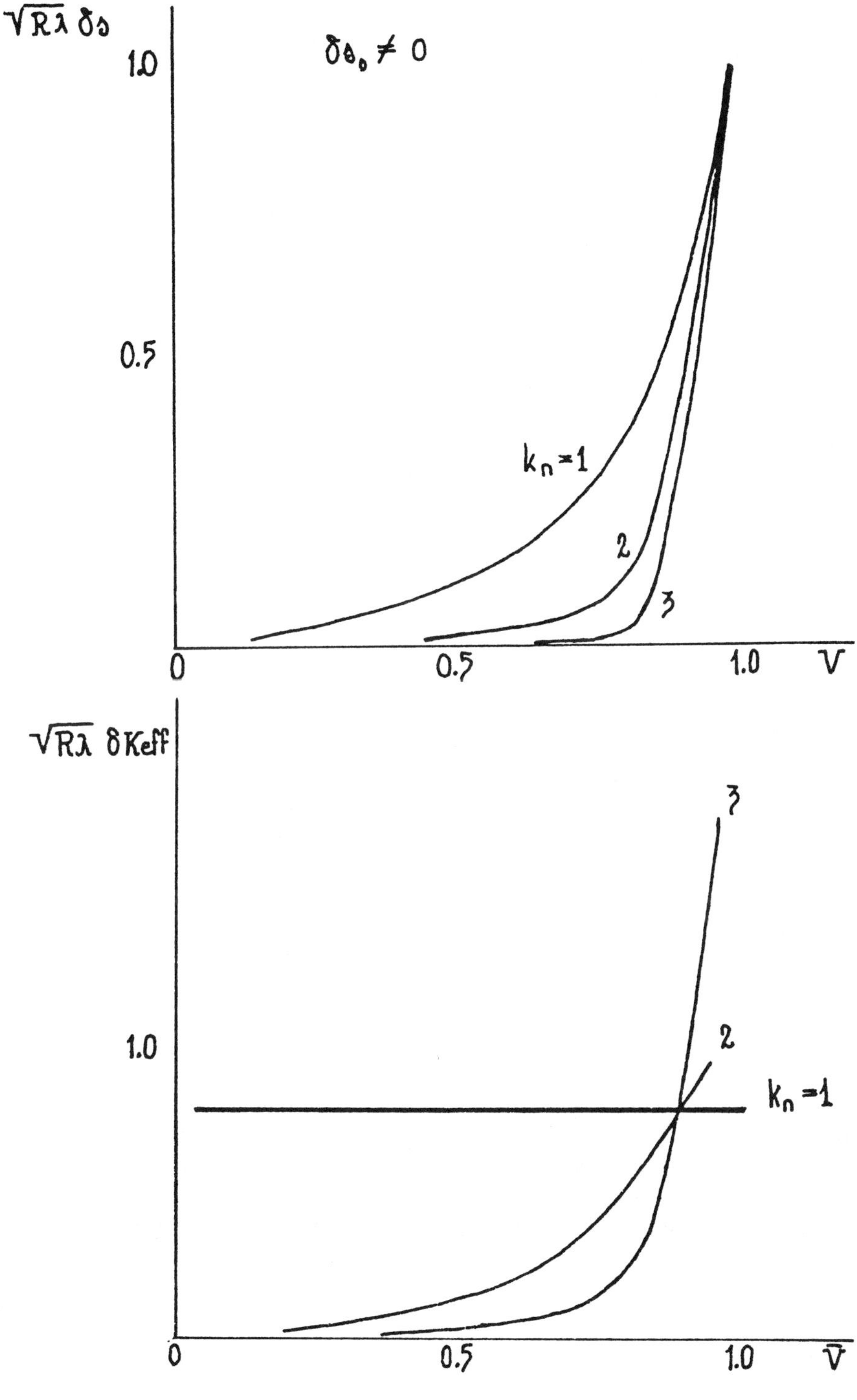

$\sqrt{R\lambda}\,\delta s$
$\delta s_0 \neq 0$
1.0
0.5
$k_n = 1$
2
3
0
0.5
1.0
V
$\sqrt{R\lambda}\,\delta K_{eff}$
3
1.0
2
$k_n = 1$
0
0.5
1.0
$\bar{V}$

Figure 14

The control of the longitudinal range continues to be the most important and most complex problem in the field of spacecraft entry control.

An approach most widely used for its solution is the method of sensitivity function control. The starting point of this is the nominal path corresponding to a certain average nominal value, γ_{nom}, of the roll angle. According to this path one calculates, in terms of the velocity V, the nominal variation regimes for, say, overload $n_{x\,nom}$ and the vertical velocity $V_{y\,nom}$. The next step is to calculate the expected deviation of the angular range Δs_{exp} (under the assumption that the quantity γ is maintained at a constant value, γ_{nom}) in the linear combination form

$$\Delta s_{exp} = \frac{\partial s_{exp}}{\partial n_x}\,[n_x - n_{x\,nom}(V)] + \frac{\partial s_{exp}}{\partial V_y}\,[V_y - V_{y\,nom}(V)]$$

where $\partial s_{exp}/\partial n_x$ and $\partial s_{exp}/\partial V_y$ are the influence functions, to be determined either numerically or analytically [19]. Following the same line of argument, one finds the sensitivity function of the effective lift to drag ratio $\Delta K_{eff} = \Delta(K \cos \gamma)$ on the final range $\partial s_{exp}/\partial K_{eff}$. This done, it is possible to evaluate the required increment ΔK_{eff} which must be maintained on the remaining portion of the path for eliminating possible final miss with respect to the longitudinal range; that is

$$\Delta K_{eff} = -\frac{\Delta s_{exp}}{\partial s_{exp}/\partial K_{eff}} \tag{9}$$

In this case, it has been shown by both numerical and analytical calculations, that the deviations of the initial conditions and other possible deviations of this type can be smoothly compensated for during the flight, see Fig. 14. If, on the other hand, some kind of "systematic" disturbance is possible, such as the deviation of lift to drag ratio, equation (9) does no longer represent a satisfactory control, because the required deviation ΔK_{eff} increases unboundedly toward the end of the flight (Fig. 15).

According to Wingrove [20], this can be remedied by rewriting (9) in the form

$$\Delta K_{eff} = -k_n \frac{\Delta s_{exp}}{\partial s_{exp}/\partial K_{eff}} \tag{10}$$

where the "overload coefficient" k_n is greater than unity. It has been shown by calculations that the acceptable values of k_n lie in the range between 1 and 3, see Figs. 14 and 15.

If the nominal path is of the quasi-stationary glide type, the variation of Δs_{exp} along the path can be shown to be approximately proportional to

$$\ln\left(\frac{1}{1-\bar{V}^2}\right)^{k_n}$$

for deviated initial conditions, and to

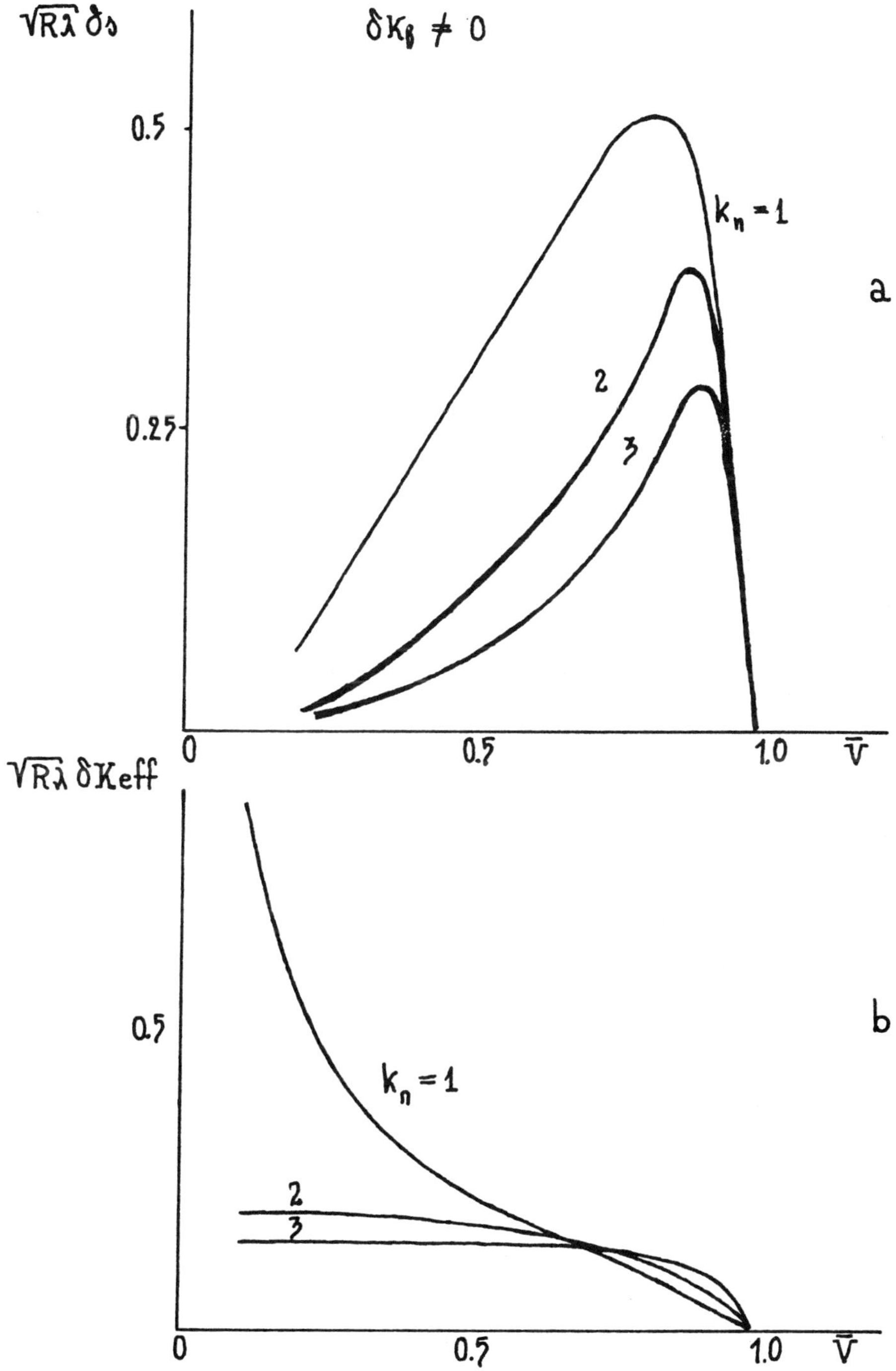

Figure 15

$$\frac{1}{k_\Pi - 1}\left[\ln\frac{1}{1-\bar{V}^2} - \frac{\left(\ln\frac{1}{1-\bar{V}^2}\right)^{k_n}}{\left(\ln\frac{1}{1-\bar{V}_0^2}\right)^{k_n-1}}\right]$$

for deviated lift to drag ratio. Accordingly, the deviation ΔK_{eff} is proportional to

$$k_\Pi\left(\ln\frac{1}{1-\bar{V}^2}\right)^{k_n-1} \quad \text{or} \quad \frac{k_\Pi}{k_\Pi - 1}\left[1 - \left(\frac{\ln\frac{1}{1-\bar{V}^2}}{\ln\frac{1}{1-\bar{V}_0^2}}\right)^{k_n-1}\right]$$

respectively.

The law (10) may be considered acceptable if the current path does not differ too much from the nominal one. In the more general case, when wide-range variations of entry conditions are possible, inflight computer calculations, numerical or analytical, are needed to periodically form the expected path.

For a discussion of various numerical computational methods now being used, the reader will be referred to [10].

When using analytical methods, it is helpful to address the solution of the inverse problem of flight dynamics [21]. It proves convenient to define the expected path analytically in the ($\bar{V}$, n_x)-plane and then to use equation (2) for calculating the required effective aerodynamic efficiency and the required vertical velocity. To make this clearer, note that since $n_x = (R\lambda)^{1/2}\, y\exp(-2x)$, the choice of a nominal program $n_x(\bar{V})$ is in fact equivalent to specifying $y(x)$, viz.,

$$y = \frac{n_x e^{2x}}{\sqrt{R\lambda}} = \frac{n_x(\bar{V})}{\sqrt{R\lambda}\,\bar{V}^2}$$

which gives

$$\left\{\begin{aligned} V_y &= V\sin\Theta = -\frac{1}{\sqrt{R\lambda}}\frac{dy}{dx}\bar{V}\sqrt{Rg} \\ &= \sqrt{\frac{g}{\lambda}}\,\bar{V}^2\frac{dy}{d\bar{V}} = \sqrt{\frac{g}{R\lambda^2}}\,\bar{V}^2\frac{d}{d\bar{V}}\left[\frac{n\times(\bar{V})}{\bar{V}^2}\right] \\ K_{\text{eff}} &= K\cos\gamma = \frac{1}{\sqrt{R\lambda}}\left[-\frac{d^2y}{dx^2} + \frac{\frac{1}{\bar{V}^2} - 1}{y}\right] \\ &= \frac{1}{R\lambda}\bar{V}\frac{d}{d\bar{V}}\left\{\bar{V}\frac{d}{d\bar{V}}\left[\frac{n\times(\bar{V})}{\bar{V}}\right]\right\} + \frac{1-\bar{V}^2}{n_x(\bar{V})} \end{aligned}\right. \tag{11}$$

after using (2) and the relation $dx = -d\bar{V}/\bar{V}$.

One possible way of forming a control algorithm is as follows [21]. First, a program function $n_x(a, \bar{V})$, a being a free parameter, is defined in an analytical or a piecewise analytical form for the remainder of the path. The parameter a is selected such as to make the desired flight range s_p (angle range to the target point) coincide with the expected flight range as determined by the expression

$$s_{\exp}(a) \approx \frac{1}{R}\int_{t_0}^{t} V\,dt \approx -\int_{V_0}^{V} \frac{V\,dV}{Rgn_x} = \int_{\bar{V}}^{\bar{V}_0} \frac{\bar{V}\,d\bar{V}}{n_x(a, \bar{V})}$$

The parameter a having been found, the nominal values of $V_{y\,\mathrm{nom}}$ and $K_{\mathrm{eff.nom}}$ are calculated by (11) giving

$$K_{\mathrm{eff}} = K_{\mathrm{eff.nom}}(V) + c_n[n_x - n_{x\,\mathrm{nom}}(V)] - c_{V_y}[V_y - V_{y\mathrm{nom}}(V)]$$

for the control law.

The values of c_x and c_{V_y} are chosen such as to secure a sufficiently rapid transition to the nominal path computed. This done, the representative point in the $(\bar{V}, n_x)$-plane moves—in the absense of disturbances—along the curve $n_{x\,\mathrm{nom}}$ $(\bar{V}, a)$ specified by the value a = const. (see Fig. 16). We thus conclude that the choice of the family of curves $n_{x\,\mathrm{nom}}(\bar{V}, a)$ determines the nature of, and can impart desired properties to, the transient processes involved in the controlled spacecraft motion.

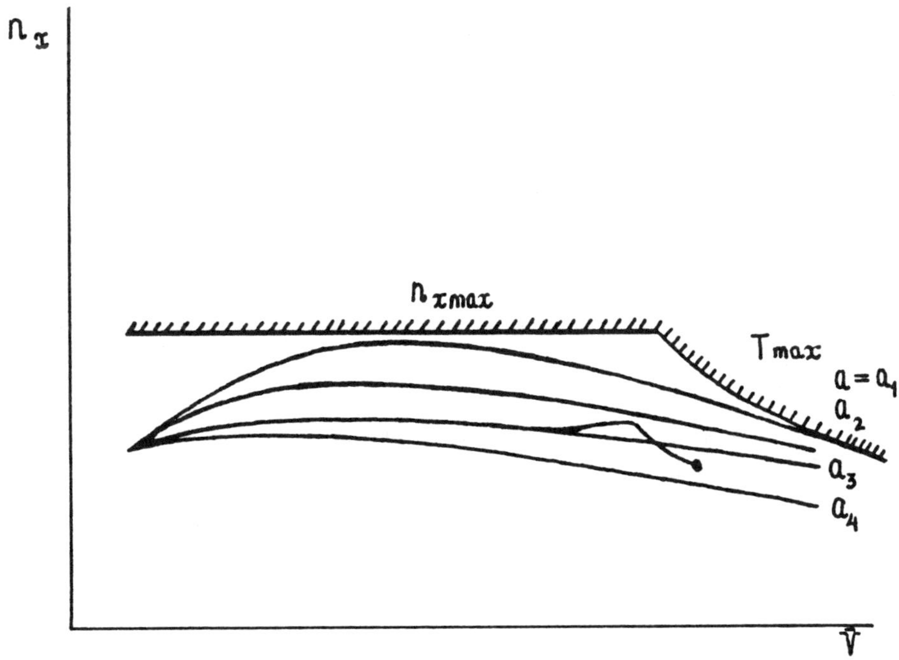

Figure 16

REFERENCES

1. Allen, H. I. and Eggers, A. I. A study of the motion and aerodynamic heating of ballistic missiles entering the earth's atmosphere at high supersonic speeds. NASA Rep., 1958, Vol. 1381.
2. Eggers, A. I. The possibility of a safe landing. Space Technology. Wiley, N.Y., 1959.
3. Chapman, D. K. An approximate analytical method for studying entry into planetary atmospheres, Washington, 1959.
4. Chapman, D. R. An analysis of the corridor and guidance requirements for supercircular entry into planetary atmospheres. NASA Techn. Rep., 1959, Vol. 55.
5. Yaroshevskii, V. A. An approximate method for calculating entry into planetary atmospheres. Kosm. Issled, 1964, Vol. 2, No. 4, pp. 407–531; No. 5, pp. 679–697.
6. Ilyin, V. A. and Kuzmak, G. E. Optimum spacecraft flights. Moscow, Nauka, 1978 (in Russian).
7. Shilov, A. A. and Zhelnin, Yu. N. On the minimum of the maximum overload of a spacecraft decelerated by an atmosphere. Kosm. Issled., 1966, Vol. 4, No. 4.
8. Shkadov, L. M. Bukhanova, R. S. Illarionov, V. F. and Plokhikh, V. P. Three-dimensional motion optimization for a spacecraft in an atmosphere. Moscow, Mashinostroyeniye, 1972 (in Russian).
9. Ivanov, N. M. and Martynov, A. I. Motion of spacecrafts in planetary atmospheres. Moscow, Nauka, 1985 (in Russian).
10. Okhotsimskii, D. E. Golubev, Yu. F. and Sikharulidze, Yu. G. Spacecraft entry control algorithms. Moscow, Nauka, 1975 (in Russian).
11. Vinh, N. X., Busemann, A. and Culp, R. D. Hypersonic and planetary entry-flight mechanics. Ann Arbor, Univ. Mich. Press, 1980.
12. Parysheva, G. V. and Yaroshevskii, V. A. Approximate analysis of quasi-stationary glide paths. Kosm. Issled., 1981, Vol. 19, No. 2, pp. 191–199.
13. Parysheva, G. V., Smirnov, S. A. and Yaroshevskii, V. A. Determination of the accessibility region of a thrustless spacecraft. Uch. Zap. TsAGI, 1981, Vol. 12, No. 5, pp. 72–83.
14. Chernous'ko, F. L. and Kolmanovskii, V. B. Optimal control for random perturbations. Moscow, Nauka, 1978 (in Russian).
15. Kolosov, G. E. Optimum automatic systems for random perturbations. Moscow, Nauka, 1984.
16. Lebedev, A. A., Krasilshchikov, M. I. and Malyshev, V. V. Optimal control of spacecraft motion. Moscow, Mashinostroyeniye, 1974 (in Russian).
17. Deyst, I. I. Optimal control at a reentry vehicle in the presence of measurement uncertainties. IEEE Trans. Automat. Contr., Apr. 1968.
18. Dudar, E. N. and Yaroshevskii, V. A. Lateral motion control for a spacecraft in an atmosphere. Kosm. Issled., 1984, Vol. 22, No. 2, pp. 189–200.
19. Rosenbaum, R. Longitudinal control for a lifting vehicle entering a planetary atmosphere. ARS Pap., 1961, No. 1911–1961.
20. Wingrove, R. C. A survey of atmosphere reentry guidance and control methods. IAS Pap., 1963, No. 63–86.
21. Harpold, I. C. and Graves, G. A. Shuttle entry guidance. J. Astronaut. Sci., 1979, Vol. 27, No. 3, pp. 239–268.

ON THE PRINCIPAL LAWS OF CLASSICAL MECHANICS

V. V. Rumyantsev

The year 1987 marks the tercentenary of the first publication of Newton's "Mathematical Principles of Natural Philosophy," the book which laid the foundations of classical mechanics in particular and classical physics in general. Although Newton's axioms and Newton's laws of motion have frequently been a subject of criticism in the past three hundred years, undoubtedly they remain a perfectly reliable basis for both theoretical and applied mechanics. As regards their interpretation, however, there has actually never been a full, if any, consensus among different writers. In the 19th century, attempts were made to deduce the second law from the law of independent forces. Today, also, we are witnessing continual disputes over the fundamentals of mechanics.

In his paper, Newton's definitions and laws are discussed. The topics on which main emphasis is placed are the modern formulations on Newton's laws, analysis of D'Alembert's principle and of the axiom of separability, and the laws of momentum, angular momentum and kinetic energy.

1. What invariably catches the imagination of the reader is the scope of the Principles, the enormous range of the physical phenomena considered, and the immense number of mathematical results, expressed as theorems, lemmas, and corollaries and often confirmed by experiments devised and performed by the author himself. Newton considered the Principles as the mathematical foundation

of natural philosophy, i.e. physics, and the purpose of physics, in his view, is "this: from the phenomena of motions to investigate the forces of nature, and then from these forces to demonstrate the other phenomena" [1]. This was the task Newton performed in the most impressive manner by creating the unified system of world on the basis of the universal law of gravitation he himself discovered.

For Lagrange [3], the Principles are "the most spectacular achievement of the human mind," a masterpiece that "shaped mechanics into an entirely new science." In S. I. Vavilov's words, "no other event in the history of science has been as significant as the appearance of Newton's Principles" [4].

2. In Newton's opinion, theoretical or "rational mechanics will be the science of motions resulting from any forces whatsoever and of the forces required to produce any motions, accurately proposed and demonstrated" [1].

By motion Newton means what we now call a mechanical motion, the continuous change in the mutual position of either different material bodies or of the different parts of the same body.

As a characteristic of the materiality of a body, Newton introduces an entirely new concept into scientific usage, the mass of a body:

"Definition 1. The quantity of matter is the measure of the same, arising from its density and bulk conjointly . . . And the same is known by the weight of each body, for it is proportional to the weight, as I have found by experiments on pendulums, very accurately made . . .." As Krylov points out, "of Newton's definitions, no other has caused so much criticism and so many interpretations as this one has . . . Before Newton, the concept of mass was not used in science, and only that of weight was known . . . and, following modern terminology, density was not defined as the mass of a unit volume, but rather was considered to be proportional to the weight of the body and inversely proportional to its volume" [2]. For Newton, mass does not change and possesses the additivity property.

As a measure of motion, Newton takes the quantity of motion, or the momentum:

"Definition II. The quantity of motion is the measure of the same, arising from the velocity and quantity of matter conjointly.

Definition III. The *vis insita,* or innate force of matter, is a power of resisting by which every body, as much as in it lies, continues in its present state, whether it be of rest or of moving uniformly forward in a right line.

This force is always proportional to the body whose force it is and does not differ from the inactivity of the mass, except in our manner of conceiving it. A body, from the inert nature of matter, is not without difficulty put out of its state of rest or motion. Upon which account, this *vis insita* may, by a most significant name, be called "inertia" (*vis inertiae*) or "force of inactivity." But a body only exerts this force when another force, impressed upon it, endeavors to change its condition; and the exercise of this force may be considered as both resistance and impulse; it is resistance so far as the body, for maintaining its present state, opposes the force impressed; it is impulse so far as the body, by not easily giving way to the impressed force of another, endeavors to change the state of that other.

Definition IV. An impressed force is an action exerted upon a body in order to change its state, either of rest or of uniform motion in a right line.

This force consists in the action only, and remains no longer in the body when the action is over. For a body maintains every new state it acquires by its inertia only".

Note that Newton distinguishes between *vis inertiae* (force of inertia) and the impressed (or accelerating) force; whereas the latter changes the state of motion of a given body, the former opposes the acting force in that it tends simultaneously to change the motion of the body acting on the given. To put it another way, the force of inertia does not fall in the category of the forces being applied to the body.

It is perhaps appropriate here to recall Euler's objection to the Newtonian concept of *vis inertiae*. In Euler's words, "if the force is defined as a certain influence affecting the state of a body, then it is in some other sense that it should be understood here; the effects of inertia are different in a high degree from those . . . of ordinary forces. To avoid confusion in this respect, we will therefore omit the word "force" and leave the word "inertia" alone to characterize the above property" [5].

In the Scholium following his eight definitions Newton insists on distinguishing time and space "into absolute and relative, true and apparent, mathematical and common.

I. Absolute, true, and mathematical time, of itself and from its own nature, flows equably without relation to anything external, and by another name is called "duration";

II. Absolute space, in its own nature, without relation to anything external, remains always similar and immovable."

Newton's comments on these definitions are definitely worth quoting here. "It may be," he writes, "that there is no such thing as an equable motion whereby time may be accurately measured . it may be that there is no body really at rest to which the place and motions of others may be referred." And later on, "true motion is neither generated nor altered but by some force impressed upon the body moved, but relative motion may be generated or altered without any force impressed upon the body. For it is sufficient only to impress some force on other bodies with which the former is compared that, by their giving way, that relation may be changed in which the relative rest or motion of this other body did consist . . ."

The definitions and the Scholium are followed by Newton's famous axioms, or laws of motion:

"Law I. Every body continues in its state of rest or of uniform motion in a right line unless it is compelled to change that state by forces impressed upon it.

Law II. The change of motion is proportional to the motive force impressed and is made in the direction of the right line in which that force is impressed.

Law III. To every action there is always imposed an equal reaction; or, the mutual actions of two bodies upon each other are always equal and directed to contrary parts.

Corollary I. A body, acted on by two forces simultaneously, will describe the diagonal of a parallelogram in the same time as it would describe the sides by those forces separately."

These laws are given by Newton as postulates summarizing the total of human experience. Spacial cases of the first two laws and of the Corollary had already been formulated before Newton by some scientists, notably by Galileo and Huygens; the third law is due to Newton exclusively. Having actually derived the law of inertia from his famous inclined-plane experiments, Galileo failed to appreciate the significance of his discovery. Newton regards inertia as the general property of matter and makes the law of inertia the corner-stone of his mechanics.

Based on his axioms, and applying mathematical apparatus, Newton considers in his treatise a great many of problems aimed at obtaining "the true motions from their causes, effects, and apparent differences, and the converse," which, in modern language, refers of course to the direct and inverse problems of mechanics.

Using Kepler's laws, Newton establishes his celebrated "**Law of universal gravitation.** Any two pieces of matter attract one another with a force which is proportional to the product of their masses and inversely proportional to the square of their distance apart."

It is this discovery which enabled Newton to create his System of World.

Newton's mechanics were far from being accepted universally and at once, and won its recognition only after many years of fierce fighting with Cartesians and Leibnizians, as Cotes indicates in his introduction to the second edition of the Principles in 1713.

In the years, decades, and centuries after the publication of the Principles, numerous authors attempted to modify Newton's axioms or to replace them by other principles—but to no avail [6].

3. We are going to discuss now the mathematical expression and modern interpretation of Newton's law.

Theoretical mechanics, as indeed any other natural science based on mathematical methods, considers ideal models representing real space, time, and material bodies. Real bodies are idealized by material points, systems of material points, perfectly rigid bodies, and continuous media.

In present-day interpretation, the term "body" involved in Newton's laws is a free material point of mass m defined as

$$m = P/g \tag{1}$$

where P is the weight of the body and g is the free fall acceleration. A free material point is understood as a geometrical point whose position and velocity are not subject to any conditions. The constant m defined by (1) is called a gravitational mass of a body represented by a material point.

Real space and real time, which are forms of the existence of matter as philosophy teaches us, are modeled by Newton in the simplest manner possible, by postulating, namely, the absolute space and the absolute time independent both of matter and of each other.

Today, the absolute space is understood as the ordinary, fixed, three-dimensional Euclidean space, isotropic and homogenous. For relatively small parts of the physical space, experience teaches us that this model is a reasonably accurate one.

The absolute time is thought of as flowing by itself from the past to the future and being the same at any point of space and for any observer, whatever his position or the mode of motion.

The determination of the position of the absolute frame of reference in the system of world and the development of methods for measuring absolute time are the subjects of astronomy and physics.

As a matter of fact, the absolute space involved in Newton's planetary studies is associated with the center of gravity of the solar system and with "fixed stars," whereas as the unity of the absolute time an average solar day is taken.

The development of science has shown that Newtonian space and time serve as quite satisfactory models of their real counterparts, and still now these models retain their significance in physics and mechanics.

Experience and observations show that material bodies influence one another mechanically, either affecting each other's motion or deforming each other.

The quantitative measure of the mechanical effect of one body on another (or of a field on a body) is called a force and characterizes the magnitude and direction of the effect. The most usual notation for this (vector) quantity is **F.** The physical nature of the force is a subject that lies outside the scope of mechanics.

It should be emphasized that Newton's axioms are valid in the absolute frame of reference and that the existence of such a frame is postulated by Newton's first law, namely: there exists a frame of reference relative to which a material point is at rest or performs a uniform straight-line motion, the point in question being either isolated (that is, not interacting with any other material bodies) or, if not isolated, subjected to a set of mutually balanced forces. Rest and uniform straight-line motion turn out to be on equal footing and undistinguishable, and represent the natural state of a body. It is the ability of a body to continue in these natural states which is the inertia of the body.

Referring to the absolute frame of reference, it is possible to find out whether the body is or is not subjected to a force or a set of forces. If not accelerated, the body is free from the action of forces; acceleration indicates that the body does experience some force or, in other words, that it interacts with other bodies.

The term "momentum" involved in the second law refers to the product $m\boldsymbol{v}$ of the mass of a body, m, and its velocity, $\boldsymbol{v} = \dot{\mathbf{r}} = d\mathbf{r}/dt$. Under the action of the force $\mathbf{F}$ a material point acquires an acceleration $\mathbf{w} = \ddot{\mathbf{r}} = d\boldsymbol{v}/dt$ directed along the force. This means that the force $\mathbf{F}$, the measure of mechanical interaction between bodies, is an accelerating force in Newtonian mechanics. In Lagrange's definition, "dynamics is the science concerning accelerating and decelerating forces and the resulting variable motions" [2].

Mathematically, Newton's second law is expressed as

$$\frac{d}{dt}(m\boldsymbol{v}) = \mathbf{F} \tag{2}$$

which is the basic equation of the dynamics of a free point. If $m =$ const., this becomes

$$m\mathbf{w} = \mathbf{F} \tag{3}$$

indicating that the acceleration of a point is proportional to the applied force and inversely proportional to the mass of the point, that is, the greater the mass the smaller the acceleration; mass thus characterizes the inertia of the point, its ability to oppose changes in its motion.

If the applied force **F** and the acceleration w are known, the inertial mass of a body is found to be inversely proportional to the acceleration produced by the given force

$$m = \frac{\mathbf{F}}{w} \tag{4}$$

The gravitational mass given by (1) is proportional to the weight of a body. Experience shows that the gravitational mass as defined by (1) is numerically equal (to an accuracy of 10^{-12}) to the inertial mass as given by (4). Mass thus represents both inertial and gravitational properties of matter and is considered today as one of its most fundamental characteristics. It is recognized at the same time that the mass of a body should be distinguished from the quantity of substance. In the SI system, the quantity of substance is the number of mechanically indivisible structural units of a substance; the unit of quantity of a substance is a mole, equal to 0.012 kg of carbon ^{12}C in the ground state at normal temperature.

It would be noted that using equation (2), a force may be defined dynamically through the resulting acceleration; a force is a vector quantity **F** applied to a material point and equal to the product of the mass m of the point and its acceleration w.

The second law thus provides an equation which relates the mass and acceleration of a material point to the force acting on it and makes it possible to determine one of these from independent measurements of the two others. The same equation enables the solution of the two major problems of mechanics, that is, given the force **F** to find the motion of a material point, or, conversely, from the known law of motion to predict the forces acting on the body point. Since for $\mathbf{F} = 0$ we have

$$w = 0, \quad \text{or} \quad \boldsymbol{v} = \text{const.}$$

from equation (3), the first law is sometimes interpreted as a special case of the second law, an opinion which should of course be rejected as incorrect in light of the preceding discussion.

The third law of motion indicates the source of the force acting on a given body and is in fact the key principle in the mechanics of systems of material bodies. The source of the force is a certain other body (or a field) which is, in its turn, under the action of the first; in other words, any change in the motion of a given body can only result from its interaction with another body or with a field. If the influence $\mathbf{F}_{12}$ of the second body on the first is called action, and the influence $\mathbf{F}_{21}$ of the first on the second reaction, then according to the third law

$$\mathbf{F}_{12} = -\mathbf{F}_{21} \tag{5}$$

If there are two interacting points, m_1 and m_2, we have

$$m_1 w_1 = \mathbf{F}_{12}, \quad m_2 w_2 = \mathbf{F}_{21}$$

by equation (3), so that equation (5) may by rewritten as

$$\mathbf{F}_{21} - m_1 w_1 \tag{6}$$

showing that $\mathbf{F}_{21}$, the force exerted by the first body on the second, is in fact that opposing changes in the motion of the first body and is therefore its force of inertia [1], in full agreement with Newton's corollary to Definition III.

Using (6), equation (5) may also be written as

$$\mathbf{F}_{12} - m_1 w_1 = 0 \tag{7}$$

with the message that at any instant of time there exists equilibrium between the force $\mathbf{F}_{12}$ acting on the first body and the body's force of inertia, $-m_1 w_1$, applied to the second body.

Surely bodies may interact under a diversity of physical conditions, and the resulting accelerations may therefore vary in wide limits with the position, motion, and the physical state of the interacting bodies. This notwithstanding, the ratio of the accelerations of both bodies remains the same for whatever experiments are performed on the same bodies

$$\frac{w_1}{w_2} = -\frac{m_2}{m_1} \tag{8}$$

This ratio is a constant quantity characterizing the given two bodies and may be used for determining the mass of one of them provided that of the other is known.

We thus conclude that the second and third axioms of Newtonian mechanics are not only laws but also very practical recipes for measuring masses and forces. The forces satisfying the second and third laws are known as Newtonian forces [7].

Today Corollary I to Newton's three laws of motion is often referred to as the fourth law. By this law, if a material point is acted upon simultaneously by two forces $\mathbf{F}_1$ and $\mathbf{F}_2$, directed along the sides *AB* and *AC* of the parallelogram *ABCD* drawn on these forces, then it is possible to replace these forces by their resultant, equal to the diagonal *AD* of the parallelogram; conversely, any force directed along *AD* may be resolved into the components *AB* and *BD* (Corollary II). It thus follows that the force addition law, or the parallelogram law, also expresses the independence of forces acting on one and the same material point.

It is worth remarking at this point that this law is generally incompatible with the functional dependence of forces on points' accelerations [7].

Apart from the absolute reference frame, theoretical mechanics also considers moving coordinate systems associated with reference bodies, it being assumed that the course of time and the dimensions of rigid bodies in such systems are the same as in the absolute system. In the finite analysis, it is the notion of the time equally running in different frames of reference which leads in classical mechanics to the possibility of instantaneous transfer of interactions or signals from one spatial point to another.

According to Coriolis' theorem, the acceleration w, a point which becomes in its motion relative to the absolute frame of reference, is considered to be the sum of three accelerations

$$w = w_r + w_e + w_c \tag{9}$$

where w_r is the acceleration of the point in its motion relative the moving coordinate system; w_e—the translational acceleration, i.e., the acceleration of that point in the moving system which coincides with the point under study at a given instant of time; w_c—the Coriolis (complementary) acceleration depending on the angular velocity of the moving system and on the relative velocity of the point.

Making use of (9), equation (3) becomes

$$mw_r = \mathbf{F} - mw_e - mw_c \tag{10}$$

which describes the motion of the point relative the moving coordinate system. In contrast to equation (3) of absolute motion, equation (10) has the terms $-mw_e$ and $-mw_c$ in its right-hand side, which have the dimension of force and are called, respectively, the force of inertia of the translation motion and the Coriolis force of inertia. The relative motion in the moving coordinate system may thus be considered as occurring in a fixed coordinate system, provided the translational and Coriolis forces are added to the forces acting on the point. These two forces are frequently called fictitious because they originate not in the interaction with other material bodies, as with real forces, but in the motion of the moving coordinate system. These forces may change considerably in transition from one moving system to another and do not obey Newton's third law.

If the moving coordinate system is in a translational, uniform, and straight-line motion, then $w_e = w_c = 0$, $w = w_r$, and equation (10) is of the same form as equation (3). The physical interpretation of this mathematical fact is that given time invariance, all laws of mechanics have the same form in all coordinate systems which are in a translational, rectilinear, and uniform motion relative the absolute system; in other words, all these inertial (or Galilean) reference frames are equivalent to one another as far as the laws of mechanics are concerned. This principle of Galileo is often called the relativity principle of classical mechanics and leads, as already remarked, to the renunciation of Newton's absolute space while leaving the absolute time notion intact.

Mathematically, a transition from one to another inertial system is expressed in the form

$$r = r' + vt, \quad t = t' \tag{11}$$

where r and t are, respectively, the radius vector of a point and time in one coordinate system, and primed letters denote the same quantities in the other system, moving with a constant velocity v relative to the first. These Galilean transformations leave equation (3) unchanged or, to put it another way, the fundamental equation of mechanics is covariant with respect to Galilean transformations.

Moving reference frames are called non-inertial if their motion is other than a translational, rectilinear and uniform one. In non-inertial systems the fundamental equation (3) must be replaced by equation (10).

For a system of N free material points, equations of motion of the form (3) are written for each point in the form

$$m_i w_i = \mathbf{F}_i + \sum_{j=1}^{N} \mathbf{F}_{ij} \; (i = 1, \ldots, N) \tag{12}$$

Here $\mathbf{F}_i$ is the resultant of the external forces exerted on the i-th point by bodies outside the given material system; the $\mathbf{F}_{ij}$ are the internal forces arising from the interaction between the points of the system and satisfying Newton's third law of motion

$$\mathbf{F}_{ij} = -\mathbf{F}_{ji}, \quad \mathbf{F}_{ii} = 0 \; (i, j = 1, \ldots, N) \tag{13}$$

By the third law, the internal forces mutually balance each other to give a zero resultant

$$\sum_{i \neq j} \mathbf{F}_{ij} = 0, \quad \sum_{ij} \bar{r}_i \times \mathbf{F}_{ij} = 0 \tag{14}$$

4. Despite popular opinion, Newton did not restrict himself to free body dynamics; the forced motion of a material point is a subject of many of the problems he deals with in his Principles.

One example of this kind is the theorem concerning a cycloidal pendulum oscillating under the action of a centripetal force proportional to the distance from the center (Proposition LI, Theorem XVIII). According to this theorem, the time of a swing of such a device is independent of the magnitude of the swing. The way Newton treated this problem is by resolving the force acting on the body (and proportional to CT, see Fig. 92 in the Principles) into the forces **F** directed along CX and TX.

Mathematically, the differential equations of motion in this problem may obviously be written in the form

$$mw = \mathbf{F} + N \tag{15}$$

where **F** is a given force and N is the normal reaction of the constraint. This equation expresses in fact the principle of separability, which was actually formulated years after Newton's death and will be discussed in some detail later on this paper.

Everyone familiar with the history of mechanics is well aware of the difficulties encountered in changing from the motions of material points to those of bodies linked together by, say, rods or strings, or by some other means. Such bodies cannot follow unhindered the applied forces and develop interactions between themselves, not known *a priori* and changing the nature of their motions.

Of problems of this kind, that of the center of oscillations of the physical pendulum is the simplest in nature and was historically the first to be considered; the problem attracted the attention of Descartes, Huygens, Bernoulli and Hermann, among others, and, accordingly, various methods of solution were proposed. The methods suggested by Johann Bernoulli, 1703, and Hermann, 1716, proved to be the most interesting. According to Bernoulli, the motions performed by the component bodies of a pendulum may be thought of as consisting of motions actually imparted to the bodies, plus motions which are added to or subtracted from the first and must balance them so as to ensure the equilibrium of the device (we shall see below that this approach contains in itself an embryo of the D'Alembert principle). Hermann, for his part, argued that the forces that must determine the motion of pendulum components are equivalent to those resulting from their weight, so that the former, when reversed in direction, must be in equilibrium with the latter. This approach was later developed by Euler and utilized by Lagrange in his formulation of D'Alembert's principle.

A general method by which any mechanics problem can be expressed in the form of equations was proposed by D'Alembert in 1743 [9]. The basis for deriving this method is D'Alembert's principle, perhaps the most controversial one in the whole of dynamics. D'Alembert formulated his principle when considering a system of bodies pulling each other by means of threads or rigid rods and colliding in an arbitrary manner with one another. In D'Alembert's words his purpose was "to show that all dynamical problems can be treated by a general, simple, and direct method formulated as a combination of equilibrium principles and of those governing the addition of motions" [9]. It is worthwhile reproducing here a concise formulation given by Lagrange [3] to D'Alembert's original interpretation of the principle. "If several bodies are given motions which they are forced to change on account of interaction between them, then . . . one may consider these motions as combined of motions actually imparted to the bodies, and of others, which cancel each other; if follows that the latter motions must be such that they would balance each other were they the only motions acting on the bodies." (Note that motion, as D'Alembert understands it, is in fact the velocity in the vector sense). Another point to be made is that in the author's formulation of the principle there is no mention either of forces applied or of inertial forces.

Considering the case of continuous motions, let $w_\nu^* = \mathbf{F}_\nu/m$ represent those accelerations which the points of a constrained system would receive were they free and acted upon by the forces $\mathbf{F}_\nu$; let w_ν be the actual accelerations of the points. It then follows that in the cancelled (or "lost") motions, the accelerations imparted to the points are given by the difference of the type $w_\nu - w_\nu$. These "lost" accelerations may be caused by the "lost" forces

$$P_\nu = m_\nu(w_\nu^* - w_\nu) = \mathbf{F}_\nu - m_\nu w_\nu \tag{16}$$

D'Alembert's principle. Whatever the motion of a system, its lost forces are at all times balanced by means of the constraints acting in system [10]. We thus see that the motions performed by constrained bodies under the action of given forces are governed by the laws reducible to those governing the equilibrium such bodies adopt under the action of the "lost" forces. However, contrary to popular opinion, this does not mean that the D'Alembert principle reduces problems dynamics to those of statics.

In the course of time, the D'Alembert principle started to be regarded as complementary to Newton's laws as far as the dynamics of constrained bodies is concerned. A number of alternative formulations of the principle have been proposed, apart from that given, and some of them will be discussed later on in the paper. To do this properly a definition of constraints and a formulation of the principle of separability are needed.

By definition, the constraints imposed on a mechanical system are certain conditions limiting in some respects the motions and positions of the points of the system.

Since the points of a constrained system behave differently under the action of given forces $\mathbf{F}_\nu$ than free material points would behave, it follows that, apart from the forces $\mathbf{F}_\nu$, the points of the system are also subjected to certain additional forces, R_ν, whose role is to fulfil the conditions imposed by the constraints. These forces have their origin in the fact of there being constraints in the system and are usually referred to as reactions of constraints. This view does not disagree with Newton's third law, because constraints are always realized, in this way or another, by means of some material devices [11] representing the physical sources of the reactions. The force with which a material point acts on a constraint is equal and directly opposite to the reaction of the constraint. In distinction to the given force $\mathbf{F}_\nu$, also called active forces, the constraint reactions R_ν are not known beforehand and are often called passive forces, to show that they depend not only on the constraint producing mechanisms, but on the active forces as well. The determination of the reactions depends not only on the form of the equations governing the constraints but also on the constraint physical characteristics, to be specified at the formulation stage of the problem.

Separability principle. The effect of constraints on a material point is completely specified by their reactions.

According to this principle, material points subjected to constraints may be considered free if, in addition to the forces given, constraint reactions are applied to the points. In accordance with Newton's second and third laws, the equations of motion of the non-free points may then be written in the form

$$m_\nu w_\nu = \mathbf{F}_\nu + R_\nu \; (\nu = 1, \ldots, N) \tag{17}$$

where R_ν is the resultant of the reactions applied to the ν-th point of the system. Clearly this coincides in form with equations (15) which were actually employed by Newton.

Comparing (16) and (17) yields

$$P_\nu + R_\nu = 0 \; (\nu = 1, \ldots, N) \tag{18}$$

allowing the following reformulation of the D'Alembert principle.

During the motion of a constrained material system the lost forces are at any time balanced by the reactions of the constraints.

If (18) is rewritten in the form of equation (7)

$$\mathbf{F}_\nu + \mathbf{S}_\nu + R_\nu = 0 \ (\nu = 1, \ldots, N) \tag{19}$$

where $\mathbf{S}_\nu = -m_\nu w_\nu$ is the Newtonian inertial force of a point. One further formulation of the D'Alembert principle suggests itself.

If, apart from the given forces, the inertial forces $\mathbf{S}_\nu$ are applied to the points of a constrained system, the set of these forces is balanced by the constraint reactions R_ν.

This formulation first appeared in Delaunay's course of mechanics in 1856 and can be found, with few exceptions, in all text-books that have since been written. Indeed it is this formulation that underlies the subject of kineto-statics, a division of technical mechanics in which methods of statics are applied for determining the dynamic reactions in systems with known motions.

In Delaunay's own opinion, his idea of applying inertial forces to the points of a mathematical system was of no more than a symbolic character [12]. This point of view was shared by many authors. Chaplygin, for example, believed that the force of inertia of a material point "does not actually exist and is merely a mechanical quantity introduced as a very convenient notion to simplify the development of the differential equations of motion." [12].

D'Alembert's principle is "an epoch in the mechanics of a system because it is owing to this principle that any problem of dynamics can be reduced to a purely mathematical investigation." [13]. But, as Sir G. Airy correctly points out, "The principle itself is not a new physical principle, nor any addition to existing physical principles; but is a convenient principle of combination of mechanical considerations, which results in a comprehensive process of great elegance." [14] It will be noted that all the problems of theoretical mechanics treated by the D'Alembert principle may actually be solved by other methods—using, for example, equations of the type (17) based on the separability principle. In this sense the D'Alembert principle is equivalent to the separability principle [12], when traces its origin back to Newton.

Equation (17) for a system of constrained material points may also be represented in the form of equations (12) if all the forces applied to the points of the system, including the reactions of constraints, are divided into internal and external forces. Assuming that the internal forces obey Newton's third law of motion, cf. equation (14), and using equations (12), it is an easy matter to obtain

$$\frac{d\mathbf{Q}}{dt} = \sum_i \mathbf{F}_i, \quad \mathbf{Q} = \sum_i m_i \boldsymbol{v}_i = M\boldsymbol{v}_c, \quad M = \sum_i m_i \tag{20}$$

$$\frac{d\mathbf{G}}{dt} = \sum_i \mathbf{r}_i \times \mathbf{F}_i, \quad \mathbf{G} = \sum_i \mathbf{r}_i \times m_i \boldsymbol{v}_i \tag{21}$$

$$dT = \sum_i \left(\mathbf{F}_i + \sum_j \mathbf{F}_{ij}\right) \cdot d\mathbf{r}_i, \quad T = \frac{1}{2} \sum_i m_i \boldsymbol{v}_i^2 \tag{22}$$

representing, respectively, the theorem of momentum; the theorem of the moment of momentum relative to a certain fixed point 0 taken as the origin for a radius-vector r_i; and the theorem of kinetic energy. Note that equations (20) and (21) contain only external forces, whereas equation (22) contains both external and internal ones (the latter disappearing in a perfectly rigid body).

The theorem (20) on the motion of the mass center seems to suggest [15] that internal forces have no effect on the motion of this point. This is true only if $\Sigma_i \mathbf{F}_i = 0$ because otherwise the external forces, even when absent from equation (20), may influence this motion indirectly through the external forces $\mathbf{F}_i$ [12].

A cautionary remark should be made regarding invariable systems of material points linked by rigid "weightless" bodies. The actions of constraints on the points of such systems are sometimes represented as forces of interaction between the material points. This approach, based on equation of the type (13) and (15), and purely mathematical in character, ignores those physical properties of constraints that secure the rigidity of the system, and may lead sometimes to serious difficulties and contradictions as the examples given below will demonstrate. These examples will be easier to understand, however, if we first discuss reactions of "weightless" constraints, bearing in mind in doing so the Newtonian view that the materiality of a body gives rise to a force on another body. In real life, all material bodies have nonzero masses, so that the case $m = 0$ should be considered as an idealization, a limiting case of $m \to 0$, which may be of relevance when the mass of the given body is so small as to be neglected in comparison with the masses of the other bodies involved in the problem. It is not permissible to actually take this limit unless equilibrium conditions for forces acting on a weightless body are satisfied.

Let a constraint be realized, for example, by a certain rigid body of mass m, which has a central tensor of inertia Θ and is subjected to given forces $\mathbf{F}_i$, and acts by means of the reactions R_i on the points of the system. By the separability principle, the equations of motion of the body are

$$m d\boldsymbol{v}_c/dt = \sum_i \mathbf{F}_i - \sum_i \mathbf{R}_i, \quad d\mathbf{G}/dt = \sum_i \mathbf{r}_i \times \mathbf{F}_i - \sum \mathbf{r}_i \times \mathbf{R}_i$$

giving, in the $m = 0$ limit, the equations

$$\sum_i \mathbf{R}_i = \sum_i \mathbf{F}_i, \quad \sum_i \mathbf{r}_i \times \mathbf{R}_i = \sum_i \mathbf{r}_i \times \mathbf{F}_i \tag{23}$$

as conditions for neglecting the mass of the body.

In other words, a "weightless" body realizing a constraint may be thought of as acting on a component body of the system by certain reactions $R_i \neq 0$ such that their principal vector and principal moment are equal to those of the forces applied to the body. Given the conditions (23), it makes no difference in principle, whether we solve the problem for all the m_ν finite and then let them approach zero, or take some of them at zero already in formulating the problem.

Example 1. Peiffer and Rouche [17] consider a rigid body formed by three points X_1, X_2 and X_3, each of mass m, situated on a weightless straight line. The points are subjected to the forces $\mathbf{F}_1 = \mathbf{F}_3 = ke_2$, $\mathbf{F}_2 = -2ke_2$ $(k > 0)$. The problem is to find the motion assuming that the points are initially at rest on the axis oe_1 and that $\mathbf{X}_1\mathbf{X}_2 = \mathbf{X}_2\mathbf{X}_3 = le_1$, $e_1 \perp e_2$.

Equations of motion will be written in the form (cf. (12))

$$m\ddot{r}_1 = ke_2 + \mathbf{F}_{12} + \mathbf{F}_{13}, \quad m\ddot{r}_2 = -2ke_2 + \mathbf{F}_{21} + \mathbf{F}_{23}, \quad m\ddot{r}_3 = ke_2 + \mathbf{F}_{31} + \mathbf{F}_{32}$$

Under the initial conditions assumed, it follows from the theorems (20) and (21) that $\boldsymbol{v}_c = \mathbf{G} = 0$ throughout the motion, that is, the system is at rest and the equilibruim conditions

$$\mathbf{F}_{12} + \mathbf{F}_{13} = -ke_2, \quad \mathbf{F}_{21} + \mathbf{F}_{23} = 2ke_2, \quad \mathbf{F}_{31} + \mathbf{F}_{32} = -ke_2$$

are satisfied. By the third law, the internal forces $\mathbf{F}_{ij}$ must be of the type $\mathbf{F}_{ij} = \lambda_{ij}e_1$ and clearly will not satisfy the equilibrium conditions.

Peiffer and Rouche conclude therefore that equations (14) cannot be satisfied simultaneously if we wish to fulfil the conditions of invariability in considering the mechanics of an N-point system. The alternative they suggest is to take the equations

$$\sum_{ij} \mathbf{F}_{ij} = 0, \quad \sum_{i,j} \mathbf{r}_i \times \mathbf{F}_{ij} = \gamma \neq 0$$

as axioms and to consider system as acted upon by a pair γ of internal forces to be determined, to quote, "by performing many and repeated experiments under the given physical conditions."

The problem can be solved in a much simpler way, however, because there is actually no need to deal with the internal forces $\mathbf{F}_{ij}$ between the i-th and the j-th points which obviously realize their interaction by means of the solid straight line they lie on. If, instead, we consider the reactions R_i of the line on the points, the equations of motion

$$m\,\ddot{r}_1 = ke_2 + R_1, \quad m\ddot{r}_2 = -\,2ke_2 + R_2, \quad m\ddot{r}_3 = ke_2 + R_3$$

provide equilibrium under the given initial conditions. We have

$$R_1 = -ke_2, \quad R_2 = 2ke_2, \quad R_3 = -ke_2$$

and conditions (23) are easily shown to be satisfied identically.

Example 2. Stadler [18] considers the motion of a rigid double pendulum containing three particles P_s with masses $m_s (s = 0, 1, 2)$ connected by a single weightless rod. The rod is suspended from a point 0 in such a way that the system may move as a pendulum in vertical plane with the particle P_0 remaining at rest. The external forces are the weights $m_s g$ of the particles and the reaction $\mathbf{F}_0$ at point 0. The problem is to find the angle of deviation Θ of the rod as a function of time.

The author starts by introducing internal inter-particle forces $\mathbf{F}_{ij}$ which describe, in his opinion, the action of constraints on material points in an ideal case. He then uses the laws II to IV and the geometrical conditions of the problem to write down the equations of motion and solves these latter for $\mathbf{F}_0$, $R_1 = \mathbf{F}_{10} + \mathbf{F}_{12}$ and $R_2 = -\mathbf{F}_{12} + \mathbf{F}_{20}$ to obtain

$$\mathbf{F}_0 = -\left[\sum_{i=1}^{2} m_i a_i \dot{\Theta}^2 + \sum_{s=0}^{2} m_s g \cos\Theta\right] e_r + \left[\sum_{i=1}^{2} m_i a_i \ddot{\Theta} + \sum_{s=0}^{2} m_s g \sin\Theta\right] e_\Theta \tag{24}$$

$$R_i = -m_i(a_i\dot{\Theta}^2 + g\cos\Theta)e_r + m_i(a_i\ddot{\Theta} + g\sin\Theta)e_\Theta \; (i = 1, 2)$$

Since $a_1 \neq a_2$, it follows that the e_Θ-components of the forces R_1 and R_2 cannot vanish simultaneously, that is, the internal forces R_i are not central ones in the sense of being directed along the line connecting the points. The theorem of the moment of momentum is thus invalidated, indicating, in Stadler's opinion, that the problem admits of no solution in the framework of traditional Newtonian dynamics. He therefore abandons the assumption of central internal forces and postulates an independent law for the change of the moment of momentum. The equation of oscillations he obtains is

$$\ddot{\Theta} = -\frac{m_1 a_1 + m_2 a_2}{m_1 a_1^2 + m_2 a_2^2} g \sin\Theta \tag{25}$$

Note that expression (24) for the R_i does not contain m_0, thus allowing one to consider the point P as not existing at all (this is precisely the formulation employed by Routh [13, Chap. II] in solving this problem by the application of the D'Alembert principle). Furthermore, Stadler writes that he included P_0 only to confirm Newton's idea of forces acting on masses (overlooking the fact that this is already secured by the presence of the fixed point 0). But if we assume that $m_0 = 0$, then $\mathbf{F}_{10} = \mathbf{F}_{20} = 0$, $R_1 = \mathbf{F}_{12}$, $R_2 = -\mathbf{F}_{12}$, and it is seen from (24) that $R_1 \neq -R_2$, that is, $\mathbf{F}_{12} \neq -\mathbf{F}_{21}$(!!) thus indicating the erroneous nature of Stadler's solution.

The same problem may be solved using equations (17) if we assign a mass m_3 to the rod and introduce reaction R_i from the rod to the particles $P_i (i = 0, 1, 2)$ instead of considering the internal forces $\mathbf{F}_{ij}$ of interaction between the particles.

The equations of motion of the component bodies are

$$m_s \ddot{r}_s = m_s g + R_s \quad (s = 0, 1, 2), \quad m_3 \ddot{r}_3 = m_3 g - \sum_{s=0}^{2} R_s + \mathbf{F}_0$$
$$J_3 \frac{d^2\Theta}{dt^2} = -\frac{m_3 a_2}{2} g \sin\Theta - \sum_{i=1}^{2} r_i \times R_i \tag{26}$$

where $\mathbf{r}_3$ is the radius vector of the center of gravity of the rod (distant $l = a_2/2$ from point 0) and J_3 is the moment of inertia of the rod relative point 0. Noting that $\mathbf{r}_0 = 0$, $\mathbf{r}_i = a_i e_\mathbf{r} (i = 1, 2)$, $\mathbf{r}_3 = l e_\mathbf{r}$, the first four of these equations yield

$$\mathbf{F}_0 = -\sum_{i=1}^{2} m_i[(a_i\dot{\Theta}^2 + g\cos\Theta)e_r - (a_i\ddot{\Theta}$$
$$+ g\sin\Theta)e_\Theta] - m_0 g(\cos\Theta e_r - \sin\Theta e_\Theta)$$
$$- m_3[(l\dot{\Theta}^2 + g\cos\Theta)e_r - (l\ddot{\Theta} + g\sin\Theta)e_\Theta], \quad R_0 = -m_0 g$$

together with expressions of the type (24) for the $R_i (i = 1, 2)$. Substitution for the R_i into the last of equations (26) then gives the equation of motion of the pendulum

$$\ddot{\Theta} = -\frac{\sum_{i=1}^{2} m_i a_i + m_3 l}{\sum_{i=1}^{2} m_i a_i^2 + J_3} g \sin \Theta$$

If $m_3 = J_3 = 0$, this simplifies to (25) and the expression for $\mathbf{F}_0$ takes the form (24), and it can be easily verified that equations (23) are satisfied in this case. If m_0 is also set to zero, expressions (24) for the R_i remain unchanged, and terms in m_0 will disappear in the expression for $\mathbf{F}_0$, indicating that the only effect of the particle P_0 is on the reaction of the fixed point 0. It follows that equations (14) in [18] are only valid if the symbols $R_i (i = 1, 2)$ in these equations denote the reactions of the rod on the particles P_i, and not the resultants of the interaction forces between these particles. This is quite understandable, because the particles interact only by means of the rod, with which they are rigidly linked: the particle P_1 moving along the arc of a smaller radius a_1 accelerates the motion of the particle P_2 while being decelerated by the slower motion of this latter [14].

The above two examples serve well to demonstrate that in considering constrained material points it is generally incorrect to replace the constraints only by the internal forces of interparticle interactions and to write down the equations of the constraints. The physical properties of the constraints must also be taken into account. If constraints are realized by rigid bodies, it should be kept in mind that a material point rigidly linked to a body experiences the action of the molecular forces from other points of the body, and the direction of the reaction is therefore not known *a priori*.

Newton's laws together with the separability principle (or D'Alembert's principle) provide the basis for the so-called vector mechanics, whose purpose is to identify all the forces acting on every material point of a system and then to investigate the motion of the points.

The alternative direction in classical mechanics, analytical mechanics, has variational principles as its basis, from which all the equations of motion and all the theorems of mechanics are obtained by the deductive method [19]. Although originated in Newton's laws and in the separability principle, the variational principles have long exceeded the limits on Newtonian mechanics.

Based on Newton's laws, the methods of classical mechanics are amply confirmed by practice and provide a firm and reliable basis for scientific progress and technological advances in many diverse fields.

Newton's laws have stood the test of time and have had a long history, even when a complete revision of our space and time concepts rendered them only a special case of relativity theory. To quote from Einstein, "May nobody think that Newton's great creation can be overthrown by the theory of relativity or by any other theory for that matter. Newton's broad and clear ideas will retain forever their significance as the foundation of our physical views" [20].

REFERENCES

1. Newton, I. Philoshophiae Naturalis Principia Mathematica. Londini, 1687.
2. Krylov, A. N. Collection of works. Moscow, USSR Acad. Sci., 1936 (in Russian).
3. Lagrange, J. L. Méchanique analytique. Paris, 1811.
4. Vavilov, S. I. Isaac Newton. Moscow, USSR Acad. Sci., 1943 (in Russian).
5. Euler, L. Theoria motus corporum solidorum neu rigidorum. Rostock, Greifswald, 1765.
6. Thomson, W. and Tait, P. G. Treatise on natural philosophy. Cambridge, 1879, Vol. 1; 1883, Vol. 2.
7. Pars, L. A. Treatise on analytical dynamics. Wiley, N.Y., 1965.
8. Kittel, Ch. Knight, W. and Ruderman, M. Berkeley physics course. Mechanics, Berkeley, Cal., 1965, Vol. 1.
9. D'Alembert, J. Traité de dynamique. Paris, 1743.
10. Lyapunov, A. M. Lectures on theoretical mechanics. Kiev, Naukova Dumka, 1982 (in Russian).
11. Suslov, G. K. Theoretical mechanics. Moscow, Gostekhizdat, 1944 (in Russian).
12. Geronimus, Ya. L. Theoretical mechanics. Moscow, Nauka, 1973 (in Russian).
13. Chaplygin, S. A. Collection of works. Gostekhizdat, 1949, Vol. 1 (in Russian).
14. Routh, E. J. Dynamics of a system of rigid bodies. London, McMillan and Co., 1930, Vol. 1.
15. Goldstein, H. Classical mechanics. Cambridge, 1950.
16. Appel, P. Traité de méchanique rationalle. Paris, Gauthier-Villars, 1926.
17. Peiffer, K. and Rouche, N. Méchanique générale. Luvain-La-Neuve, Cabay, 1984, Parts 1, 2.
18. Stadler, W. Inadequancy of the usual Newtonian formulation for certain problems in particle mechanics. Amer. J. Phys., 1982, Vol. 50, No. 7.
19. Rumyantsev, V. V. On the basic laws and variational principles of classical mechanics. Moscow Prepr. USSR Acad. Sci. Inst. Mechanics, Computer Center USSR Acad. Sci., 1985, No. 257.
20. Einstein, A. Ideas and opinions. Wiley, N.Y., 1954.

NEWTON'S *PRINCIPIA* (TO THE TERCENTENARY OF THE FIRST EDITION)

G. K. Mikhailov

On July 5, 1687, a parcel with sixty leather-bound copies of a newly printed book was sent from London to Cambridge. The sender was Edmond Halley, a prominent scientist of the time and a future Royal Astronomer, famous today primarily for the comet named after him. The addressee, to whom a third of the copies was intended, was the author, Isaac Newton, then 44 years old and not yet Sir, and the book was titled *Principia Mathematica Philosophiae Naturalis* [1], or *Mathematical Principles of Natural Philosophy*. The book, a crown-quarto volume of about 500 pages, had been printed under the auspices of the London Royal Society but, as estimable as this corporation was, it was in a difficult budgetary situation at the time, and it was actually the same Halley who financed and indeed initiated and organized the whole venture. Since that summer day of 1687, *Principia* has become a scientific classic and is now known worldwide as one of the glories of the human intellect, a founding document of our modern exact science. In S. I. Vavilov's words "no other event in the history of science has been as significant as the appearance of Newton's *Principia*" [2].

Given the enormity of the world Newtoniana and the availability of many excellent and fundamental studies on the subject, it would be no less than presumptuous to attempt here a new analysis of *Principia* and its background. We hope, however, that some aspects of its pre-history and certain untraditional views adopted here concerning Newtonian thought will be of interest to the reader who, we note parenthetically, should be more or less acquainted with Newton's life and work and with the socioeconomic characteristics of the late 17th century.

The first edition of *Principia* contained an official dedication, preface, two introductory parts titled "Definitiones" and "Axiomata sive Leges Motus," and three main parts traditionally called "libri" (books), namely "De motu corporum liber primus" (later referred to as Book I), "De motu corporum liber secundus" (Book II), and "De mundi systemate" (Book III). Book I deals with the motions a mass point performs under the action of central forces, including conic and mobile orbits, and provides in fact the basis for celestial mechanics, even though no single mention of the Solar system is ever made. In Book II, Newton is concerned with motions of bodies in resistant media and discusses fluid motions and wave propagation processes. Most of Book II is in fact a Summa Contra Cartesian physics, although without the explicit reference to Descartes' work. Book III gives a general description of the Copernican picture of the Universe, based on the universal law of gravitation, and includes certain elements of the theory of lunar and comet motions.

Although it is recognized universally that *Principia* is the starting point of rational mechanics, hardly any scientist can be found in our time whose knowledge of this treatise would extend further than just a few statements from the introductory chapters and something very general concerning Newton's justification of the Copernican system. What makes the great book so very inaccessible to the modern reader is—sometimes—the incomprehensibility of Newton's approach to the subject matter and—throughout the book—the exceptional difficulty of Newton's obsolete mathematical language for everyone used to the simple Leibnizian symbolics of the modern mathematical analysis. This, of course, if we ignore the fact that *Principia* was originally written in Latin. Undoubtedly, the study of Newton's *Principia* is greatly facilitated by translations into new languages, of which the Russian translation by A. M. Krylov [3] deserves perhaps special mention as a translation immaculate in its literary style and annotated most exhaustively from the mathematical point of view. It should be kept in mind, however, that hardly any translator can help tacitly expressing his own attitude to, indeed his commentary on, the material being translated (as poetry most vividly exemplifies), and so there is actually no hope of finding a totally equivalent translation of a 17th century physicomathematical tract written in the now almost completely dead Latin.

The mathematical justification for the Copernican world system was in fact the primary reason why *Principia* was appreciated by Newton's contemporaries, scientific and not, and by many generations that followed. (As Lagrange used to say, the system of the world cannot be invented but once, an expression of admiration, but also of deep regret on the part of the great savant). Today, about two-and-a-half centuries after Newton's death, the Copernican system is a matter of fact at even the primary school level and its justification, Newton's immortal contribution to the treasury of human thought, is now regarded, not unnaturally, as of academic interest only. The question which does remain open is that of the foundations of Newtonian mechanics. The truth is that although Newton's exposition leaves much to be desired in terms of its inner logic, no one has yet succeeded in removing all the inconsistencies of this exposition. This is not to say, of course, that Newtonian mechanics rests on an unsound basis; rather, this is indicative of the profoundness and complexity of the basic principles underlying the subject of mechanics.

The three centuries of heated and often fruitless scholarly disputes over the foundations of Newtonian mechanics are perhaps the most magnificent monument to the immortal genius of the great author of *Principia*. There is no other man in the history of science whose concepts would have remained generally intact under such long, constant, and scrupulous criticism as that which Newton's ideas have undergone.

There exist three editions of *Principia* that appeared in Newton's lifetime, of which the first, the raison d'être of this article, comprised at most 300 to 400 copies by modern estimates and was sold out within four years or so. It took about a quarter of a century of concerted efforts by his scientific contemporaries to persuade the great man to publish the second edition in 1713, and then yet another 13 years passed before the third and ultimate authorized edition was undertaken in 1726, one year before Newton's death. There are a great number of changes and improvements that were made by the author in these successive editions, the most significant of which concern fluid motions (in Book II) and some philosophical aspects and concrete celestial applications of mechanics (Book III). During the past three centuries about a hundred editions of *Principia* have appeared in ten new languages in either a complete or abridged form. The complete editions have been translated into English (1729), French (1756), German (1872), Russian (1915–1916), Swedish (1927–1931), Japanese (1930), and Rumanian (1956).

Although as far back as in the late 19th century positive opinions were expressed on the need for a critical edition of *Principia,* with textual, historical, and scientific analysis, it was not until the middle of this century that such an edition was actually undertaken. In 1971, a fundamental historico-critical "Introduction" to Newton's *Principia* was published [4], to be followed, in 1972, by a complete two-volume Latin text of the *Principia,* with all the variants found in the three authorized editions and in numerous Newton's manuscripts *quae exstant* [5].[1] Plans also have been made to complement these three volumes with a critical English

1) Curiously enough, these Cambridge University editions had been prepared by an American Bernard Cohen, and a Russian-origin Frenchman Alexander Koyré, who worked in the USA for a few years.

translation, a parallel analytic presentation of the *Principia,* glossaries and other auxiliary materials, but as things stand, there is slight promise that these plans will be realized in the foreseeable future.

Apart from the *Principia,* there have been other major developments in the "Newton industry" in the last three decades or so. In the years from 1967 to 1981, D. Whiteside published 8 volumes of The "Mathematical Papers" of Isaac Newton [6], a total of about 5000 pages, of which the 6th volume (1974) contains a number of Newton's preparatory material to the *Principia* (dated in the period from 1684 to 1686) and is therefore of special value for those interested in the history of mechanics. In 1959–1977, seven volumes of Newton's "Correspondence" were published [7], about 4000 pages in total, including (along with new material) all of the earlier published letters. These two editions were based on the vast archive of Newton's manuscript material, the major part of which (the so-called Portsmouth Collection) was entrusted to the Cambridge University Library about a hundred years ago by the then Earl of Portsmouth.[2] The Collection contains an immense mass of Newton's correspondence; the copies of *Principia* with the author's corrections; the now widely known "Waste-Book" with Newton's first notes on dynamics; and numerous other papers. It should be noted, however, that the Catalogue of the Portsmouth Collection published in 1888 was too brief and inadequate to arise scholarly interest at the time. It was the edition of the "Mathematical Papers" and the "Correspondence" and the preceeding enormous work on the systematization of Newton's extant writings that put these materials into circulation and triggered a large number of fundamental studies (mostly in the English-speaking world) on Newton's life and personality and on the development of his mechanical ideas in the course of his long life (see [8–11], for example; the reader is referred to [12] for a rather complete Newtonian bibliography for the period 1672–1972). It is perhaps only natural that the first twenty years of Newton's creative work, starting from the mid-1660's, have been in the focus of attention of the most investigators.

As a result of these research activities, we now have sufficiently complete information at our disposal on the study of mechanics by young Newton in 1660's; materials (not without lacunas, though) on the circumstances surrounding the preparation of the 1687 edition of *Principia;* detailed data on the revision of the first and preparation of the second and third editions. And yet we know little or nothing about the actual process of writing the first edition of the *Principia* for, surprising as it may seem—especially considering the vast extent of Newton's manuscripts of all sorts remain—we do not have the manuscript draft of the treatise, nor are there any traces of the analytical solutions allegedly replaced by geometrical ones at the final stage of the book preparation. "What ever happened to the work-sheets of *Principia?*" B. Cohen asks. "Do they still exist in some obscure private or public collection? Was this particular set of manuscripts—alone

2) One of whose ancestors married Newton's grand-niece in 1740; for about a century-and-a-half the Collection remained an idle stock in the hereditary castle of the family.

out of all of Newton's papers—lost or mislaid, either when the Portsmouth Collection was still in Hurstbourne Castle or during the actual transfer to the University Library in Cambridge? Did such work-sheets still exist among Newton's papers at the time of his death? Or were they lost or destroyed—either by chance or design—during Newton's own lifetime? We may possibly never be certain of the answers to these questions." As things stand the penultimate phrase is by no means a rhetorical question. Considering Newton's unbalanced character and his particularly grave persecution mania and depression of the earlier half of the 1690's; considering, further, his notorious controversy with Hooke over priority in the discovery of the universal law of gravitation, and a similar controversy with Leibniz as to the validity of their respective claims to the invention of the calculus, there is evey reason to believe that the drafts of *Principia* could have been destroyed in Newton's lifetime, that is, by Newton himself, to call a spade a spade.

But let us return to the origins of Newton's dynamics. How did he come to *Principia?*

Isaac Newton was born at Woolsthorpe in Lincolnshire, on January 4, 1643 in the new (Gregorian) calendar, which corresponds to Christmas Day, December 25, 1642 in the then used in England old Julian calendar.[3] In 1661, Newton was admitted to Trinity College in Cambridge. The earlier period of his creative work is associated with the dreadful Plague years of 1665–1666, during which Newton lived and worked in his native village. It was then that this youngster of 22–23 paved new ways in optics and dynamics, and in the theory of mathematical series made fundamental contributions to mathematical analysis, and all but discovered the universal law of gravitation.

It can be deduced rather confidently from the analysis of Newton's manuscript material that there were two major factors on which Newton's mechanical works of this period were based. The first was the Galileo's $s \sim t^2$ free fall law generalized by Newton to include an arbitrary constant force (not that he had a very clear idea of this notion at the time). The second factor was the mechanics of Descartes. Even though Newton did not accept, indeed had a staunch aversion to, the Cartesian philosophy, he did adopt Descartes' idea of inertia (which was right) and his concept of centrifugal forces (which was not). There is every evidence that young Newton was on good terms with kinematics; in one of his notes of this time he defines velocity as the slope of a tangent for an arbitrary time dependence of a path, which is identical to the analytical definition $v = ds/dt$. It is a pity that Newton ignored analytical geometry completely in his later work (was it an anti-Cartesian démarche?), because the classical synthetic method of presentation he borrowed from the ancients for his *Principia* delayed for at least

3) The Gregorian calendar was adopted in England only in the autumn of 1752, and it was only then that the beginning of a year was translated from March 25th to January 1st. (It should be kept in mind therefore in reading old English literature that to avoid ambiguity it was customary in the latter half of the 17th century to give double year denomination for the first quarter of the year, namely, in both the "Continental" and "English" sense).

a half century the development of mechanical science.[4] In his study of elastic collision Newton perceived very early the erroneous nature of Descartes' approach to the problem and the necessity of employing algebraic—indeed vector—sums for the momenta of colliding bodies. It was a detailed analysis of this problem and the thorough understanding of the laws of inertia and of the notion of momentum that led Newton to the concept of force as we find it in the *Principia* (although it proves impossible to trace in full detail the evolution of this concept through Newton's manuscripts).

Newton's earlier analysis of circular motion, the simplest type of curvilinear motion, may serve well to illustrate the way in which he approached dynamical problems in the 1660's (see the recent study by T. Herivel [9]). Where Descartes considers a stone rotating on the end of a string, Newton takes straight-line motion as his departure point and analyzes first the motion of a ball in a hollow circular cylinder and its collisions with the cylinder's wall. He assumes first that the ball moving at the velocity v within the cylinder describes a four-cornered broken line as shown in Fig. 1. The change of the ball's momentum in each of the four collisions is (in absolute magnitude) $mv\sqrt{2}$ (note that Newton does not introduce the notion of mass), so that the sum of these over one revolution is $\Sigma|\Delta mv| = 4mv\sqrt{2}$. The ratio of this sum to mv, the momentum of the ball, is equal to the ratio of the length of the ball's path, $4R\sqrt{2}$, to the radius of the circle, R; and it appeared self-evident to Newton that these ratios are always equal to each other whatever regular polygon is described by the ball within the cylinder, that is, we have always that $\Sigma|\Delta mv|/mv = L/R$, L being the perimeter of the polygon. Since the trajectory of the ball tends to coincide with the circle's circumference as the number of the sides of polygons is indefinitely increased, it follows that the above ratio takes to value 2π in this limit. We are thus forced to conclude that the total momentum the ball acquires over one revolution in this sense ($T = 2\pi R/v$) is $2\pi mv$.

Although there is nothing about centrifugal force in the above argument, what Newton does next is virtually the evaluation of this force. He calculates, namely,

4) One might argue (and some people do) that the contribution of a great scientist may be assessed by the extent to which he slows down the later development of his particular field. Undoubtedly, Newton's contribution was a very substantial one according to this criterion, especially in England, where his authority prevented for about one and a half century the development in the fields of optics and mathematical analysis. It was not until 1803 that the easy-to-follow Leibnizian notation was employed in an English paper on mathematical analysis; the "Analytical Society," established by Ch. Babbage, J. Herschel and G. Peacock in 1812, made it its aim to implant the Leibnizian notation into English scientific literature; and it was only in 1819 that the first English course on mechanics (by W. Whewell) was published in this notation [13]. According to A. R. Forsyth [14], Leibnizian symbols for integral and differential were prohibited as late as in 1880's at the earliest stage of Cambridge Tripos examinations. It may be interesting to note that for about a century after Newton's death the problems in Tripos papers were closely related to those found in Newton's writings. A problem posed at the 1786 examinations sounds very familiar in our space age: "To determine the velocity with which a body must be thrown, in a direction parallel to the horizon, so as to become a secondary planet to the Earth; as also to describe a parabola, and never return" [15].

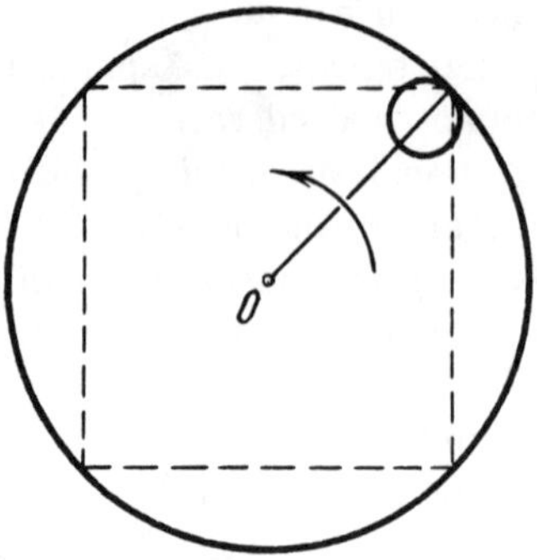

Figure 1

the distance from the center at which the ball would have been found on the passage of one full period of revolution had its motion been along a straight line from the center and caused by the change of its momentum we have just found. To do this Newton uses the formula $S = (\frac{1}{2})VT$ for uniformly accelerated motion, which had already been known to Galileo and can actually be traced back to the Merton school. He finds

$$S = \frac{1}{2}\left(\frac{2\pi mv}{m}\right)\left(\frac{2\pi R}{v}\right) = 2\pi^2 R \tag{1}$$

About the same period, Newton rederives this relationship by comparing the motion of a ball in a circle with its motion along the tangent under the influence of inertia. Referring to Fig. 2, it is known from elementary geometry that $BE/BA = BA/BD$, which allows us to write $2R/DA = DA/DB$ (2) for small arcs DA. Using Galileo's $s \sim t^2$ law we thus have

$$S/DB = \left(\frac{2\pi R}{DA}\right)^2 \tag{3}$$

for the space S travelled along the radius under the action of the centrifugal force.

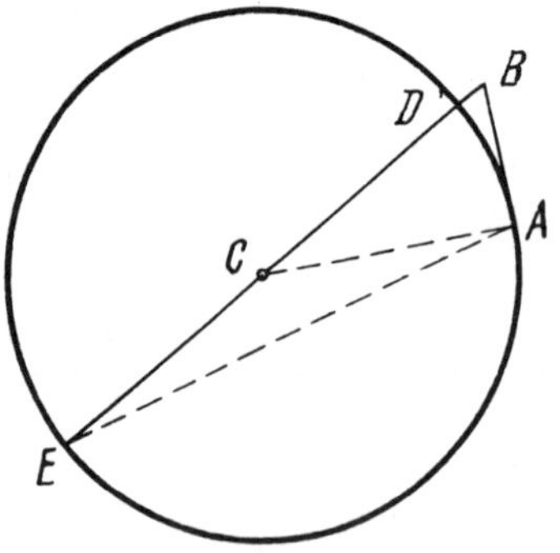

Figure 2

Combining (2) and (3), we deduce that $S = 2\pi^2 R$, which means, using the present-day terminology, that the acceleration of the ball in its imaginary motion away from the center is $a = v^2/R$ (since $a = 2S/T^2$). This rather simple argument is presented in the second half of the Scholium to Proposition IV, Book I, although it probably will not seem too simple to the modern reader when read in Newton's own language (or rather in a contemporaneous English translation):

"In any circle suppose a polygon to be inscribed of any number of sides. And if a body, moved with a given velocity along the sides of the polygon, is reflected from the circle at the several angular points, the force, with which at every reflection it strikes the circle, will be as its velocity: and therefore the sum of the forces, in a given time, will be as that velocity and the number of reflections conjunctly; that is (if the species of the polygon be given), as the length described in that given time, and increased or diminished in the ratio of the same length to the radius of the circle; that is, as the square of that length applied to the radius; and therefore the polygon, by having its sides diminished *ad infinitum,* coincides with the circle, as the square of the arc described in a given time applied to the radius. This is the centrifugal force,[5] with which the body impels the circle; and to which the contrary force, wherewith the circle continually repels the body towards the centre, is equal."[6]

Now that Newton knew how centrifugal forces could be evaluated, it is only very natural to assume that he made an attempt (in the same year 1666) to compare the force of gravity at the Earth's surface with the gravitational force necessary for keeping the Moon in its orbit (which latter force he had earlier obtained from the inverse-square law). This attempt must have failed, however, possibly because correct numerical data were not available at the time, and caused Newton's deep disappointment in his celestial studies—up to the point of totally abandoning them. Be it as it may, it was not until after 15 years later that Newton turned again to the problems of mechanics. In the decade of the 1670's his interests lay in optics, theology, but above all—alchemy.

There is a great deal of (mostly recent) evidence that alchemy played a central role in Newton's life since the late 1660's. It is known, in particular, that in 1669 Newton bought a vast amount of chemical equipment and the six-volume *Theatrum Chemicum,* the largest then known collection of alchemical texts. Newton's extreme reticence with regard to his (al)chemical studies (he did not publish one single line on the subject) may of course be explained by his jealousy for professional secrets and partly is due to his almost abnormal fear of public discussions,[7]

5) The word "centrifugal" is absent from the first edition of the *"Principia"* but does appear in the second. The concept of centrifugal forces in today's interpretation is found in Book III where the dependence of free fall on geographical latitude is discussed.

6) Here and hereafter, excerpts from the *Principia* are given in A. Motte's translation.

7) ". . . I see a man must either resolve to put out nothing new or to become a slave to defend it" (from a letter written in 1676 to Henry Oldenburg, Secretary to the Royal Society). It is not out of place here to remember that Newton published his famous *Opticks* only in 1704, a quarter century after it had actually been written—but almost immediately after the death of Robert Hooke (1635–1703), with whom he had a long and fierce debate in the 1670's over the nature of light and color.

but undoubtedly it should also be taken as an indication that he was fully aware of the importance of this work. The estimated volume of his papers on this subject is about one million words, which compares with his mathematical studies and exceeds his writings on physics and mechanics.

Newton's assistant and amanuensis (copyist) later wrote about his master's work in the latter half of 1680's: "He very rarely went to bed till two or three of the clock, sometimes not until five or six, lying about four or five hours, especially at spring and fall of the leaf, at which times he used to employ about six weeks in his laboratory, the fire scarcely going out either night or day; he sitting up one night and I another, till he had finished his chemical experiments, in the performances of which he was the most accurate, strict, exact. What his aim might be I was not able to penetrate into, but his pains, his diligence at these set times made me think he aimed at something beyond the reach of human art and industry . . . The laboratory . . . was . . . furnished with . . . crucibles . . . in which he fused his metals, . . . the transmuting of metals being his chief design . . ." [16]. It is also known from other independent sources that Newton's active interest in metal transmutation actually began in the earlier 1970's. The scholarly investigation of Newton's alchemical interests is only a newcomer in the vast field of Newtonian studies (cf. [17]) and there is every reason to believe that new facts will soon be established concerning this aspect of Newton's scientific work.

But let us return to the problems of mechanics or, more specifically, to the mechanics of the Solar system. In 1673, C. Huygens published a method for calculating a centrifugal (centripetal) force, which made it rather a simple matter to derive the gravitational inverse-square law from Kepler's third planetary law. In 1670 Robert Hooke expressed some thoughts in regard to the universal nature of the law of gravitation. In the late autumn of 1679 Hooke, by then Secretary to the London Royal Society, invited Newton to speak out on the motion of bodies under the action of gravitation and in particular, on the idea that this motion could be regarded as a combination of a centripetal component and a rectilinear inertial component. Newton declined the invitation under the dubious pretext of being too busy and unaware of Hooke's results, but was unwary enough to make a wrong statement in his answer, namely, that falling terrestrial bodies move eastward along a "spiral line"; which means, of course, that at the moment he had no proof for the ellipticity of orbits of bodies acted on by gravity. Hooke's immediate reply was to the effect that, first, falling bodies deflected not as much eastward as southward; and second, that the curve in question resembled an ellipse [7]. Newton tried—but failed—to explain away his error, and, understandably, this episode could only worsen the relations between the two scientists already marred by their controversy over optical problems. There follows a four-year lacuna in our knowledge about their relations because Hooke's diaries for this period mysteriously disappeared (as did many of his later papers), and Newton was too preoccupied with his (al)chemical experiments and spent virtually all his time in his Cambridge laboratory. There is little doubt, however, that Newton was too much annoyed with his lapse to leave the problem suspended, and that he did find some time to analyze it thoroughly.

In early 1684, R. Hooke, E. Halley, and C. Wren met together in London to discuss the inverse-square law in connection with Kepler's third law and with Huygens' formula for centrifugal forces. In retrospect, the meeting may well be considered as a historic one, for it was then that the problem of deducing elliptical planetary orbits from an inverse-square force was distinctly formulated. Of the three Royal Society colleagues, none possessed a mathematical skill sufficient to solve the problem. In August of the same year Halley visited Newton in Cambridge and learned, to his utmost surprise, that the solution had already been found—by Newton. Halley gets excited and asks for the solution, Newton fails to find it amongst his papers and asks for time, and subsequent events follow one another at a remarkable speed for those days. At the close of 1684 the Royal Society receives Newton's tract "De Motu"—only to be registered, but not intended for print (the tract was officially entered into the Register in February 1685). Under Halley's pressure Newton begins to write a major treatise on motion. In the spring of 1686, the manuscript of Book I of the *Principia* is officially presented to the Society, in May the Society adopts a resolution to undertake the printing of the whole of Newton's treatise, and on July 5, 1686, the President of the Society formally signs the order for print, although it was already in June that Newton actually received the proof of the first sheet of his book. Very soon Newton changes his mind as to the plan of the work and decides to exclude Book III of the treatise, devoted to the mathematical description of the Copernican system of the world. This decision was caused by Newton's constant fear of public discussions,[8] and Halley had to exert all his persuading and diplomatic power to make Newton change his mind. In March 1687 Newton sends to London the manuscript of Book II, and finally, on April 4, Halley receives Book III of the *Principia*. Despite Newton's instructions not to hasten things, Halley hired a second composer who began to compose simultaneously Book III (which accounts for some inconsequences in the pagination of the first edition of the *Principia*). On July 5, 1687, as already mentioned, the book was ready. The pace at which Halley did the job—and this three hundred years ago—is one that many publishers of today might do well to consider emulating: Books II and III that constitute more than a half of the whole treatise were composed, proofread, and printed (from a manu-, not a typescript!) in four months exactly.

One problem which Halley met with little success in solving in the course of the preparation of the *Principia* was to persuade Newton to acknowledge in some way or another the role of Hooke in establishing the universal law of gravitation. The maximum Halley achieved was a rather ambiguous mention of Hooke in connection with the sixth Corollary to the theorem (Proposition IV) concerning circular motions (Book I): "The case of the 6th corollary obtains in the celestial bodies (as Sir Christopher Wren, Dr. Hooke, and Dr. Halley have severally ob-

8) He writes in one of his letters in June 1686: ". . . The third I now design to suppress. Philosophy is such an impertinently litigious Lady that a man had as good be engaged in Law suits as have to do with her. I found it so formerly and now I no sooner come near her again but she gives me warning."

served).[9] Strictly speaking, Newton failed to prove that he was fully independent of Hooke's ideas in his work.

Contrary to the traditional opinion, it should be noted that Newton did not actually succeed in showing that under the action of an inverse-square force the planets should move along conic sections with a focus in the center of attraction (i.e., the Sun); nor is this opinion supported much by the fact of Newton's having derived quadratures necessary for the direct solution of the problem (in polar coordinates). What Newton did prove in his *Principia* is actually the inverse statement, and one much easier for the time, that if a body does move along a conic section under the action of a centripetal force directed toward a focus, then the force must vary inversely as the square of the distance. This shortcoming was pointed out by Johann Bernoulli already in Newton's lifetime (indeed prior to the second edition of *Principia*) and became again a subject of discussion in more recent times [18]. That Newton himself was not entirely happy about his presentation may be seen from the fact that he made an attempt to strengthen his reasoning in the second edition of *Principia*. He failed, however, to achieve anything more than to complement his proof by some verisimilar arguments in support of the truth of the direct statement.

It is perhaps worth noting in this connection that, at least as his own scientific priorities were at issue, Newton's attitude toward formal mathematical proofs was rather a peculiar one. Suffice it to say that Newton not only refused to recognize Hooke's role in formulating the universal law of gravitation and the problem of planetary motion, but went so far as to take credit for the two statements now generally known as the first two of Kepler's laws on the ground that it was he, Newton, who had proved these statements mathematically. It was only Kepler's third law which Newton was benign enough to leave to its author.[10]

Today, when we can more objectively assess the scientific events to those bygone days, it is impossible to deny Hooke's significance as a Newton forerunner en route to the gospel of new mechanics. In S. I. Vavilov's words, "In the 17th century Newton was the only person who could have written *Principia,* but it is an undisputable fact that the plan of the book was first outlined by Hooke" [2]. This sounds perhaps a little overcomplimentary with regard to Hooke, but certainly he did play a very important role in forming our views concerning the system of the world, and on this point we cannot but disagree with the great Newton.[11]

Universal gravitation, the cornerstone of the Copernican world picture, was for a long time a subject of emotional philosophical discussions. Newton himself

9) The corollary reads: "If the periodic times are in the sesquiplicate ratio of the radii, and therefore the velocities reciprocally in the subduplicate ratio of the radii, the centripetal forces will be in the duplicate ratio of the radii inversely; and the contrary."

10) This law is mentioned in *Principia* twice as merely a law discovered by Kepler. Krylov missed this point in his translation and made Newton to speak of Kepler's "planetary laws," in plural (Book III, Scholium to Prop. IV.).

11) There exists a thorough comparative analysis of the views on free fall and planetary motions as expressed in *Principia* and in Hooke's works [19].

was extremely cautious in his opinions as to the nature of gravitation and spoke more than once against the hypothesis of action at a distance, apparently admitting the existence of some "incorporeal" propagation agent for gravitation. In the *Scholium Generale* added to the second edition of the *Principia* and containing Newton's theological views on God in the framework of natural philosophy, the 70 year old author writes: "And to us it is enough that gravity does really exist, and act according to the laws which we have explained, and abundantly serves to account for all the motions of the celestial bodies, and of our sea." It is in the same *Scholium Generale* that Newton proclaimed his famous *Hypotheses non fingo,* a thesis which is frequently given a much more profound philosophical significance than the author himself seem to have had in mind.[12]

A few words should be said about Newton's exposition of the system of mechanics. The main body of *Principia* is preceded by eight definitions, of which the first and the last are:

"Definition I. The quantity of matter is the measure of the same, arising from its density and bulk conjunctly.

Definition VIII. The motive quantity of a centripetal force is the measure of the same, proportional to the motion which it generates in a given time."

The remaining six definitions refer to the quantity of motion, the force of inertia[13] (*vis insita,* or innate force), impressed force, and centripetal force with its absolute and accelerative quantities.

In the *Scholium* which follows, Newton gives detailed explanation of the concepts he is going to rely upon in his exposition, namely, absolute time, relative time, space, place, and motion. Next follow his famous laws of motion:

"Law I. Every body perseveres in its state of rest, or of uniform motion in a right line, unless it is compelled to change that state by forces impressed thereon.

Law II. The alteration of motion is ever proportional to the motive force impressed; and is made in the direction of the right line in which that force is impressed.

Law III. To every action there is always opposed an equal reaction; or the mutual actions of two bodies upon each other are always equal, and directed to contrary parts."

All the definitions and laws are accompanied by explanatory comments. The laws are complemented by six corollaries, of which we reproduce here the following three:

"Corollary I. A body by two forces conjoined will describe the diagonal of a parallelogram, in the same time that it would describe the sides, by those forces apart.

12) The word *hypothesis* was actually not totally alien to Newton's thinking (and writing). It will be noted that the separate sections "Regulae philosophandi" and "Phaenomena" that opened Book III in the second and third editions are found under the common rubric *Hypotheses* in the first edition.

13) This term of Newton's (for a body's resistance to any changes in its state of rest or of uniform rectilinear motion) is actually a revitalization of rather an unfortunate term formerly used by Kepler in very nearly the same sense. The term has been given quite a number of interpretations and has been—and indeed is now—a subject of much and mostly fruitless controversy among scientists.

Corollary IV. The common centre of gravity of two or more bodies does not alter its state of motion or rest by the actions of the bodies among themselves; and therefore the common centre of gravity of all bodies acting upon each other (excluding outward actions and impediments) is either at rest, or moves uniformly in a right line.

Corollary V. The motions of bodies included in a given space are the same among themselves, whether that space is at rest, or moves uniformly forwards in a right line without circular motion" [Galileo's relativity principle—G. M.].

The *Scholium* that closes the introductory chapters begins as follows: "Hitherto I have laid down such principles as have been received by mathematicians, and are confirmed by abundance of experiments. By the two first Laws and the first two Corollaries, Galileo discovered that the descent of bodies observed the duplicate ratio of the time, and that the motion of projectiles was in the curve of a parabola . . ." Newton then goes on to the laws of collision established by C. Wren, J. Wallis, and C. Huygens, and to the collision experiments by E. Mariotte, and the impression is that he makes no claims whatsoever to his own role in the development of the basic principles of mechanics. We are faced here with a rare case of Newton's underestimating his role in science, because prior to Newton the laws of motion had not actually been formulated by anyone as a unifed system, and to formulate them was certainly to do much more on Newton's part than merely to unite or generalize the views of his predecessors.

It is appropriate here to mention Newton's totally independent contributions to mechanical science. He was the first to have established the equivalence of inertial and gravitational masses, the principle that underlies both terrestrial and celestial mechanics. He was also the first to give a clear and inambiguous formulation for the law of action and reaction. (Newton attached great importance to this law, and it may be noted in passing that he had totally abandoned his erroneous views on centrifugal forces to the time *Principia* started to be written). And finally, we find thoughts here and there throughout *Principia,* whose profundity goes far beyond the Newtonian methodology of mechanics as it is commonly understood. To give but one example, here is what he writes in his *Scholium* to the laws of motion: "If we estimate the action of the agent from its force and velocity conjunctly, and likewise the re-action of the impediment conjunctly from the velocities of its several parts, and from the forces of resistance arising from the attrition, cohesion, weight, and acceleration of those parts, the action and re-action in the use of all sorts of machines will be found always equal to one another . . ." To quote A. N. Krylov, "these words . . . contain in themselves not only the principle of virtual displacements, with its all-embracing application to the equilibrium problems of machines, that is, of systems with a complete constraint or one degree of freedom; they also contain the idea of the principle of D'Alembert, albeit in a form so concise that it took the genius of Lagrange to express it in one mathematical formula including statics and dynamics in their entirety" [3].

Although Newton's laws and definitions undoubtedly form a basis on which the whole of classical (i.e., non-relativistic) mechanics has been erected, their formal and logical inconsistency has been a headache for many scientific generations over the past three centuries. The most serious dissatisfaction has been caused by the relationship between the first two laws of motion and by Newton's definitions of mass and force (or the motive quantity of force, in Newton's own terminology) in their relation to the basic law of motion. Two hundred years after the appearance of *Principia,* E. Mach [20] took it upon himself to carry out a constructive analysis of Newton's definitions and suggested his own definition of mass (*via* the third law) based on its inertial properties. This definition has become very popular, as exemplified by an 8th grade Russian physics textbook by I. K. and A. K. Kikoins, in which all three laws of motion are retained, but the first one is interpreted as an assertion of the existence of inertial reference systems.

Considering that Newton was extremely cautious and self-disciplined in expressing himself publicly, there is little doubt that he was fully aware of the logical difficulties in formulating the basics of mechanics and that the representation he adopted in the *Principia* was the best possible he had been able to think of, and one on which, he believed, further constructions could be confidently built up. One of such constructions, the dynamic model of the Solar system, demonstrated in the most spectacular manner the might of Newton's thought and indeed sanctified the foundations he laid for the future edifice of classic mechanics.

Newton's incontestable authority and the lack of knowledge of his scientific heritage are the reasons why it has not been possible until very recently to assess objectively his contribution to mechanics and to examine impartially what Newton achieved himself, what he borrowed from his predecessors, and what—and whose—achievements have been ascribed to him by mistake.

What impresses most about Newton's life and work is, of course, the grandeur of his *Principia,* the more so that it took less than two years to complete the preparation of its first version, and that on the background on Newton's continuous work in his alchemic laboratory! Undoubtedly, the writing of the book must have mobilized all Newton's intellectual resources. Ignoring insignificant errors excusable for a pathfinder, the structure of the *Principia* may well be called perfect, and it is no exaggeration to say that Newton's genius rose to the highest level of imagination and descended to unfathomable depths of intuition when creating this *opus magnum.*

We can consolidate with C. Truesdell [21] in subdividing in two major parts Newton's contribution into mechanics, namely, works that brought fame to him but were not entirely free from external influences; and works that remained in the shadow but were totally and completely original.

From this viewpoint, Newton's mechanics of a point mass and, in particular, his laws of motion were actually contained in the work of his predecessors, although it took Newton's genius and unfathomable perspicacity to see this and to express in the impressive form he did.[14]

14) It is amazing that sometimes Newton made mistakes in solving even elementary problems. One example of this kind, pertaining to the momentum conservation law, was recently analyzed by V. A. Fabrikant [29].

Although it was Newton who provided a mathematical foundation for the Copernican system of the world, the universal law of gravitation and indeed the necessity of the mathematical description of the Copernican world had already been voiced before Newton. But no one of his contemporaries had a mathematical skill of anywhere near the level required for such a grandiose undertaking. Although it remains unknown to what extent (if at all) he used calculus in his work, there is little doubt that the mastery of new mathematical methods did play a role in Newton's work on the subject. Note that Leibniz's followers on the Continent remained unprepared for solving similar problems until the 1700's, that is, about two decades later.

Whatever some people may think to the contrary, Newton did not build up the whole of classical mechanics, even though it was he who laid its foundations. He failed to formulate general equations of motion, he did not consider in any general manner mechanical systems of more than two mass points, he remained unaware of the equation of angular momentum, and made no attempts to develop some general approach to the mechanics of absolutely rigid body or, less still, to the mechanics of a continuous medium. It is for these reasons that Newton should more appropriately be regarded as the creator of the mechanics of a particle. A system, a rigid body, and a continuous medium are by far more complex objects, whose mechanics call for concepts that were not—and hardly could be—known to Newton.

For all this there are many results contained in the *Principia,* especially in its Book II which, although less popular and not so frequently mentioned, are undoubtedly original and bear the mark of genius even though most of them have undergone some, often very serious, modifications before actually having found their applications. These are certain results pertaining to the problem of three bodies, principles of the dynamical similitude of mechanical systems, the theory of air elasticity, the problem of a liquid issuing through a hole, wave propagation on a water surface, vibrations of a liquid in a pipe, resistance of rarified and dense media (Newton's sine law has been confirmed in our days for very high velocities),[15] sound propagation (with the adiabat replaced by the isothermal line), and internal friction in liquids ("Newtonian liquid").[16]

Newton's *Principia* thus provided the basis on which the great edifice of classic mechanics was virtually completed in the 18th century by the efforts of L. Euler, J. D'Alembert, and L. Lagrange. The 18th century first brought with it the general differential equations of motion of a particle in a fixed coordinate system; then the fundamentals of the mechanics of a system, of a rigid body, and of an ideal liquid; the beginnings of the variational approach to mechanical problems; and finally, the Lagrangian analytical mechanics. In the 19th century much work

15) Newton was also the first to explain the Robins-Magnus effect, in which a rotating body is changing its trajectory when moving in a resistant medium.

16) One and a half centuries after the appearance of the *Principia* G. Stokes found a mistake Newton had made in analyzing the rotation of a viscous liquid.

had been done, of course, on the further development of the earlier established principles but, as far as fundamentals are concerned, mention should only be made of the treatment of the stress-strained state of elastic bodies; the study of the motion of viscous (Newtonian) liquids; and the development of the theory of discontinuities in gas dynamics which was actually completed only in the 20th century when necessary thermodynamic methods had become available.

To conclude, it can rightfully be said that modern classic mechanics rests on the ideas formulated by Newton three hundred years ago in his immortal *Principia,* and it is a pity we know so little about how Newton actually wrote his great book. Using the ancient metaphor—and following Fontenelle—we can compare Newton with the Nile whose origin is jealously concealed by Nature and who does not appear but in its full Majesty to us mortals: *"Arcanum Natura caput non prodidit ulli, Nec licuit populis parvum te, Nile, videre."* Lucan. 1.x. Nature conceals thy infant Stream with care, Nor lets thee, but in Majesty appear.

REFERENCES

1. Newton, I. Philosophiae naturalis principia mathematica. London, 1687.
2. Vavilov, S. I. Isaac Newton. USSR Acad. Sci., Moscow, 1945 (in Russian).
3. Krylov, A. N. Collection of works. USSR Acad. Sci., Moscow, 1936, (in Russian).
4. Cohen, I. B. Introduction to Newton's *Principia*. Cambridge, Univ. Press, 1971.
5. Newton, I. Philosophiae Naturalis Principia Mathematica. 2 Vol., Ed. A. Koyré, I. B. Cohen, 3rd ed. with variant readings, Cambridge Univ. Press, 1972.
6. Newton, I. The mathematical papers. 8 Vol., Ed. D. T. Whiteside, Cambridge Univ. Press, 1967–1981.
7. Newton, I. The Correspondence, 7 Vol., Ed. H. W. Turnbull, J. F. Scott, A. R. Hall, and L. Tilling, Cambridge Univ. Press, 1959–1977.
8. Newton, I. Unpublished scientific papers. A selection from the Portsmouth Collection. Chosen, Ed., and Transl. by A. R. Hall and M. B. Hall, Cambridge Univ. Press, 1962.
9. Herivel, J. The background to Newton's *Principia*. Oxford, Clarendon Press, 1965.
10. Westfall, R. S. Force in Newton's physics. The science of dynamics in the 17th century. London, MacDonald, 1971.
11. Westfall, R. S. Never at rest. A biography of I. Newton. Cambridge Univ. Press, 1980.
12. Wallis, P. and Wallis, R. Newton and Newtoniana. 1672–1975. A bibliography. Folkestone (Kent), Dawson, 1977.
13. Dubbey, J. M. The introduction of the differential notation to Great Britain. Ann. Sci., 1963, Vol. 19, No. 1, pp. 37–48.
14. Forsyth, A. R. Old tripos days at Cambridge. Math. Gazette, pp. 162–179.
15. Ball, W. W. R. A History of the study of mathematics at Cambridge. Cambridge Univ. Press, 1889.
16. Brewster, D. Mémoires of the life, writings, and discoveries of Sir Isaac Newton. 2 Vol., Edinburgh, Constable, 1885.
17. Dobbs, B. J. T. The foundations of Newton's alchemy or "The hunting of the Greene Lyon." Cambridge Univ. Press, 1975.
18. Weinstock, R. Dismantling a centuries-old myth: Newton's *Principia* and inverse-square orbits. Amer. J. Phys., 1982, Vol. 50, No. 7, pp. 610–617.
19. Lohne, J. Hook versus Newton. An analysis of the documents in the case on free fall and planetary motion. Centaurus, 1960, Vol. 7, No. 1, pp. 6–52.

20. Mach, E. Die Mechanik in ihrer Entwickelung Historisch-Kritisch Dargestellt. Leipzig, Brockhaus, 1883.
21. Truesdell, G. Essays in the history of mechanics. Berlin, Springer, 1968.
22. Fabrikant, V. A. Newton, Bernoulli and momentum conservation law. Uspekhi Fiz. Nauk, 1960, Vol. 70, No. 3, pp. 575–580.

INDEX